Michał Baczyński and Balasubramaniam Jayaram

Fuzzy Implications

Studies in Fuzziness and Soft Computing, Volume 231

Editor-in-Chief

Prof. Janusz Kacprzyk
Systems Research Institute
Polish Academy of Sciences
ul. Newelska 6
01-447 Warsaw
Poland
E-mail: kacprzyk@ibspan.waw.pl

Further volumes of this series can be found on our homepage: springer.com

Vol. 215. Paul P. Wang, Da Ruan, Etienne E. Kerre (Eds.)
Fuzzy Logic, 2007
ISBN 978-3-540-71257-2

Vol. 216. Rudolf Seising
The Fuzzification of Systems, 2007
ISBN 978-3-540-71794-2

Vol. 217. Masoud Nikravesh, Janusz Kacprzyk, Lofti A. Zadeh (Eds.)
Forging New Frontiers: Fuzzy Pioneers I, 2007
ISBN 978-3-540-73181-8

Vol. 218. Masoud Nikravesh, Janusz Kacprzyk, Lofti A. Zadeh (Eds.)
Forging New Frontiers: Fuzzy Pioneers II, 2007
ISBN 978-3-540-73184-9

Vol. 219. Roland R. Yager, Liping Liu (Eds.)
Classic Works of the Dempster-Shafer Theory of Belief Functions, 2007
ISBN 978-3-540-25381-5

Vol. 220. Humberto Bustince, Francisco Herrera, Javier Montero (Eds.)
Fuzzy Sets and Their Extensions: Representation, Aggregation and Models, 2007
ISBN 978-3-540-73722-3

Vol. 221. Gleb Beliakov, Tomasa Calvo, Ana Pradera
Aggregation Functions: A Guide for Practitioners, 2007
ISBN 978-3-540-73720-9

Vol. 222. James J. Buckley, Leonard J. Jowers
Monte Carlo Methods in Fuzzy Optimization, 2008
ISBN 978-3-540-76289-8

Vol. 223. Oscar Castillo, Patricia Melin
Type-2 Fuzzy Logic: Theory and Applications, 2008
ISBN 978-3-540-76283-6

Vol. 224. Rafael Bello, Rafael Falcón, Witold Pedrycz, Janusz Kacprzyk (Eds.)
Contributions to Fuzzy and Rough Sets Theories and Their Applications, 2008
ISBN 978-3-540-76972-9

Vol. 225. Terry D. Clark, Jennifer M. Larson, John N. Mordeson, Joshua D. Potter, Mark J. Wierman
Applying Fuzzy Mathematics to Formal Models in Comparative Politics, 2008
ISBN 978-3-540-77460-0

Vol. 226. Bhanu Prasad (Ed.)
Soft Computing Applications in Industry, 2008
ISBN 978-3-540-77464-8

Vol. 227. Eugene Roventa, Tiberiu Spircu
Management of Knowledge Imperfection in Building Intelligent Systems, 2008
ISBN 978-3-540-77462-4

Vol. 228. Adam Kasperski
Discrete Optimization with Interval Data, 2008
ISBN 978-3-540-78483-8

Vol. 229. Sadaaki Miyamoto, Hidetomo Ichihashi, Katsuhiro Honda
Algorithms for Fuzzy Clustering, 2008
ISBN 978-3-540-78736-5

Vol. 230. Bhanu Prasad (Ed.)
Soft Computing Applications in Business, 2008
ISBN 978-3-540-79004-4

Vol. 231. Michał Baczyński, Balasubramaniam Jayaram
Fuzzy Implications, 2008
ISBN 978-3-540-69080-1

Michał Baczyński and
Balasubramaniam Jayaram

Fuzzy Implications

Springer

Authors

Dr. Michał Baczyński
University of Silesia
Institute of Mathematics
ul. Bankowa 14
40-007 Katowice
Poland
E-mail: michal.baczynski@us.edu.pl

Dr. Balasubramaniam Jayaram
Department of Mathematics and Computer Sciences
Sri Sathya Sai University
Prasanthi Nilayam
Anantpur District
Andhra Pradesh - 515134
India
E-mail: jbala@ieee.org

ISBN 978-3-642-08861-2 e-ISBN 978-3-540-69082-5

DOI 10.1007/978-3-540-69082-5

Studies in Fuzziness and Soft Computing ISSN 1434-9922

Printed in acid-free paper

9 8 7 6 5 4 3 2 1

springer.com

To my wife Agnieszka and
our daughters Dominika and Justyna
with love.

Michał

To Bhagwan Sri Sathya Sai Baba
and His University.

Balasubramaniam

Preface

Nothing new had been done in Logic since Aristotle!
– Kurt Gödel (1906-1978)

Fuzzy implications are one of the main operations in fuzzy logic. They generalize the classical implication, which takes values in $\{0,1\}$, to fuzzy logic, where the truth values belong to the unit interval $[0,1]$. In classical logic the implication can be defined in different ways. Three of these have come to assume greater theoretical importance, viz. the usual material implication from the Kleene algebra, the implication obtained as the residuum of the conjunction in Heyting algebra (also called pseudo-Boolean algebra) in the intuitionistic logic framework and the implication (also called as 'Sasaki arrow') in the setting of quantum logic. Interestingly, despite their differing definitions, their truth tables are identical in classical case. However, the natural generalizations of the above definitions in the fuzzy logic framework are not identical. This diversity is more a boon than a bane and has led to some intensive research on fuzzy implications for close to three decades. It will be our endeavor to cover the various works churned out in this period to sufficient depth and allowable breadth in this treatise.

In the foreword to Klir and Yuan's book [147], Professor Lotfi A. Zadeh states the following:

> "The problem is that the term 'fuzzy logic' has two different meanings. More specifically, in a narrow sense, fuzzy logic, FL_n, is a logical system which may be viewed as an extension and generalization of classical multivalued logics. But in a wider sense, fuzzy logic, FL_w, is almost synonymous with the theory of Fuzzy Sets. In this context, what is important to recognize is that:
> a) FL_w is much broader than FL_n and subsumes FL_n as one of its branches;
> b) the agenda of FL_n is very different from the agendas of classical multivalued logics; and
> c) at this juncture, the term fuzzy logic is usually used in its wide rather than the narrow sense, effectively equating Fuzzy Logic with FL_w."

Since a fuzzy implication is a generalization of the classical (Boolean) implication to the unit interval $[0,1]$, its study predominantly belongs to the domain of FL_n. Keeping the above concern and the subsequent distinction, we have made a sincere attempt to discuss fuzzy implications in the context of FL_w too, vis-á-vis the role of fuzzy implications in potential applications. Moreover, the organization of the book is made to reflect the above philosophy. This treatise has been divided into the following 3 parts:

Part I Analytical Study of Fuzzy Implications;
Part II Algebraic Study of Fuzzy Implications;
Part III Applicational Study of Fuzzy Implications.

Chap. 1 introduces the readers to fuzzy implications. After defining a fuzzy implication, we present some of the most desirable properties that a fuzzy implication can possess and discuss their interrelationships. Based on this study we list out, what we consider, as 9 basic fuzzy implications.

Part I, consisting of Chaps. 2-5, deals with the analytical aspects of fuzzy implications, viz., different approaches to generating fuzzy implications from other fuzzy logic operations, the families obtained through such processes, their properties, representations, characterizations and the overlaps that exist among these families.

Part II, consisting of Chaps. 6 and 7, deals with the algebraic aspects of fuzzy implications, viz., different approaches to generating fuzzy implications from existing fuzzy implications, the closure of such processes, the mathematical structures that can be imposed on different classes or families of fuzzy implications and the functional equations/tautologies they satisfy.

While Parts I and II can be seen as dealing with fuzzy implications in the setting of FL_n, Part III, consisting of Chap. 8, is an 'applicational' study of fuzzy implications, wherein we deal with the effecting role that fuzzy implications play in approximate reasoning (AR) and fuzzy control (FC) which occupy the center stage in the framework of FL_w. Towards helping the readers better appreciate this facet of fuzzy implications, a basic overwiew of AR is given, with particular emphasis on two of the most important inference schemes therein, namely, Compositional Rule of Inference and Similarity Based Reasoning. Following this, we investigate how different families of fuzzy implications with the myriad properties they possess and tautologies they satisfy influence both the performance and functioning of the above schemes of inference.

This is probably the first comprehensive treatise on fuzzy implications. The terms *first* and *comprehensive* may seem contradictory and mutually exclusive. We should confess that *it was* for the most part during the writing of the book. The challenge of achieving this twin, but seemingly conflicting goals, has spurred us to put in our sincere efforts in making this up-to-date with all the latest relevant results, complete with proofs, examples and counter-examples to highlight the concepts and plots of different operations for better illustration. Each chapter is preceded by a short summary and ends with some bibliographical remarks on the contents of the chapter, viz., historical details, the context in which the operations were proposed and their subsequent line of analysis.

However, it should be noted that in this treatise we have dealt only with the classical approaches to fuzzy implications, owing mainly to keep the size of this monograph to a manageable extent. For example, studies on *interval-valued and intuitionistic fuzzy implications* (see BUSTINCE et al. [44], CORNELIS et al. [61, 64], DESCHRIJVER and KERRE [74, 75]) and those that deal with fuzzy implications where the underlying set is any *discrete* and/or *finite* (see MAS et al. [172, 173, 177], MAYOR and TORRENS [178]) have not been included in detail in this treatise. From the applicational aspects of fuzzy implications, in Part III, we have covered only their influence in approximate reasoning, while there are many other areas wherein they wield a dominant influence - see, for example, the excellent survey of MAS et al. [176]. Another notable omission is the role of fuzzy implications in fuzzy logics, in the sense of FL_n, viz., the Basic-Logic (BL) of HÁJEK [119] which was shown to be the logic of continuous t-norms and their residua by CIGNOLI et al. [53], the Monoidal T-norm Logic (MTL) of ESTEVA and GODO [97] - which was shown to be the logic of left-continuous t-norms and their residua - and later on extended to Involutive MTL, WNML and NML in ESTEVA et al. [98, 96], etc. In this case, the predominant reason being the availability of many excellent treatises on such topics, e.g., CIGNOLI et al. [55], GOTTWALD [113], HÁJEK [120], TURUNEN [243].

This book is intended for any researcher in fuzzy logic operations and it can serve as an auxiliary textbook for different courses on fuzzy logic or fuzzy control. It does not assume any special background in fuzzy set theory, but the reader should have some knowledge in analysis, algebra and classical logic (on the level of a graduate student).

Parts of this book may also be of interest to practitioners in fuzzy logic applications, especially in approximate reasoning or fuzzy control, where fuzzy implications play a central role.

Acknowledgments
It is said that "No book is written in isolation". This book is an exception to this rule, in that, most parts of this book was written when one of the authors was in Poland while the other was in India - surely isolation cannot get any farther than this. It is immediately obvious that to surmount this part-challenge-part-obstacle of geographical distance the help and benevolence of many were involved.

First on our list are our respective supervisors, Prof. Józef Drewniak and Prof. C. Jagan Mohan Rao, who magnanimously gave their time, energy and knowledge that have opened many vistas, plugged many gaps in the various topics, shaped the subsequent organization and contents of this treatise. Following them are our colleagues and fellow researchers Dr. Paweł Drygaś, Dr. Urszula Dudziak, Mgr Anna Król and Dr. Martin Štěpnička who have patiently read many parts of the various versions of our draft and whose suggestions have made a remarkable difference in the final outcome. Thank you, all.

We thank the constant support extended to us by our host universities, University of Silesia, Poland and Sri Sathya Sai University, India, respectively, and the colleagues at our respective departments for the excellent ambience they provided, without which a venture of this sort would have been difficult to undertake.

While the gaps in the academic content were narrowed with the help of the above, the geographical distance was bridged by Google. The authors would like to gratefully acknowledge the excellent voice-chat tool 'Google Talk', which brought us to hearing-distance with each other. The countless hours of discussion we have had online with the help of GTalk is a testimony to the power of the Internet in conquering distances.

This geographical gap finally vanished in the final stages of this venture, thanks to the SAIA scholarship under the National Scholarship Program of the Slovak Republic, which made it possible for Balasubramaniam Jayaram to travel to Bratislava, Slovakia where many important decisions were made and the book finalized in this current form. Balasubramaniam Jayaram sincerely thanks SAIA and Prof. Radko Mesiar of DMDG, FCE, Slovak University of Technology, Bratislava for graciously hosting him during this period.

While writing this monograph we have also employed and gained familiarity with a number of fantastic tools. This text was typeset entirely in LaTeX2e and all the 3-d plots were made in Matlab 7.0. We have still not ceased to wonder at these magical tools and marvel at how the lifeless characters that we type suddenly metamorphose into such breath-taking symbols and plots. Is this where science meets art?

No acknowledgement section in a book is deemed complete without the authors' part-apologetic, part-warning but nevertheless proud proclamation of the errors that they have managed to retain in their book. We would like to part with this tradition with the supreme knowledge that errors are endowed with a life-expectancy that survives any editorial vetting and have an uncanny sense of direction to creep seemingly innocuously, in the least obtrusive way, into that part of the text where it lies hidden only to leap into the full view of all the readers, especially the reviewer and except the writer, causing maximum catastrophe. To err may not be the right of the authors, but to ignore it is left to the readers.

Katowice, Puttaparthi
June 2008

Michał Baczyński
Balasubramaniam Jayaram

Notations and Some Preliminaries

Mathematics is written for mathematicians.
– Nicholaus Copernicus (1473-1543)

In this book we have tried to employ notations that are generally used in the mathematical literature. For the classical logical operations like conjunction, disjunction, negation and implication we write $\wedge$, $\vee$, $\neg$ and $\rightarrow$, respectively. If A is a subset of B, where $A = B$ is possible, we denote this by $A \subseteq B$ or $B \supseteq A$. If A is a subset of, but not equal to, B, then A is called a proper subset of B, written $A \subsetneq B$ or $B \supsetneq A$. The union, the intersection and the difference of two sets A and B are denoted by $A \cup B$, $A \cap B$ and $A \setminus B$, respectively. The complement of a subset $A \subset X$ is denoted by A'. The Cartesian product of two sets A and B is denoted by $A \times B$. The cardinality of a set A is denoted by card A. The symbol $\mathbb{N}$ denotes the set of positive integers, $\mathbb{Z}$ denotes the set of all integers and $\mathbb{R}$ stands for the set of all real numbers. Moreover $\mathbb{N}_0 = \mathbb{N} \cup \{0\}$.

If $f\colon X \rightarrow Y$ is a function, then X is called the domain of f denoted by $\mathrm{Dom}(f)$ and the set Y is called the codomain. The range of f is given by $\mathrm{Ran}(f) = \{f(x) \mid x \in X\}$. If $A \subseteq \mathrm{Dom}(f)$, then $f(A)$ is the subset of the range consisting of all images of elements of A, i.e., $f(A) = \{f(x) \in Y \mid x \in A\}$. The preimage (or inverse image) of a subset B of the codomain Y under a function f is the subset of the domain X defined by $f^{-1}(B) = \{x \in X | f(x) \in B\}$. The composition of two functions $f\colon X \rightarrow Y$ and $g\colon Y \rightarrow Z$ is given by $(g \circ f)(x) = g(f(x))$ for all $x \in X$. The identity function $\mathrm{id}_X : X \rightarrow X$ is defined by $\mathrm{id}_X(x) = x$ for all $x \in X$. If f is a function from X to Y and A is any subset of X, then the restriction of f to A is denoted by $f|_A = f \circ \mathrm{id}_A$. Let $F\colon X \times Y \rightarrow Z$ be a function of two variables. Then for each fixed $x_0 \in X$, the vertical section of F is denoted by $F(x_0, \,\cdot\,)$. Similarly, for each fixed $y_0 \in Y$, the horizontal section of F is denoted by $F(\,\cdot\,, y_0)$.

For a closed interval we write $[a, b]$, for an open interval (a, b) and for half-open intervals $[a, b)$ or $(a, b]$. The set $\mathbb{R} \cup \{-\infty, +\infty\}$ is denoted by $[-\infty, +\infty]$ and the set $[a, b] \times [a, b]$ is denoted by $[a, b]^2$. The conventions when arithmetic operations are done on ∞ and $-\infty$ (e.g., the symbols $\infty + (-\infty)$ or $\infty \cdot (-\infty)$) will be explained in the context.

Let A be a subset of $\mathbb{R}$, and let $f\colon A \rightarrow \mathbb{R}$ be a function. Then we say that

- f is increasing, if $x \leq y$ implies that $f(x) \leq f(y)$ for all $x, y \in A$;
- f is strictly increasing, if $x < y$ implies that $f(x) < f(y)$ for all $x, y \in A$;

- f is decreasing, if $x \le y$ implies that $f(x) \ge f(y)$ for all $x, y \in A$;
- f is strictly decreasing, if $x < y$ implies that $f(x) > f(y)$ for all $x, y \in A$;
- f is monotone, if f is either increasing or decreasing;
- f is strictly monotone, if f is either strictly increasing or strictly decreasing.

If A is a nonempty set, then in the family of all real functions from A to $\mathbb{R}$ we can consider the partial order induced from the standard order on $\mathbb{R}$, i.e., if $f_1, f_2 \colon A \to \mathbb{R}$, then

$$f_1 \le f_2 :\Longleftrightarrow f_1(x) \le f_2(x) , \qquad x \in A .$$

If $f_1 \le f_2$ and $f_1 \ne f_2$, then we will write $f_1 < f_2$.

If a function $f \colon X \to Y$ is a bijection, then $f^{-1} \colon Y \to X$ denotes its inverse function, i.e., $f^{-1} \circ f = \mathrm{id}_X$ and $f \circ f^{-1} = \mathrm{id}_Y$. If $[a, b]$ and $[c, d]$ are two closed subintervals of $[-\infty, +\infty]$ and $f \colon [a, b] \to [c, d]$ is a monotone function, then it is well-known that the set of discontinuity points of f is a countable subset of $[a, b]$. In this case we will use the pseudo-inverse $f^{(-1)} \colon [c, d] \to [a, b]$ of f defined by

$$f^{(-1)}(y) := \sup\{x \in [a, b] \mid (f(x) - y)(f(b) - f(a)) < 0\} , \qquad y \in [c, d] .$$

For a decreasing and non-constant function $f \colon [a, b] \to [c, d]$, the pseudo-inverse of f is defined by

$$f^{(-1)}(y) = \sup\{x \in [a, b] \mid f(x) > y\} , \qquad y \in [c, d] .$$

Since $([a, b], \le)$ is a lattice, with $[a, b] \subseteq [-\infty, +\infty]$ and classical order $\le$ on the (extended) real line, $\inf \emptyset = b$ and $\sup \emptyset = a$. More results connected with the pseudo-inverses of monotone functions can be found in Schweizer and Sklar [222] and Klement et al. [146].

By Φ we denote the family of all increasing bijections from $[0, 1]$ to $[0, 1]$. We say that functions $f, g \colon [0, 1]^n \to [0, 1]$ are Φ-conjugate (see Kuczma [150], p. 156), if there exists a $\varphi \in \Phi$ such that $g = f_\varphi$, where

$$f_\varphi(x_1, \ldots, x_n) := \varphi^{-1}(f(\varphi(x_1), \ldots, \varphi(x_n))) , \qquad x_1, \ldots, x_n \in [0, 1] .$$

If F is an associative binary operation on $[a, b]$ with the neutral element e, then the power notation $x_F^{[n]}$, where $n \in \mathbb{N}_0$, is defined by

$$x_F^{[n]} := \begin{cases} e, & \text{if } n = 0 , \\ x, & \text{if } n = 1 , \\ F(x, x_F^{[n-1]}), & \text{if } n > 1 . \end{cases}$$

Let X be non-empty set. By $\mathcal{F}(X)$ we denote the fuzzy power set of X, i.e., $\mathcal{F}(X) = \{A | A \colon X \to [0, 1]\}$.

Finally, we recall the definition of a convex set. A subset X of a linear space is convex over $\mathbb{R}$ if, for any two points $x, y \in X$, X also contains the line segment between x and y, i.e., for all $\lambda \in [0, 1]$, we have

$$z = \lambda x + (1 - \lambda) y \in X .$$

The definitions, theorems, etc. are numbered triplewise: they all have a common counter with a chapter and section prefix. Thus there exist Definition 1.1.1 and Definition 2.1.1 (the first in Chap. 1, the second in Chap. 2), but there is no Theorem 1.1.1. The numbering of formulae is independent of the numbering of definitions, etc. and is doublewise. The fact that the book is divided into three parts has no reflection in the enumeration.

Contents

Part III: Applicational Study of Fuzzy Implications

Appendix

1 An Introduction to Fuzzy Implications

Thus, be it understood, to demonstrate a theorem,
it is neither necessary nor even advantageous to know
what it means.
– Jules Henri Poincaré (1854-1912)

The *implication operator* ($\rightarrow$) plays a significant role in the classical two-valued logic. Firstly, from the classical implication one can obtain all other basic logical connectives of the binary logic, viz., the binary operators - *and* ($\wedge$), *or* ($\vee$) - and the unary *negation* operator ($\neg$). Secondly, the implication operator holds the center stage in the inference mechanisms of any logic, like *modus ponens*, *modus tollens*, *hypothetical syllogism* in classical logic. The truth table for the classical implication is given in Table 1.1.

Table 1.1. Truth table for the classical implication

p	q	$p \rightarrow q$
0	0	1
0	1	1
1	0	0
1	1	1

A fuzzy implication is a generalization of the classical one to fuzzy logic, much the same way as a t-norm and a t-conorm are generalizations of the classical conjunction and disjunction, respectively. In this chapter we give the basic definitions and examples of fuzzy implications. Next, we introduce some main properties of fuzzy implications and their inter-dependencies. Following this, we discuss the relationship between fuzzy negations and fuzzy implications and deal with the laws of contraposition. In the last section some bibliographical remarks are given.

1.1 Definition and Basic Examples

In the literature, especially in the beginning, we can find several different definitions of fuzzy implications. In this book we will use the following one, which is equivalent to the definition proposed by Kitainik [141] (see also Fodor and Roubens [105]).

M. Baczyński and B. Jayaram: Fuzzy Implications, STUDFUZZ 231, pp. 1–35, 2008.
springerlink.com

Definition 1.1.1. *A function* $I\colon [0,1]^2 \to [0,1]$ *is called a* fuzzy implication *if it satisfies, for all* $x, x_1, x_2, y, y_1, y_2 \in [0,1]$, *the following conditions:*

$$\text{if } x_1 \le x_2, \text{ then } I(x_1, y) \ge I(x_2, y)\ , \text{ i.e., } I(\,\cdot\,, y) \text{ is decreasing}\ , \tag{I1}$$

$$\text{if } y_1 \le y_2, \text{ then } I(x, y_1) \le I(x, y_2)\ , \text{ i.e., } I(x, \,\cdot\,) \text{ is increasing}\ , \tag{I2}$$

$$I(0,0) = 1\ , \tag{I3}$$

$$I(1,1) = 1\ , \tag{I4}$$

$$I(1,0) = 0\ . \tag{I5}$$

The set of all fuzzy implications will be denoted by $\mathcal{FI}$.

The property (I1) is the *left antitonicity* of the function I that gives a fuzzy implication its unique flavor. It captures the idea that a decrease in the truth value of the antecedent increases its efficacy to state more about the truth value of its consequent. In fact, (I1) strongly encourages the truth-functional realization of an implication, since the truth values of both the antecedent and the consequent independently affect the overall truth value of the statement itself. The axiom (I2) captures the *right isotonicity* of the overall truth value as a direct function of the consequent. Axioms (I3) – (I5) comes from the basic properties of the classical implication. Therefore, we can also use the alternative name 'monotonic implications' as the characteristic property of the family $\mathcal{FI}$.

Example 1.1.2. The following examples of functions in Table 1.2 show that the axiom set from the above definition is mutually independent. Note that we denote the absence (presence, respectively) of a property by a × (✓, respectively) in the appropriate column. This convention will be followed in the rest of the book.

Remark 1.1.3. Directly from Definition 1.1.1 we can deduce that each fuzzy implication I is constant for $x = 0$ and for $y = 1$, i.e., I satisfies the following properties, called left and right boundary condition, respectively:

$$I(0, y) = 1\ , \qquad y \in [0,1]\ , \tag{LB}$$

$$I(x, 1) = 1\ , \qquad x \in [0,1]\ . \tag{RB}$$

Table 1.2. The mutual independence of the axioms from Definition 1.1.1

Function from $[0,1]^2$ to $[0,1]$	(I1)	(I2)	(I3)	(I4)	(I5)
$I_{-1}(x,y) = \max(1-x, \min(x,y))$	×	✓	✓	✓	✓
$I_{-2}(x,y) = \max(y, \min(1-x, 1-y))$	✓	×	✓	✓	✓
$I_{-3}(x,y) = \begin{cases} 0, & \text{if } y < 1 \\ 1, & \text{if } y = 1 \end{cases}$	✓	✓	×	✓	✓
$I_{-4}(x,y) = \begin{cases} 1, & \text{if } x = 0 \\ 0, & \text{if } x > 0 \end{cases}$	✓	✓	✓	×	✓
$I_{-5}(x,y) = 1$	✓	✓	✓	✓	×

Indeed, (LB) follows from (I2) and (I3), since $I(0,y) \geq I(0,0) = 1$. Similarly, because of (I1) and (I4) we get $I(x,1) \geq I(1,1) = 1$, i.e., I satisfies (RB). Therefore, I satisfies also the normality condition:

$$I(0,1) = 1 . \tag{NC}$$

Consequently, every fuzzy implication restricted to the set $\{0,1\}^2$ coincides with the classical implication, and hence fulfils the binary implication truth table provided in Table 1.1. Note that (LB) means that falsity implies anything, while (RB) expresses that tautology is implied by anything.

Example 1.1.4. Since there exist uncountably many fuzzy implications, the following Table 1.3 lists a few basic fuzzy implications, while Figs. 1.1a, 1.1b give the plots of these operations (the plot of $I_{\mathbf{WB}}$ is given in Fig. 1.3). They are usually listed with the corresponding author's name. Note that the Łukasiewicz implication can also be written as

$$I_{\mathbf{LK}}(x,y) = \begin{cases} 1, & \text{if } x \leq y , \\ 1 - x + y, & \text{if } x > y , \end{cases} \qquad x, y \in [0,1] ,$$

and the Goguen implication as

$$I_{\mathbf{GG}}(x,y) = \begin{cases} 1, & \text{if } x = 0 , \\ \min\left(1, \frac{y}{x}\right), & \text{if } x > 0 , \end{cases} \qquad x, y \in [0,1] .$$

Remark 1.1.5. We must also underline that some formulas which are considered in the scientific literature as multi-valued implications do not belong to the family $\mathcal{FI}$. For example, the function I_{-1}, also called in the literature as the Zadeh implication and noted as (I7) in [59], is not monotonic with respect to the first variable, as shown in Table 1.2. Consider also the following function

$$F(x,y) = \max(0, y - x) , \qquad x, y \in [0,1] ,$$

noted as (I12) in [59]. It is monotonic with respect to both variables, but does not satisfy (I3) and (I4), since $F(0,0) = F(1,1) = 0$.

According to the terminology introduced earlier, in the family of all fuzzy implications we can consider the partial order induced from the unit interval $[0,1]$. However, it can be easily seen that not all fuzzy implications are comparable. For our 9 basic fuzzy implications we have the ordering given in Example 1.1.6, a graphical representation of which is given in Fig. 6.1.

Example 1.1.6. Fuzzy implications from Table 1.3 form the following chains:

$$I_{\mathbf{KD}} < I_{\mathbf{RC}} < I_{\mathbf{LK}} < I_{\mathbf{WB}} , \tag{1.1}$$

$$I_{\mathbf{RS}} < I_{\mathbf{GD}} < I_{\mathbf{GG}} < I_{\mathbf{LK}} < I_{\mathbf{WB}} , \tag{1.2}$$

$$I_{\mathbf{YG}} < I_{\mathbf{RC}} < I_{\mathbf{LK}} < I_{\mathbf{WB}} , \tag{1.3}$$

$$I_{\mathbf{KD}} < I_{\mathbf{FD}} < I_{\mathbf{LK}} < I_{\mathbf{WB}} , \tag{1.4}$$

$$I_{\mathbf{RS}} < I_{\mathbf{GD}} < I_{\mathbf{FD}} < I_{\mathbf{LK}} < I_{\mathbf{WB}} . \tag{1.5}$$

Table 1.3. Examples of basic fuzzy implications

Name	Year	Formula
Łukasiewicz	1923, [157]	$I_{\mathbf{LK}}(x,y) = \min(1, 1-x+y)$
Gödel	1932, [117]	$I_{\mathbf{GD}}(x,y) = \begin{cases} 1, & \text{if } x \leq y \\ y, & \text{if } x > y \end{cases}$
Reichenbach	1935, [209]	$I_{\mathbf{RC}}(x,y) = 1 - x + xy$
Kleene-Dienes	1938, [143]; 1949, [77]	$I_{\mathbf{KD}}(x,y) = \max(1-x, y)$
Goguen	1969, [111]	$I_{\mathbf{GG}}(x,y) = \begin{cases} 1, & \text{if } x \leq y \\ \frac{y}{x}, & \text{if } x > y \end{cases}$
Rescher	1969, [210]	$I_{\mathbf{RS}}(x,y) = \begin{cases} 1, & \text{if } x \leq y \\ 0, & \text{if } x > y \end{cases}$
Yager	1980, [258]	$I_{\mathbf{YG}}(x,y) = \begin{cases} 1, & \text{if } x = 0 \text{ and } y = 0 \\ y^x, & \text{if } x > 0 \text{ or } y > 0 \end{cases}$
Weber	1983, [255]	$I_{\mathbf{WB}}(x,y) = \begin{cases} 1, & \text{if } x < 1 \\ y, & \text{if } x = 1 \end{cases}$
Fodor	1993, [102]	$I_{\mathbf{FD}}(x,y) = \begin{cases} 1, & \text{if } x \leq y \\ \max(1-x, y), & \text{if } x > y \end{cases}$

As examples, we will consider the first inequality in (1.1), the third inequality in (1.2) and the first inequality in (1.3). The other inequalities are the consequence of formulas of fuzzy implications and can be proven in a similar way.

Let us fix arbitrarily $x, y \in [0,1]$. Obviously, $1 - x \leq 1 - x + xy$. Moreover, $1-x+xy = xy+(1-x)1 \in [y,1]$ as a convex combination of y and 1. Therefore, $y \leq 1 - x + xy$ and $\max(1-x, y) \leq 1 - x + xy$, which proves the first inequality in (1.1).

Now, let $x, y \in [0,1]$ and assume that $x > y$. From the convex combination of y and x we have $xy + (1-x)x \in [y, x]$, so we see that $xy + x - x^2 \geq y$, which is equivalent to $1 - x + y \geq \frac{y}{x}$. This completes the proof of the third inequality in (1.2).

Finally, we consider the first inequality in (1.3). Since both the Reichenbach and the Yager implications satisfy (LB) and $I_{\mathbf{RC}}(1,y) = I_{\mathbf{YG}}(1,y) = y$, it is enough to show this inequality for $x \in (0,1)$ and $y \in [0,1]$. Therefore, let us fix arbitrarily $x \in (0,1)$ and consider the function $H_x(y) := y^x - 1 + x - xy$ for $y \in [0,1]$. Observe that $H_x(0) = x - 1 < 0$ and $H_x(1) = 0$. Since the derivative $(H_x)'(y) = x\left(y^{x-1} - 1\right)$, we get that $(H_x)'(y) > 0$ if and only if $y^{x-1} - 1 > 0$, but this is true for all $x, y \in (0,1)$. This implies that H_x is a continuous and increasing function such that $H_x(1) = 0$, so $H_x(y) \leq 0$. Therefore $y^x \leq 1-x+xy$ for all $x, y \in [0,1]$, i.e., we have proven the first inequality in (1.3).

The following result is easily established (see Kitainik [141], Baczyński and Drewniak [12], Dubois and Prade [94]).

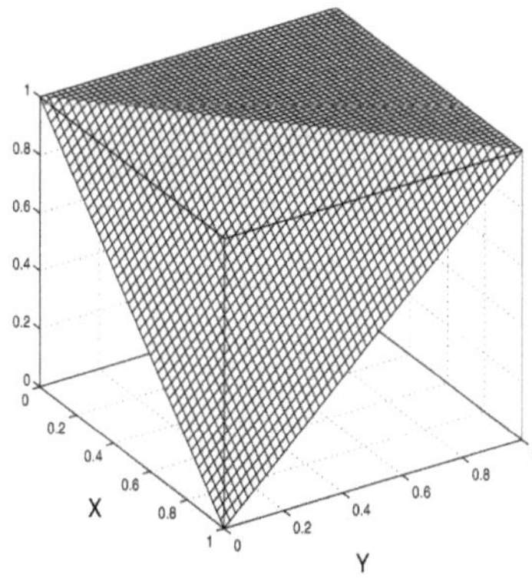

(a) Łukasiewicz implication $I_{\mathbf{LK}}$

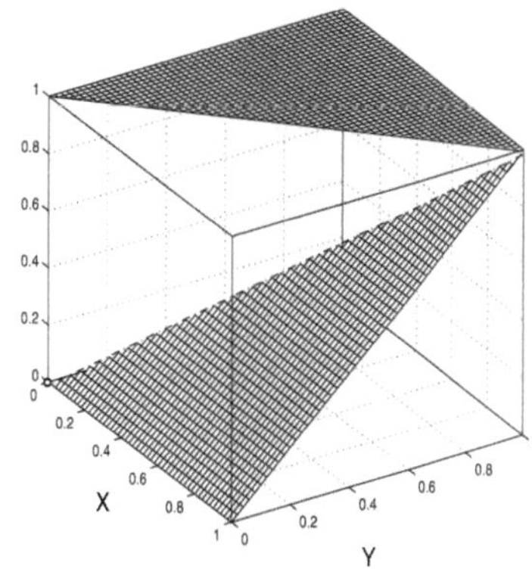

(b) Gödel implication $I_{\mathbf{GD}}$

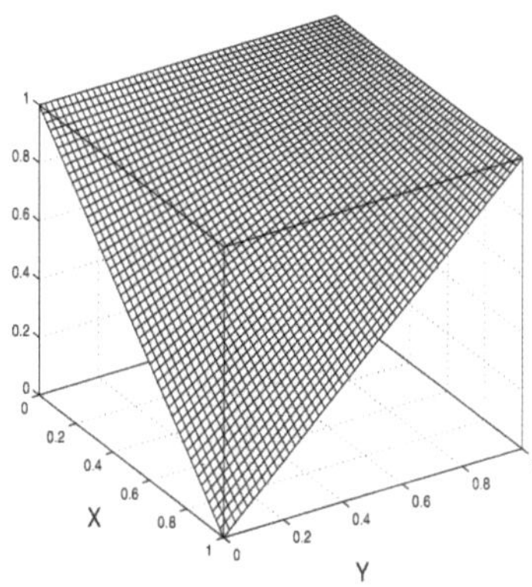

(c) Reichenbach implication $I_{\mathbf{RC}}$

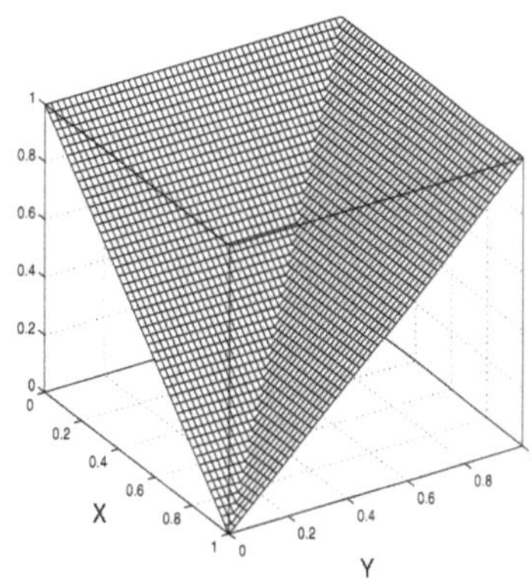

(d) Kleene-Dienes implication $I_{\mathbf{KD}}$

Fig. 1.1a. Plots of $I_{\mathbf{LK}}$, $I_{\mathbf{GD}}$, $I_{\mathbf{RC}}$ and $I_{\mathbf{KD}}$ fuzzy implications

Proposition 1.1.7. *The family $\mathcal{FI}$ has the least fuzzy implication*

$$I_0(x,y) = \begin{cases} 1, & \text{if } x = 0 \text{ or } y = 1\,, \\ 0, & \text{if } x > 0 \text{ and } y < 1\,, \end{cases} \qquad x, y \in [0,1]\,,$$

and the greatest fuzzy implication

$$I_1(x,y) = \begin{cases} 1, & \text{if } x < 1 \text{ or } y > 0\,, \\ 0, & \text{if } x = 1 \text{ and } y = 0\,, \end{cases} \qquad x, y \in [0,1]\,.$$

The next result illustrates a first method for generating new fuzzy implications from a given fuzzy implication. This method will be discussed in more detail in Part II (cf. p. XII).

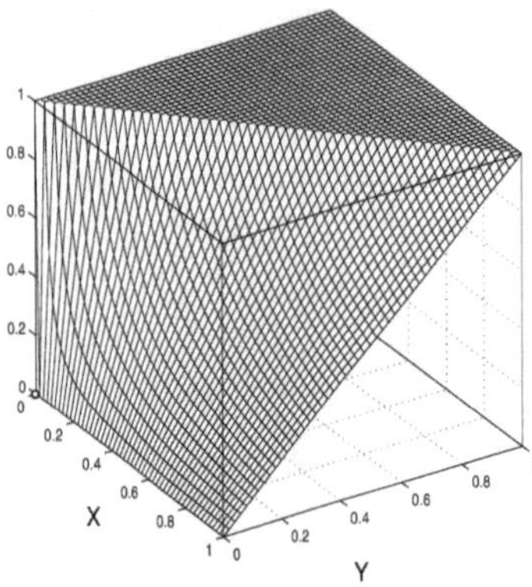

(a) Goguen implication $I_{\mathbf{GG}}$

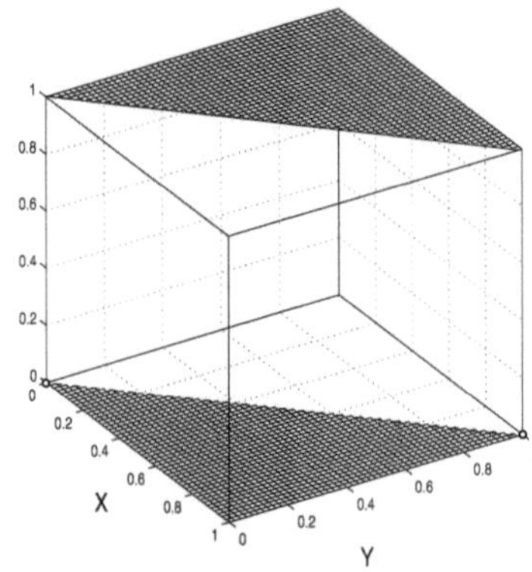

(b) Rescher implication $I_{\mathbf{RS}}$

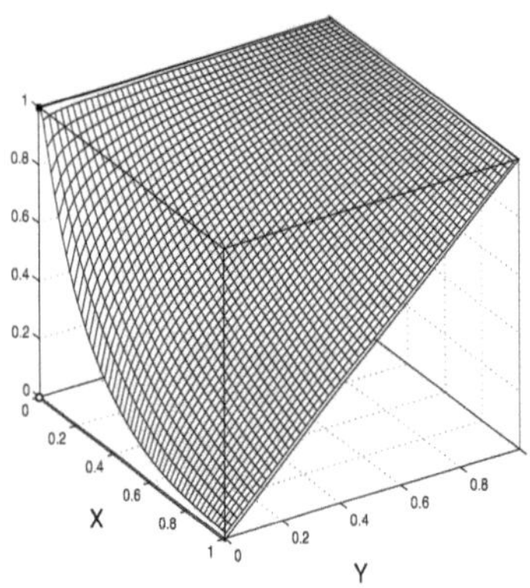

(c) Yager implication $I_{\mathbf{YG}}$

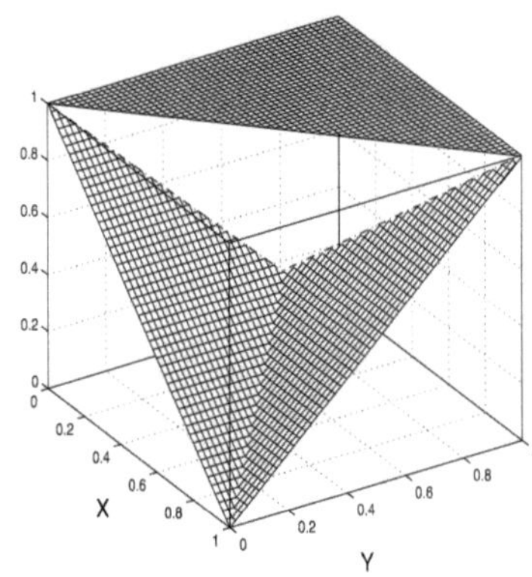

(d) Fodor implication $I_{\mathbf{FD}}$

Fig. 1.1b. Plots of $I_{\mathbf{GG}}$, $I_{\mathbf{RS}}$, $I_{\mathbf{YG}}$ and $I_{\mathbf{FD}}$ fuzzy implications

Proposition 1.1.8. *If $I \in \mathcal{FI}$ and $\varphi \in \Phi$, then $I_\varphi \in \mathcal{FI}$.*

Proof. Let $I \in \mathcal{FI}$, $\varphi \in \Phi$ and $x_1, x_2, y \in [0,1]$. If $x_1 \leq x_2$, then $\varphi(x_1) \leq \varphi(x_2)$, thus

$$I_\varphi(x_1, y) = \varphi^{-1}(I(\varphi(x_1), \varphi(y))) \geq \varphi^{-1}(I(\varphi(x_2), \varphi(y))) = I_\varphi(x_2, y) .$$

If $x, y_1, y_2 \in [0,1]$ and $y_1 \leq y_2$, then, similarly,

$$I_\varphi(x, y_1) = \varphi^{-1}(I(\varphi(x), \varphi(y_1))) \leq \varphi^{-1}(I(\varphi(x), \varphi(y_2))) = I_\varphi(x, y_2) .$$

Therefore, I_φ fulfils axioms (I1) and (I2). The other three conditions from Definition 1.1.1 can be easily verified, since $\varphi(0) = 0$ and $\varphi(1) = 1$. □

We will show in Sect. 6.3 that none of the 9 basic fuzzy implications from Example 1.1.4 is a Φ-conjugate of another.

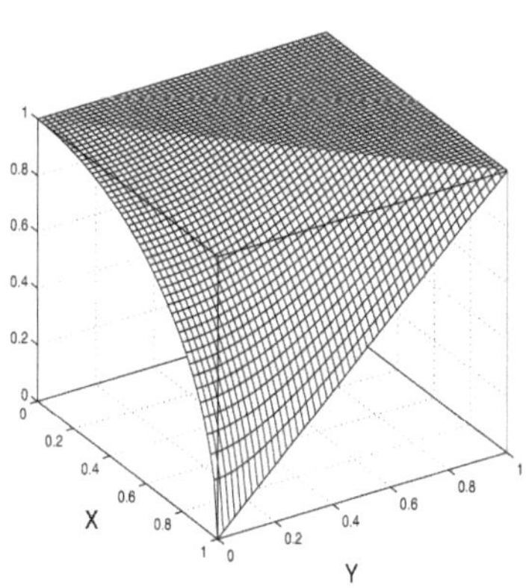

(a) $(I_{\mathbf{LK}})_\varphi$ with $\varphi(x) = x^2$

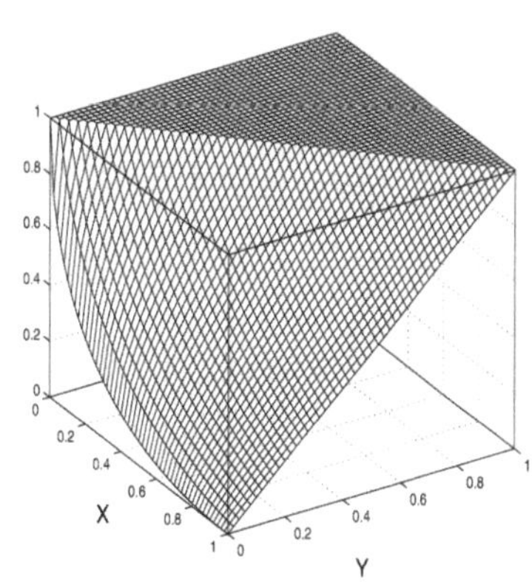

(b) $(I_{\mathbf{LK}})_\varphi$ with $\varphi(x) = \sqrt{x}$

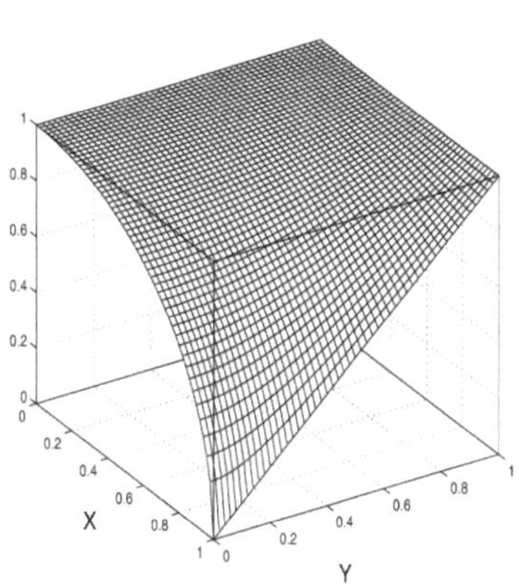

(c) $(I_{\mathbf{RC}})_\varphi$ with $\varphi(x) = x^2$

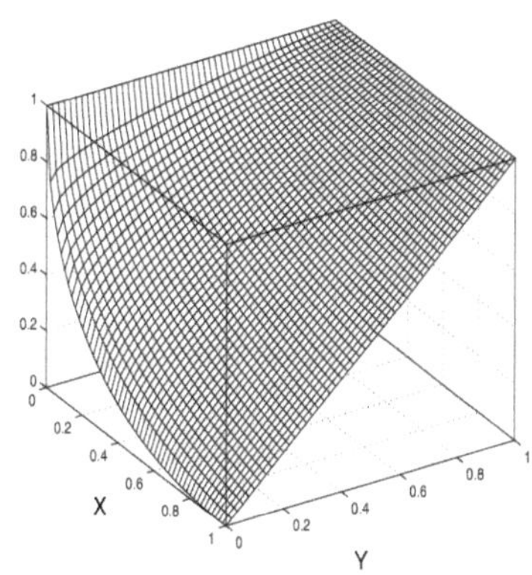

(d) $(I_{\mathbf{RC}})_\varphi$ with $\varphi(x) = \sqrt{x}$

Fig. 1.2. Plots of Φ-conjugates of $I_{\mathbf{LK}}$ and $I_{\mathbf{RC}}$ for different $\varphi \in \Phi$

Example 1.1.9. In Fig. 1.2 we present plots of some fuzzy implications which are Φ-conjugate with $I_{\mathbf{LK}}$ and $I_{\mathbf{RC}}$ for different $\varphi \in \Phi$.

1.2 Continuity of Fuzzy Implications

The continuity of fuzzy implications is an important property, since it ensures that small changes in the truth value of the antecedent and/or the consequent does not cause large changes in the overall truth value. It is natural that a fuzzy implication I, as a real function of two variables, is continuous at the point $(x_0, y_0) \in [0,1]^2$ if and only if for all convergent sequences $(x_n)_{n\in\mathbb{N}}$, $(y_n)_{n\in\mathbb{N}}$, where $x_n, y_n \in [0,1]$ for all $n \in \mathbb{N}$, we have

$$I(\lim_{n\to\infty} x_n, \lim_{n\to\infty} y_n) = \lim_{n\to\infty} I(x_n, y_n) .$$

It is also well-known that a real function of two variables can be continuous in each component without being continuous. As an example consider the following function

$$F(x,y) = \begin{cases} \frac{xy}{x^2+y^2}, & \text{if } x^2+y^2 \neq 0 , \\ 0, & \text{if } x=0 \text{ and } y=0 , \end{cases} \qquad x,y \in \mathbb{R} .$$

This function is not continuous at the point $(0,0)$, but it is continuous in each variable at this point. Interestingly, for monotone real functions we have the following very important result.

Theorem 1.2.1. *For a function $F\colon [0,1]^2 \to [0,1]$ which is monotonic with respect to one variable the following statements are equivalent:*

(i) F is continuous.
(ii) F is continuous in each variable.

For a proof see Appendix A, Theorem A.0.4

Since any fuzzy implication satisfies (I2), i.e., it is increasing with respect to the second variable, we get

Corollary 1.2.2. *For $I \in \mathcal{FI}$ the following statements are equivalent:*

(i) I is continuous.
(ii) I is continuous in each variable.

Example 1.2.3. Considering our 9 basic fuzzy implications from Example 1.1.4 it can be easily verified that $I_{\mathbf{LK}}$, $I_{\mathbf{RC}}$ and $I_{\mathbf{KD}}$ are continuous fuzzy implications, while $I_{\mathbf{GD}}$, $I_{\mathbf{GG}}$, $I_{\mathbf{RS}}$, $I_{\mathbf{YG}}$, $I_{\mathbf{FD}}$ and $I_{\mathbf{WB}}$ are not continuous. In fact $I_{\mathbf{GD}}$, $I_{\mathbf{GG}}$, $I_{\mathbf{RS}}$ and $I_{\mathbf{FD}}$ are left-continuous with respect to the first variable and right-continuous with respect to the second variable, $I_{\mathbf{YG}}$ is non-continuous at the point $(0,0)$, while $I_{\mathbf{WB}}$ is continuous in the second variable.

In the sequel the family of all continuous fuzzy implications will be denoted by $\mathcal{CFI}$. Since a composition of continuous functions is continuous, from Proposition 1.1.8 we get

Proposition 1.2.4. *If $I \in \mathcal{CFI}$ and $\varphi \in \Phi$, then $I_\varphi \in \mathcal{CFI}$.*

The analytical property related to continuity is the following Lipschitz property (see [146], Definition 1.25).

Definition 1.2.5. *A function $F\colon [0,1]^2 \to [0,1]$ is said to satisfy the* Lipschitz property, *if there exists a constant $c \in (0,\infty)$ such that*

$$|F(x_1,y_1) - F(x_2,y_2)| \leq c \cdot (|x_1 - x_2| + |y_1 - y_2|) , \tag{1.6}$$

for all $x_1, x_2, y_1, y_2 \in [0,1]$.

We only remark that, in general, the Lipschitz property of a real function F implies the uniform conntinuity and, therefore, the continuity of F. Thus any fuzzy implication I that satisfies the Lipschitz property (1.6) is also continuous.

1.3 Basic Properties of Fuzzy Implications

Additional properties of fuzzy implications were postulated in many works (see TRILLAS and VALVERDE [239], DUBOIS and PRADE [92], SMETS and MAGREZ [226], FODOR and ROUBENS [105], GOTTWALD [113]). The most important of them are presented below.

Definition 1.3.1. *A fuzzy implication I is said to satisfy*

(i) the left neutrality property *if*

$$I(1,y) = y\,, \qquad y \in [0,1]\,; \tag{NP}$$

(ii) the exchange principle, *if*

$$I(x, I(y,z)) = I(y, I(x,z))\,, \qquad x,y,z \in [0,1]\,; \tag{EP}$$

(iii) the identity principle, *if*

$$I(x,x) = 1\,, \qquad x \in [0,1]\,; \tag{IP}$$

(iv) the ordering property, *if*

$$I(x,y) = 1 \iff x \le y\,, \qquad x,y \in [0,1]\,. \tag{OP}$$

The left neutrality property (NP) captures the notion that a tautology allows the truth value of the consequent to be assigned as the overall truth value of the statement and can be seen as the generalization of the following law:

$$(1 \to p) \equiv p\,.$$

The exchange property (EP) is the generalization of the classical tautology known as the exchange principle:

$$p \to (q \to r) \equiv q \to (p \to r)\,.$$

The identity principle (IP) states that the overall truth value should be 1 when the truth values of the antecedent and the consequents are equal and can be seen as the generalization of the following tautology from the classical logic:

$$p \to p\,.$$

The ordering property (OP), called also the degree ranking property, imposes an ordering on the underlying set.

Example 1.3.2. Table 1.4 lists the 9 basic fuzzy implications from Example 1.1.4 along with the properties they satisfy from Definition 1.3.1.

The following results illustrate some interdependence between the axioms of fuzzy implications - that form part of the definition of a fuzzy implication - and the properties given in Definition 1.3.1. The proof of the first lemma can be easily verified.

Lemma 1.3.3. *Let $I\colon [0,1]^2 \to [0,1]$ be any function.*

(i) If I satisfies (LB), *then I satisfies* (I3) *and* (NC).
(ii) If I satisfies (RB), *then I satisfies* (I4) *and* (NC).
(iii) If I satisfies (NP), *then I satisfies* (I4) *and* (I5).
(iv) If I satisfies (IP), *then I satisfies* (I3) *and* (I4).
(v) If I satisfies (OP), *then I satisfies* (I3), (I4), (NC), (LB), (RB) *and* (IP).

The second lemma shows that the exchange principle (EP) and the ordering property (OP), together, are strong conditions. This result will be very important in the sequel.

Lemma 1.3.4. *If a function $I\colon [0,1]^2 \to [0,1]$ satisfies* (EP) *and* (OP), *then I satisfies* (I1), (I3), (I4), (I5), (LB), (RB), (NC), (NP) *and* (IP).

Proof. From Lemma 1.3.3 we know that if I satisfies (OP), then it satisfies (I3), (I4), (NC), (LB), (RB) and (IP).

Now we show that I satisfies (I1). If $x_1, x_2, y \in [0,1]$ and $x_1 \le x_2$, then because of (EP) and (IP) we get

$$I(x_2, I(I(x_2,y),y)) = I(I(x_2,y), I(x_2,y)) = 1\,,$$

thus, by (OP), we obtain $x_2 \le I(I(x_2,y),y)$. Since $x_1 \le x_2$ we have also $x_1 \le I(I(x_2,y),y)$. Therefore, (OP) and (EP) imply

$$1 = I(x_1, I(I(x_2,y),y)) = I(I(x_2,y), I(x_1,y))\,,$$

i.e., $I(x_1,y) \ge I(x_2,y)$ by (OP).

Finally, I satisfies (NP), since from (EP) and (IP) we have

$$I(x, I(1,x)) = I(1, I(x,x)) = I(1,1) = 1\,,$$

Table 1.4. Basic fuzzy implications and their main properties

Fuzzy implication	(NP)	(EP)	(IP)	(OP)	continuity
$I_{\mathbf{LK}}$	✓	✓	✓	✓	✓
$I_{\mathbf{GD}}$	✓	✓	✓	✓	×
$I_{\mathbf{RC}}$	✓	✓	×	×	✓
$I_{\mathbf{KD}}$	✓	✓	×	×	✓
$I_{\mathbf{GG}}$	✓	✓	✓	✓	×
$I_{\mathbf{RS}}$	×	×	✓	✓	×
$I_{\mathbf{YG}}$	✓	✓	×	×	×
$I_{\mathbf{WB}}$	✓	✓	✓	×	×
$I_{\mathbf{FD}}$	✓	✓	✓	✓	×

for all $x \in [0,1]$. Thus, because of (OP), we get $x \le I(1,x)$. On the other side,

$$I(1, I(I(1,x),x)) = I(I(1,x), I(1,x)) = 1 .$$

Again, by (OP), we get $I(I(1,x),x) = 1$, which proves that $I(1,x) \le x$. Therefore, $I(1,x) = x$ for all $x \in [0,1]$. (I5) is now obvious. □

Remark 1.3.5. The fact that the converse of the above lemma does not hold can be easily shown.

(i) $I_{\mathbf{WB}}$ is a fuzzy implication satisfying (NP), (IP) and (EP), but it does not satisfy (OP), while the function

$$I(x,y) = \begin{cases} 1, & \text{if } x \le y , \\ \max\left(\frac{y}{x}, \frac{1-x}{1-y}\right), & \text{otherwise} , \end{cases} \qquad x, y \in [0,1] ,$$

is a fuzzy implication (see Fodor [103]) satisfying (NP), (IP) and (OP), but does not satisfy (EP), because for $x = 0.9$, $y = 0.2$ and $z = 0.1$ we obtain $I(x, I(y,z)) = \frac{80}{81} \neq \frac{9}{10} = I(y, I(x,z))$.

(ii) In a recent article Smogura [227] showed that (EP) and (OP) do not imply the right isotonicity (I2). As an example we can consider the following function, defined for all $x, y \in [0,1]$,

$$F(x,y) = \begin{cases} 1, & \text{if } x \le y , \\ 0.75, & \text{if } x \in (0.25, 0.75] \text{ and } y = 0.25 , \\ y, & \text{otherwise} . \end{cases}$$

It is interesting to note that with the additional assumption that I is continuous, (EP) and (OP), in fact, imply (I2) (see Corollary 1.5.12).

(iii) In Table 1.5 we show that other dependencies between basic properties of fuzzy implications, without any other additional assumptions like continuity, are not possible. The formulas for some fuzzy implications used in this example are the following:

$$I_{\mathbf{SQ}}(x,y) = \begin{cases} 1, & \text{if } (x,y) \in \{(0,0),(1,1)\} , \\ \max(y^x, (1-x)^{\sqrt{1-y}}), & \text{otherwise} , \end{cases}$$

$$I_{\mathbf{MN}}(x,y) = \begin{cases} \min(1-x, y), & \text{if } \max(1-x,y) \le 0.5 , \\ \max(1-x, y), & \text{otherwise} , \end{cases}$$

$$I_{\mathbf{NI}}(x,y) = \begin{cases} \min(1, 1-x^2+y), & \text{if } y > 0 , \\ 1, & \text{if } x \in [0, 0.25) \text{ and } y = 0 , \\ 0.1, & \text{if } x \in [0.25, 0.75) \text{ and } y = 0 , \\ 0, & \text{otherwise} , \end{cases}$$

$$I_{\mathbf{BZ}}(x,y) = \min\left(\max\left(0.5, \min(1-x+y, 1)\right), 2-2x+2y\right) ,$$

$$I_{\mathbf{MM}}(x,y) = \min\left(1, \max\left(\frac{y}{x}, \frac{1-x}{1-y}\right)\right) ,$$

Table 1.5. Examples of fuzzy implications for Remark 1.3.5(iii)

Fuzzy implication	(NP)	(EP)	(IP)	(OP)
$I(x,y) = \begin{cases} 0, & \text{if } (x,y) \in [0.7,1] \times [0,0.6] \\ 0.5, & \text{if } (x,y) \in [0.4,0.7) \times [0,0.6] \\ 1, & \text{otherwise} \end{cases}$	×	×	×	×
$I_{\mathbf{SQ}}$	✓	×	×	×
$I_{\mathbf{MN}}$	×	✓	×	×
$I(x,y) = \begin{cases} 1, & \text{if } x < 1 \text{ and } y > 0 \\ \sqrt{y}, & \text{if } x = 1 \text{ and } y > 0 \\ 1 - x^2, & \text{otherwise} \end{cases}$	×	×	✓	×
$I_{\mathbf{RC}}$	✓	✓	×	×
$I_{\mathbf{NI}}$	✓	×	✓	×
$I(x,y) = \begin{cases} 0, & \text{if } x > 0 \text{ and } y = 0 \\ 1, & \text{otherwise} \end{cases}$	×	✓	✓	×
$I_{\mathbf{BZ}}$	×	×	✓	✓
$I_{\mathbf{WB}}$	✓	✓	✓	×
$I_{\mathbf{MM}}$	✓	×	✓	✓
$I_{\mathbf{LK}}$	✓	✓	✓	✓

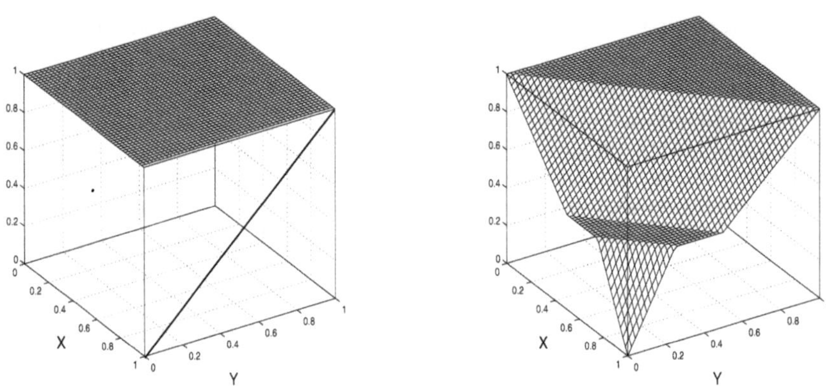

(a) Weber implication $I_{\mathbf{WB}}$ (b) Baczyński implication $I_{\mathbf{BZ}}$

Fig. 1.3. Plots of $I_{\mathbf{WB}}$ and $I_{\mathbf{BZ}}$ fuzzy implications

where $x, y \in [0,1]$. Plots of the Weber implication $I_{\mathbf{WB}}$ and the Baczyński implication $I_{\mathbf{BZ}}$ (see BACZYŃSKI [9]) are given in Fig. 1.3, while the plot of $I_{\mathbf{MN}}$ is given in Sect. 5.3 in Fig. 5.1(c).

We can easily prove that the Φ-conjugation preserves all the properties given in Definition 1.3.1.

Proposition 1.3.6. *Let $\varphi \in \Phi$. If a function $I\colon [0,1]^2 \to [0,1]$ satisfies* (NP) *(*(EP), (IP), (OP)*), then I_φ also satisfies* (NP) *(*(EP), (IP), (OP)*).*

Proof. Assume that $\varphi \in \Phi$. If I satisfies (NP), then for all $y \in [0,1]$ we have

$$I_\varphi(1,y) = \varphi^{-1}(I(\varphi(1),\varphi(y))) = \varphi^{-1}(I(1,\varphi(y))) = \varphi^{-1}(\varphi(y)) = y\,.$$

If I satisfies (EP), then for any $x, y, z \in [0,1]$ we get

$$\begin{aligned} I_\varphi(x, I_\varphi(y,z)) &= \varphi^{-1}(I(\varphi(x), \varphi(\varphi^{-1}(I(\varphi(y),\varphi(z)))))) \\ &= \varphi^{-1}(I(\varphi(x), I(\varphi(y),\varphi(z)))) \\ &= \varphi^{-1}(I(\varphi(y), I(\varphi(x),\varphi(z)))) \\ &= \varphi^{-1}(I(\varphi(y), \varphi(\varphi^{-1}(I(\varphi(x),\varphi(z)))))) \\ &= I_\varphi(y, I_\varphi(x,z))\,. \end{aligned}$$

If I satisfies (IP), then for all $x \in [0,1]$ we have

$$I_\varphi(x,x) = \varphi^{-1}(I(\varphi(x),\varphi(x))) = \varphi^{-1}(1) = 1\,.$$

If I satisfies (OP), then for any $x, y \in [0,1]$ the following equivalences are true:

$$\begin{aligned} I_\varphi(x,y) = 1 &\iff \varphi^{-1}(I(\varphi(x),\varphi(y))) = 1 \iff I(\varphi(x),\varphi(y)) = \varphi(1) \\ &\iff I(\varphi(x),\varphi(y)) = 1 \iff \varphi(x) \le \varphi(y) \iff x \le y\,. \end{aligned}$$

□

1.4 Negations from Fuzzy Implications

A fuzzy negation N is a generalization of the classical complement or negation $\neg$, whose truth table consists of the two conditions: $\neg 0 = 1$ and $\neg 1 = 0$. In this section, we first give in brief the definitions, some examples and characterizations of fuzzy negations, before we see how to obtain negations from fuzzy implications.

1.4.1 Fuzzy Negations

The following definitions and basic results, with proofs, can be found in any introductory text book on fuzzy logic (see, for example, Fodor and Roubens [105], Klir and Yuan [147], Nguyen and Walker [188]).

Definition 1.4.1. *A function $N\colon [0,1] \to [0,1]$ is called a* fuzzy negation *if*

$$N(0) = 1\,, \qquad N(1) = 0\,, \tag{N1}$$

$$N \text{ is decreasing.} \tag{N2}$$

The property (N1) is a generalization of the boundary conditions of the classical negation, while (N2) captures the notion that as the belongingness of an element to one set increases, its belongingness to the complement of the set decreases. Since the class of functions satisfying conditions in Definition 1.4.1 is wide, in the literature other axioms have been introduced.

Definition 1.4.2. *(i) A fuzzy negation N is called* strict *if, in addition,*

$$N \text{ is strictly decreasing,} \tag{N3}$$

$$N \text{ is continuous.} \tag{N4}$$

(ii) A fuzzy negation N is called strong *if it is an involution, i.e.,*

$$N(N(x)) = x\,, \qquad x \in [0,1]\,. \tag{N5}$$

(iii) A fuzzy negation N is said to be non-vanishing if $N(x) = 0 \Longleftrightarrow x = 1$.
(iv) A fuzzy negation N is said to be non-filling if $N(x) = 1 \Longleftrightarrow x = 0$.

Note that (N5) is a generalization of the classical law of double negation:

$$\neg(\neg p) \equiv p\,. \tag{1.7}$$

In the family of all fuzzy negations we can consider the partial order induced from the unit interval $[0,1]$.

Theorem 1.4.3. *(i) The family of all fuzzy negations is a complete, completely distributive lattice with lattice operations*

$$(N_1 \vee N_2)(x) := \max(N_1(x), N_2(x))\,, \qquad x \in [0,1]\,,$$
$$(N_1 \wedge N_2)(x) := \min(N_1(x), N_2(x))\,, \qquad x \in [0,1]\,.$$

(ii) The family of all strict negations is a distributive lattice.
(iii) The family of all strong negations is a distributive lattice.

For a proof see Appendix A, Theorem A.0.5.

Example 1.4.4. Table 1.6 lists a few fuzzy negations with properties they satisfy. Parameters t, λ, w take values in their respective domains and thus define parametric families of negations, viz., N^t, N^λ, N^w. The classical (standard) fuzzy complement

$$N_{\mathbf{C}}(x) = 1 - x\,, \qquad x \in [0,1]\,,$$

can be obtained by setting $\lambda = 0$ in N^λ or $w = 1$ in N^w. Considering the threshold class of negations N^t, in the limiting cases, namely, when $t = 0$ we obtain the least fuzzy negation

$$N_{\mathbf{D1}}(x) = \begin{cases} 1, & \text{if } x = 0\,, \\ 0, & \text{if } x \in (0,1]\,, \end{cases}$$

while for $t = 1$ we obtain the greatest fuzzy negation

$$N_{\mathbf{D2}}(x) = \begin{cases} 0, & \text{if } x = 1 , \\ 1, & \text{if } x \in [0,1) . \end{cases}$$

These negations are also called the intuitionistic negation (or the Gödel negation) and the dual intuitionistic negation (or the dual Gödel negation), respectively (cf. [105], pp. 2-4 and [146], pp. 231-232). Observe that there does not exist any least or greatest strong (continuous) fuzzy negation.

Theorem 1.4.5. *If a function $N\colon [0,1] \to [0,1]$ satisfies* (N2) *and* (N5)*, then it also satisfies* (N1) *and* (N4)*. Moreover, N is a bijection, i.e., it satisfies* (N3).

Corollary 1.4.6. *Every strong negation is strict.*

Theorem 1.4.7. *Every continuous fuzzy negation N has a unique fixed point, i.e., there exists an $e \in (0,1)$ such that $N(e) = e$.*

Similar to Propositions 1.1.8 and 1.3.6, the following fact can be easily shown.

Proposition 1.4.8. *If $\varphi \in \Phi$ and N is a fuzzy (strict, strong) negation, then N_φ is also a fuzzy (strict, strong) negation.*

The next two facts will be used later in the book. For a proof of Lemma 1.4.10 see Appendix A, Lemma A.0.6

Lemma 1.4.9. *If N_1, N_2 are fuzzy negations such that $N_1 \circ N_2 = \mathrm{id}_{[0,1]}$, then*

(i) N_1 is a continuous fuzzy negation,

Table 1.6. Examples of fuzzy negations and their properties

Name	Formula	Properties
threshold class	$N^t(x) = \begin{cases} 1, & \text{if } x < t \\ 1 \text{ or } 0, & \text{if } x = t \\ 0, & \text{if } x > t \end{cases}, \quad t \in (0,1)$	(N1), (N2)
—	$N(x) = \begin{cases} 1-x, & \text{if } x \in [0,0.5) \\ 0.8(1-x), & \text{if } x \in [0.5,1] \end{cases}$	(N1), (N2), (N3)
—	$N(x) = \begin{cases} 1-x, & \text{if } x \in [0,0.5) \\ 0.5, & \text{if } x \in [0.5,0.8] \\ 2.5(1-x), & \text{if } x \in [0.8,1] \end{cases}$	(N1), (N2), (N4)
—	$N_{\mathbf{K}}(x) = 1 - x^2$	(N1) – (N4), strict
—	$N_{\mathbf{R}}(x) = 1 - \sqrt{x}$	(N1) – (N4), strict
Sugeno class	$N^\lambda(x) = \dfrac{1-x}{1+\lambda x}, \quad \lambda \in (-1,\infty)$	(N1) – (N5), strong
Yager class	$N^w(x) = (1-x^w)^{\frac{1}{w}}, \quad w \in (0,\infty)$	(N1) – (N5), strong

(ii) N_2 *is a strictly decreasing fuzzy negation,*
(iii) N_2 *is a continuous fuzzy negation if and only if* N_1 *is a strictly decreasing fuzzy negation. In both cases* $N_1 = N_2^{-1}$.

Proof. (i) It is obvious that N_1 is onto $[0,1]$. Additionally, since N_1 is decreasing it is continuous. Indeed, let us assume that N_1 is not continuous at some point $x_0 \in [0,1]$. From the monotonicity of N_1 there exist constants

$$(N_1)_-(x_0) = \begin{cases} \lim\limits_{x \to x_0^-} N_1(x), & \text{if } x_0 > 0\,, \\ 1, & \text{if } x_0 = 0\,, \end{cases}$$

$$(N_1)_+(x_0) = \begin{cases} \lim\limits_{x \to x_0^+} N_1(x), & \text{if } x_0 < 1\,, \\ 0, & \text{if } x_0 = 1\,. \end{cases}$$

From the discontinuity we have $(N_1)_-(x_0) < (N_1)_+(x_0)$, so N_1 does not take values between $(N_1)_-(x_0)$ and $(N_1)_+(x_0)$, a contradiction to the fact that N_1 is onto $[0,1]$.

(ii) Let us assume that $N_2(x) = N_2(y)$ for some $x, y \in [0,1]$. Given that $(N_1 \circ N_2)(x) = x$ for all $x \in [0,1]$, we get $x = N_1(N_2(x)) = N_1(N_2(y)) = y$, so N_2 is one-to-one and, consequently, a strictly decreasing fuzzy negation.

(iii) Let us assume, firstly, that N_2 is a continuous fuzzy negation. From point (ii) above we know that N_2 is one-to-one, so N_2 is a decreasing bijection on $[0,1]$, i.e., it is a strict negation. Hence N_2^{-1} is also a strict negation and $N_1 = N_2^{-1}$. In particular, N_1 is a strict negation.
Let us assume now that N_1 is a strictly decreasing negation. From point (i) above we know that N_1 is continuous, so N_1 is a strict negation. Hence N_1^{-1} is also a strict negation and $N_2 = N_1^{-1}$. In particular, N_2 is strict. □

Lemma 1.4.10. *If* N *is a continuous fuzzy negation, then the function* $\mathfrak{N}\colon [0,1] \to [0,1]$ *defined by*

$$\mathfrak{N}(x) = \begin{cases} N^{(-1)}(x), & \text{if } x \in (0,1]\,, \\ 1, & \text{if } x = 0\,, \end{cases} \tag{1.8}$$

is a strictly decreasing fuzzy negation. Moreover

$$\mathfrak{N}^{(-1)} = N\,, \tag{1.9}$$

$$N \circ \mathfrak{N} = \mathrm{id}_{[0,1]}\,, \tag{1.10}$$

$$\mathfrak{N} \circ N\,|_{\mathrm{Ran}(\mathfrak{N})} = \mathrm{id}_{\mathrm{Ran}(\mathfrak{N})}\,. \tag{1.11}$$

Remark 1.4.11. Note that, even if N is a continuous negation, the pseudo-inverse $N^{(-1)}$ need not be a fuzzy negation. For instance, for the following continuous fuzzy negation

$$N(x) = \begin{cases} -2x + 1, & \text{if } x \in [0, 0.5]\,, \\ 0, & \text{if } x \in (0.5, 1]\,, \end{cases}$$

the $N^{(-1)}(x) = -0.5x + 1$ is a strictly decreasing function, but not a fuzzy negation.

In the rest of this section we present theorems related to the representation of strict and strong negations, which will be employed in the proofs of theorems in the sequel. The first theorem was originally presented by FODOR [101]. The second theorem was firstly obtained by TRILLAS [232] (for the proofs see also KLIR and YUAN [147] or GOTTWALD [113]).

Theorem 1.4.12. *For a function $N\colon [0,1] \to [0,1]$ the following statements are equivalent:*

(i) N is a strict negation.
(ii) There exist $\varphi, \psi \in \Phi$ such that

$$N(x) = \psi^{-1}(1 - \varphi(x)) , \qquad x \in [0,1] . \tag{1.12}$$

Theorem 1.4.13. *For a function $N\colon [0,1] \to [0,1]$ the following statements are equivalent:*

(i) N is a strong negation.
(ii) N is Φ-conjugate with the classical negation $N_{\mathbf{C}}$, i.e., there exists $\varphi \in \Phi$ such that

$$N(x) = (N_{\mathbf{C}})_{\varphi}(x) = \varphi^{-1}(1 - \varphi(x)) , \qquad x \in [0,1] .$$

(iii) There exists a strictly increasing, continuous function $g\colon [0,1] \to [0,\infty)$ such that $g(0) = 0$ and

$$N(x) = g^{-1}(g(1) - g(x)) , \qquad x \in [0,1] .$$

(iv) There exists a strictly decreasing, continuous function $f\colon [0,1] \to [0,\infty)$ such that $f(1) = 0$ and

$$N(x) = f^{-1}(f(0) - f(x)) , \qquad x \in [0,1] .$$

It should be pointed out that the increasing bijection φ of the unit interval in Theorem 1.4.13(ii) is not determined uniquely (see [105], Proposition 1.1).

1.4.2 Natural Negations of Fuzzy Implications

From the truth table for the classical implication operator (see Table 1.1) we see that whenever the truth value of the consequent is 'False', the overall truth value of the statement assumes the value that is opposite of the truth value of the antecedent. This gives us a cue to define a negation from a boolean implication. In this section, we discuss the natural negations of fuzzy implications and the different properties related to them.

Lemma 1.4.14. *If a function* $I\colon [0,1]^2 \to [0,1]$ *satisfies* (I1), (I3) *and* (I5), *then the function* $N_I\colon [0,1] \to [0,1]$ *defined by*

$$N_I(x) := I(x,0)\ , \qquad x \in [0,1]\ , \tag{1.13}$$

is a fuzzy negation.

Proof. By (I3) we get $N_I(0) = I(0,0) = 1$ and from (I5) we have $N_I(1) = I(1,0) = 0$. Moreover, (I1) implies that N_I is decreasing. □

This fact implies that the following terminology is valid.

Definition 1.4.15. *Let* $I \in \mathcal{FI}$. *The function* N_I *defined by* (1.13) *is called the* natural negation *of* I *or the* negation induced *by* I.

Example 1.4.16. In Table 1.7 we list the natural negations of 9 basic fuzzy implications from Example 1.1.4. Note that, in general, N_I is neither strong, strict nor even continuous.

Proposition 1.4.17. *If a function* $I\colon [0,1]^2 \to [0,1]$ *satisfies* (EP) *and* (OP), *then*

(i) N_I *is a fuzzy negation,*
(ii) $x \le N_I(N_I(x))$ *for all* $x \in [0,1]$,
(iii) $N_I \circ N_I \circ N_I = N_I$.

Proof. (i) From Lemmas 1.3.4 and 1.4.14 we easily deduce that N_I is a fuzzy negation.
(ii) Lemma 1.3.4 implies that I satisfies (IP). Now, also using (EP), we have

$$I(x, I(I(x,0),0)) = I(I(x,0), I(x,0)) = 1\ , \qquad x \in [0,1]\ ,$$

hence, by (OP), we get $x \le I(I(x,0),0) = N_I(N_I(x))$ for all $x \in [0,1]$.

Table 1.7. Basic fuzzy implications and their natural negations

Fuzzy implication I	Natural negation N_I
$I_{\mathbf{LK}}$	$N_{\mathbf{C}}$
$I_{\mathbf{GD}}$	$N_{\mathbf{D1}}$
$I_{\mathbf{RC}}$	$N_{\mathbf{C}}$
$I_{\mathbf{KD}}$	$N_{\mathbf{C}}$
$I_{\mathbf{GG}}$	$N_{\mathbf{D1}}$
$I_{\mathbf{RS}}$	$N_{\mathbf{D1}}$
$I_{\mathbf{YG}}$	$N_{\mathbf{D1}}$
$I_{\mathbf{WB}}$	$N_{\mathbf{D2}}$
$I_{\mathbf{FD}}$	$N_{\mathbf{C}}$

(iii) Since the negation N_I is decreasing, from point (ii) above we get

$$N_I(N_I(N_I(x))) \leq N_I(x) , \qquad x \in [0,1] . \tag{1.14}$$

From the other side, for all $x \in [0,1]$ we have

$$I(I(x,0), I(I(I(x,0),0),0)) = I(I(I(x,0),0), I(I(x,0),0)) = 1 ,$$

by (EP) and (IP). Thus, by (OP), $I(x,0) \leq I(I(I(x,0),0),0)$, i.e.,

$$N_I(x) \leq N_I(N_I(N_I(x))) , \qquad x \in [0,1] . \tag{1.15}$$

From the inequalities (1.14) and (1.15) we obtain the claim. □

Proposition 1.4.18. *If a function* $I\colon [0,1]^2 \to [0,1]$ *satisfies* (EP) *and* (OP), *then the following statements are equivalent:*

(i) N_I *is a continuous function.*
(ii) N_I *is a strong negation.*

Proof. Firstly, observe that from the previous result N_I is a fuzzy negation.
(i) $\Longrightarrow$ *(ii)* If N_I is a continuous negation, then it satisfies the Darboux property (see Appendix A), so for any $x \in [0,1]$ there exists $y \in [0,1]$ such that $N_I(y) = x$. Thus,

$$N_I(N_I(x)) = N_I(N_I(N_I(y))) = N_I(y) = x , \qquad x \in [0,1] ,$$

from Proposition 1.4.17(iii), so N_I is strong.
(ii) $\Longrightarrow$ *(i)* If N_I is a strong negation, then by virtue of Theorem 1.4.5 it is always continuous. □

Corollary 1.4.19. *If a function* $I\colon [0,1]^2 \to [0,1]$ *satisfies* (EP) *and* (OP), *then* N_I *is either a strong negation or a discontinuous negation.*

Example 1.4.20. Note that both the conditions in Proposition 1.4.18 are important.

(i) Consider the Baczyński implication $I_{\mathbf{BZ}}$ from Remark 1.3.5(iii). It satisfies (OP) but not (EP) and its natural negation given by

$$N_{I_{\mathbf{BZ}}}(x) = \begin{cases} 1-x, & \text{if } 0 \leq x \leq 0.5 , \\ 0.5, & \text{if } 0.5 \leq x \leq 0.75 , \\ 2(1-x), & \text{if } 0.75 \leq x \leq 1 , \end{cases}$$

although continuous, is not strong.
(ii) Consider the following function

$$I(x,y) = \max\left(\sqrt{1-x}, y\right) , \qquad x,y \in [0,1] .$$

It is a fuzzy implication (cf. Sect. 2.4), which satisfies (EP) but not (OP) and its natural negation is strict but not strong.

Finally, we have

Theorem 1.4.21. *If* $I \in \mathcal{FI}$ *and* $\varphi \in \Phi$, *then* $(N_I)_\varphi = N_{I_\varphi}$.

Proof. By Proposition 1.1.8, we know that I_φ is a fuzzy implication, thus N_{I_φ} is a fuzzy negation. Now, for every $x \in [0,1]$, we get

$$\begin{aligned}(N_I)_\varphi(x) &= \varphi^{-1}(N_I(\varphi(x))) = \varphi^{-1}(I(\varphi(x),0)) = \varphi^{-1}(I(\varphi(x),\varphi(0))) \\ &= I_\varphi(x,0) = N_{I_\varphi}(x) \,.\end{aligned}$$

□

1.5 Laws of Contraposition

One of the most important tautologies in the classical two-valued logic is the law of contraposition:

$$p \to q \equiv \neg q \to \neg p \,.$$

Its natural generalization to fuzzy logic is based on fuzzy negations and fuzzy implications. In fuzzy logic, contrapositive symmetry of a fuzzy implication I with respect to a fuzzy negation N plays an important role in the applications of fuzzy implications, for instance, *approximate reasoning, deductive systems, decision support systems, formal methods of proof.* (cf. [103] and [133]). Since the classical negation satisfies the law of double negation (1.7), the following laws are also tautologies in the classical logic

$$\begin{aligned}\neg p \to q &\equiv \neg q \to p \,, \\ p \to \neg q &\equiv q \to \neg p \,.\end{aligned}$$

Consequently, we can consider different laws of contraposition in fuzzy logic.

Definition 1.5.1. *Let* $I \in \mathcal{FI}$ *and* N *be a fuzzy negation.* I *is said to satisfy the*

(i) law of contraposition *(or in other words, the* contrapositive symmetry*) with respect to* N, *if*

$$I(x,y) = I(N(y),N(x)) \,, \qquad x,y \in [0,1] \,. \tag{CP}$$

(ii) law of left contraposition *with respect to* N, *if*

$$I(N(x),y) = I(N(y),x) \,, \qquad x,y \in [0,1] \,. \tag{L-CP}$$

(iii) law of right contraposition *with respect to* N, *if*

$$I(x,N(y)) = I(y,N(x)) \,, \qquad x,y \in [0,1] \,. \tag{R-CP}$$

If I *satisfies the (left, right) contrapositive symmetry with respect to* N, *then we also denote this by CP(*N*) (respectively, by L-CP(*N*), R-CP(*N*)).*

Firstly, we can easily observe that all the three properties are equivalent in some cases.

Proposition 1.5.2. *If $I\colon [0,1]^2 \to [0,1]$ is any function and N is a strict negation, then the following statements are equivalent:*

(i) I satisfies (L-CP) *with respect to N.*
(ii) I satisfies (R-CP) *with respect to N^{-1}.*

Proof. (*i*) $\Longrightarrow$ (*ii*) If I satisfies the law of left contraposition with respect to a strict negation N, then for all $x, y \in [0,1]$ we have

$$\begin{aligned} I(x, N^{-1}(y)) &= I((N \circ N^{-1})(x), N^{-1}(y)) \\ &= I((N \circ N^{-1})(y), N^{-1}(x)) = I(y, N^{-1}(x)) . \end{aligned}$$

(*ii*) $\Longrightarrow$ (*i*) If I satisfies the law of right contraposition with respect to a strict negation N^{-1}, then for all $x, y \in [0,1]$ we have

$$\begin{aligned} I(N(x), y) &= I(N(x), (N^{-1} \circ N)(y)) \\ &= I(N(y), (N^{-1} \circ N)(x)) = I(N(y), x) . \end{aligned}$$ □

Proposition 1.5.3. *If $I\colon [0,1]^2 \to [0,1]$ is any function and N is a strong negation, then the following statements are equivalent:*

(i) I satisfies (CP) *with respect to N.*
(ii) I satisfies (L-CP) *with respect to N.*
(iii) I satisfies (R-CP) *with respect to N.*

Proof. (*i*) $\Longrightarrow$ (*ii*) Since I satisfies CP(N) we get

$$I(N(x), y) = I(N(y), (N \circ N)(x)) = I(N(y), x) , \qquad x, y \in [0,1] .$$

(*ii*) $\Longrightarrow$ (*iii*) If I satisfies L-CP(N), then for all $x, y \in [0,1]$ we have

$$I(x, N(y)) = I((N \circ N)(x), N(y)) = I((N \circ N)(y), N(x)) = I(y, N(x)) .$$

(*iii*) $\Longrightarrow$ (*i*) If I satisfies R-CP(N), then

$$I(x, y) = I(x, (N \circ N)(y)) = I(N(y), N(x)) , \qquad x, y \in [0,1] .$$ □

If a negation N is not strong, then these laws may not be equivalent. Therefore, in the rest of this section, we discuss the relationships between the above laws and other properties of fuzzy implications. Like in the previous sections we often assume that I is any function defined on $[0,1]^2$. We start with the classical contrapositive symmetry.

Lemma 1.5.4. *Let $I\colon [0,1]^2 \to [0,1]$ be any function and N a fuzzy negation.*

(i) If I satisfies (I1) *and CP(N), then I satisfies* (I2).
(ii) If I satisfies (I2) *and CP(N), then I satisfies* (I1).

(iii) If I satisfies (LB) *and CP(N), then I satisfies* (RB).
(iv) If I satisfies (RB) *and CP(N), then I satisfies* (LB).
(v) If I satisfies (NP) *and CP(N), then I satisfies* (I3), (I4), (I5) *and $N = N_I$ is a strong negation.*

Proof. Let $x, y, z \in [0,1]$.

(i) If $y \leq z$, then $N(z) \leq N(y)$, so $I(N(y), N(x)) \leq I(N(z), N(x))$ by (I1). However, from (CP) we get

$$I(x,y) = I(N(y), N(x)) \leq I(N(z), N(x)) = I(x,z) ,$$

thus I satisfies (I2).

(ii) If $x \leq y$, then $N(y) \leq N(x)$, so $I(N(z), N(y)) \leq I(N(z), N(x))$ by (I2). However, from (CP) we get

$$I(x,z) = I(N(z), N(x)) \geq I(N(z), N(y)) = I(y,z) ,$$

thus I satisfies (I1).

(iii) From (CP) we have $I(x,1) = I(0, N(x))$, which is equal, by (LB), to 1.

(iv) From (CP) we have $I(0,y) = I(N(y), 1)$, which is equal, by (RB), to 1.

(v) Since I satisfies (NP), by virtue of Lemma 1.3.3, I satisfies (I4) and (I5). Moreover, since I satisfies CP(N), we get

$$N_I(x) = I(x,0) = I(N(0), N(x)) = I(1, N(x)) = N(x) ,$$

for every $x \in [0,1]$. Further, for any $y \in [0,1]$ we have

$$y = I(1,y) = I(N_I(y), N_I(1)) = I(N_I(y), 0) = N_I(N_I(y)) ,$$

so N_I must be a strong negation. Finally, $I(0,0) = I(1,1) = 1$, so I satisfies (I3). □

Immediately, from the last point we get

Corollary 1.5.5. *Let $I\colon [0,1]^2 \to [0,1]$ be a function that satisfies* (NP). *If N_I is not a strong negation, then I does not satisfy* (CP) *with any fuzzy negation.*

Lemma 1.5.6. *Let $I\colon [0,1]^2 \to [0,1]$ be any function and N_I be a strong negation.*

(i) If I satisfies CP(N_I), then I satisfies (NP).
(ii) If I satisfies (EP), *then I satisfies* (I3), (I4), (I5), (NP) *and* (CP) *only with respect to N_I.*

Proof. (i) For every $y \in [0,1]$ we get

$$I(1,y) = I(N_I(y), N_I(1)) = I(N_I(y), 0) = N_I(N_I(y)) = y .$$

(ii) I satisfies the contrapositive symmetry with N_I, since

$$\begin{aligned} I(N_I(y), N_I(x)) &= I(N_I(y), I(x,0)) = I(x, I(N_I(y),0)) \\ &= I(x, N_I(N_I(y))) = I(x,y) , \end{aligned}$$

for all $x, y \in [0,1]$. The other parts follow from the previous point and Lemma 1.5.4(v). □

From the above facts the following corollaries can be easily proven.

Corollary 1.5.7. *Let a function $I\colon [0,1]^2 \to [0,1]$ satisfy* (CP) *with respect to a fuzzy negation N. Then*

(i) (I1) *holds if and only if* (I2) *holds.*
(ii) (LB) *holds if and only if* (RB) *holds.*
(iii) If N is strong, then (NP) *holds if and only if $N = N_I$.*

Corollary 1.5.8. *If a function $I\colon [0,1]^2 \to [0,1]$ satisfies* (I1), (NP) *and* (CP) *with respect to a fuzzy negation N, then $I \in \mathcal{FI}$ and $N = N_I$ is a strong negation.*

Corollary 1.5.9. *If $I \in \mathcal{FI}$ satisfies* (NP) *and* (EP), *then the following statements are equivalent:*

(i) N_I is a strong negation.
(ii) I satisfies (CP) *with respect to N_I.*

On the one hand, if a fuzzy implication I does not satisfy (NP), it still can satisfy the contrapositive symmetry with some fuzzy, even strong, negation.

Example 1.5.10. (i) Consider the following fuzzy implication, noted as I_3 in Drewniak [80] (see also [81]),

$$I(x,y) = \begin{cases} 0, & \text{if } x > 0 \text{ and } y = 0 , \\ 1, & \text{otherwise} , \end{cases} \qquad x, y \in [0,1] ,$$

which satisfies (EP), does not satisfy (NP) and whose natural negation is the least fuzzy negation, i.e., $N_I = N_{\mathbf{D1}}$. It can be easily verified that I satisfies CP(N_I).

(ii) Similarly, consider the fuzzy implication, noted as I_4 in [80] (see also [81]),

$$I(x,y) = \begin{cases} 0, & \text{if } x = 1 \text{ and } y < 1 , \\ 1, & \text{otherwise} , \end{cases} \qquad x, y \in [0,1] ,$$

which satisfies (EP), does not satisfy (NP) and whose natural negation is the greatest fuzzy negation, i.e., $N_I = N_{\mathbf{D2}}$. Again, I satisfies (CP) with respect to N_I.

(iii) Consider the fuzzy implication $I_{\mathbf{MN}}$ defined in Remark 1.3.5(iii). It satisfies (EP), does not satisfy (NP) and the natural negation is the discontinuous fuzzy negation

$$N_{I_{\mathbf{MN}}}(x) = \begin{cases} 0, & \text{if } x \geq 0.5\,, \\ 1-x, & \text{otherwise}\,, \end{cases} \qquad x \in [0,1]\,.$$

Interestingly, $I_{\mathbf{MN}}$ satisfies (CP) with respect to $N_{\mathbf{C}}$.

(iv) Finally, consider the Baczyński implication $I_{\mathbf{BZ}}$ (see Remark 1.3.5(iii)). It does not satisfy either (NP) or (EP) and the natural negation of $I_{\mathbf{BZ}}$ is not strong (see Example 1.4.20(i)). However, it can be easily verified that $I_{\mathbf{BZ}}$ satisfies the contrapositive symmetry with respect to $N_{\mathbf{C}}$.

On the other hand, even if a fuzzy implication I satisfies (NP) or other properties, it still may not satisfy the contrapositive symmetry with any fuzzy negation as shown in the following examples (cf. Corollary 1.5.7).

Example 1.5.11. (i) Let us consider the fuzzy implication $I_{\mathbf{SQ}}$ defined in Remark 1.3.5(iii). It can be directly verified that I satisfies (NP) but not (EP). Although the natural negation of $I_{\mathbf{SQ}}$ is strong, in fact $N_{I_{\mathbf{SQ}}} = N_{\mathbf{C}}$, it does not satisfy the contrapositive symmetry with respect to $N_{\mathbf{C}}$ and, consequently, with any fuzzy negation.

(ii) Consider the Yager implication $I_{\mathbf{YG}}$. It satisfies (NP) and (EP), but its natural negation is not strong and, consequently, it does not satisfy the contrapositive symmetry with any fuzzy negation.

(iii) Finally, consider the Weber implication $I_{\mathbf{WB}}$. It satisfies (NP), (EP) and (IP), but $N_{I_{\mathbf{WB}}} = N_{\mathbf{D2}}$ is not strong and hence $I_{\mathbf{WB}}$ does not satisfy the contrapositive symmetry with any fuzzy negation.

The above examples show that (NP) and (EP) are neither necessary nor sufficient for a fuzzy implication to possess contrapositive symmetry. Moreover, if a fuzzy implication does not satisfy (NP), then it can satisfy the contrapositive symmetry with infinitely many negations. As an example consider the greatest fuzzy implication I_1 (cf. Proposition 1.1.7). It satisfies (CP) with respect to any fuzzy negation N, which is non-vanishing and non-filling.

From Lemma 1.3.4, Proposition 1.4.18, Lemma 1.5.6 and Corollary 1.5.7 we obtain the following very important result (cf. Fodor and Roubens [105], Corollary 1.2).

Corollary 1.5.12. *If* $I\colon [0,1]^2 \to [0,1]$ *satisfies* (EP), (OP) *and* N_I *is a continuous function, then* $I \in \mathcal{FI}$ *and it satisfies* (NP), (IP) *and* (CP) *only with respect to* N_I*, which is a strong negation.*

Now we investigate the law of left contraposition (L-CP). Similarly as for (CP) we get the following facts.

Lemma 1.5.13. *Let* $I\colon [0,1]^2 \to [0,1]$ *be any function and* N *a continuous fuzzy negation.*

(i) If I *satisfies* (I1) *and L-CP(*N*), then* I *satisfies* (I2).
(ii) If I *satisfies* (I2) *and L-CP(*N*), then* I *satisfies* (I1).

(iii) If I satisfies (LB) *and L-CP(N), then I satisfies* (RB).
(iv) If I satisfies (RB) *and L-CP(N), then I satisfies* (LB).

Proof. Let $x, x_1, x_2, y, y_1, y_2 \in [0,1]$ be fixed.

(i) If $y_1 \leq y_2$, then $N(y_2) \leq N(y_1)$. Since N is continuous, there exists $x_0 \in [0,1]$ such that $N(x_0) = x$. Now, from (L-CP) and (I1) we get

$$I(x, y_1) = I(N(x_0), y_1) = I(N(y_1), x_0) \\ \leq I(N(y_2), x_0) = I(N(x_0), y_2) = I(x, y_2) \ ,$$

thus I satisfies (I2).

(ii) If $x_1 \leq x_2$, then, again by continuity, there exist $x_1' \geq x_2'$ such that $N(x_1') = x_1$ and $N(x_2') = x_2$. Therefore, from (L-CP) and (I2) we have

$$I(x_1, y) = I(N(x_1'), y) = I(N(y), x_1') \\ \geq I(N(y), x_2') = I(N(x_2'), y) = I(x_2, y) \ ,$$

thus I satisfies (I1).

(iii) By the continuity of N, there exists $x_0 \in [0,1]$ such that $N(x_0) = x$. From (L-CP) we have $I(x,1) = I(N(x_0),1) = I(N(1), x_0) = I(0, x_0)$, which is equal, by (LB), to 1.

(iv) From (L-CP) we have $I(0,y) = I(N(1), y) = I(N(y), 1)$, which is equal, by (RB), to 1. □

Lemma 1.5.14. *Let a function $I\colon [0,1]^2 \to [0,1]$ satisfy* (NP), (L-CP) *with some fuzzy negation N and let N_I be a fuzzy negation. Then*

(i) if I satisfies, in addition, (I1), (I2), *then $I \in \mathcal{FI}$,*
(ii) $N_I \circ N = \mathrm{id}_{[0,1]}$,
(iii) N_I is a continuous fuzzy negation,
(iv) N is a strictly decreasing fuzzy negation,
(v) N is a continuous fuzzy negation if and only if N_I is a strictly decreasing fuzzy negation. In both cases $N = N_I^{-1}$.

Proof. (i) Since I satisfies (NP) and (L-CP), it is obvious that it satisfies (I3), (I4) and (I5). Consequently, if I satisfies (I1), (I2), then $I \in \mathcal{FI}$.

(ii) Let us take a fixed $x \in [0,1]$. Then by (L-CP) we have

$$x = I(1,x) = I(N(0), x) = I(N(x), 0) = N_I(N(x)) \ .$$

The other three points follow from Lemma 1.4.9. □

Immediately, we get

Corollary 1.5.15. *Let $I\colon [0,1]^2 \to [0,1]$ be a function that satisfies* (NP). *If N_I is not a continuous negation, then I does not satisfy* (L-CP) *with any fuzzy negation.*

In the case I also satisfies (EP), we have the following sufficient condition.

Lemma 1.5.16. *If a function* $I\colon [0,1]^2 \to [0,1]$ *satisfies* (EP), N_I *is a continuous fuzzy negation and* N *is a strictly decreasing fuzzy negation such that* $N_I \circ N = \mathrm{id}_{[0,1]}$, *then* I *satisfies* (L-CP) *with respect to* N.

Proof. By our assumptions, we have for all $x, y \in [0,1]$,

$$\begin{aligned} I(N(x), y) &= I(N(x), (N_I \circ N)(y)) = I(N(x), I(N(y), 0)) \\ &= I(N(y), I(N(x), 0)) = I(N(y), (N_I \circ N)(x)) \\ &= I(N(y), x) . \end{aligned}$$

□

When N_I is a strict negation we get the following result.

Lemma 1.5.17. *Let* $I\colon [0,1]^2 \to [0,1]$ *be any function and* N_I *be a strict negation.*

(i) If I *satisfies L-CP(*N_I^{-1}*), then* I *satisfies* (NP).
(ii) If I *satisfies* (EP), *then* I *satisfies* (NP) *and* (L-CP) *only with* N_I^{-1}.

Proof. (i) For every $y \in [0,1]$ we get

$$I(1, y) = I(N_I^{-1}(0), y) = I(N_I^{-1}(y), 0) = N_I(N_I^{-1}(y)) = y .$$

(ii) By Lemma 1.5.16, we know that I satisfies (L-CP) with respect to N_I^{-1}. The fact that I satisfies (NP) follows from (i) above. From Lemma 1.5.14(ii), we deduce that N_I^{-1} is the only negation with which I satisfies (L-CP). □

If N_I is a continuous fuzzy negation, then by Lemmas 1.4.10 and 1.5.16 we can consider the modified pseudo-inverse $\mathfrak{N}_I$ given by

$$\mathfrak{N}_I(x) = \begin{cases} N_I^{(-1)}(x), & \text{if } x \in (0,1] , \\ 1, & \text{if } x = 0 , \end{cases} \tag{1.16}$$

as the potential candidate for the fuzzy negation in (L-CP).

Corollary 1.5.18. *Let* $I\colon [0,1]^2 \to [0,1]$ *be any function and* N_I *be a continuous fuzzy negation.*

(i) If I *satisfies L-CP(*$\mathfrak{N}_I$*), then* I *satisfies* (NP).
(ii) If I *satisfies* (EP), *then* I *satisfies L-CP(*$\mathfrak{N}_I$*) and* (NP).

Remark 1.5.19. (i) If a function $I\colon [0,1]^2 \to [0,1]$ satisfies (EP) and N_I is a strict but not strong negation, then I satisfies (L-CP) only with the strict negation N_I^{-1}, but does not satisfy (CP) with any fuzzy negation.

(ii) If a function $I\colon [0,1]^2 \to [0,1]$ satisfies (EP) and N_I is a continuous but not a strict negation, then the strictly decreasing negation with which I satisfies (L-CP) need not be unique. The negation $\mathfrak{N}_I$ is one of them. As an example consider the following function:

$$I(x, y) = \begin{cases} 1, & \text{if } x, y \in (0,1] , \\ -2x + 1, & \text{if } x \in [0, 0.25] \text{ and } y = 0 , \\ 0.5, & \text{if } x \in (0.25, 0.75) \text{ and } y = 0 , \\ -2x + 2, & \text{if } x \in [0.75, 1] \text{ and } y = 0 , \\ y, & \text{otherwise} , \end{cases} \qquad x, y \in [0,1] .$$

One can easily check that it is a fuzzy implication which satisfies (EP). Moreover,

$$N_I(x) = \begin{cases} -2x+1, & \text{if } x \in [0, 0.25] , \\ 0.5, & \text{if } x \in (0.25, 0.75) , \\ -2x+2, & \text{if } x \in [0.75, 1] , \end{cases}$$

is a continuous, but not a strict negation. From Corollary 1.5.18(ii) we have that I satisfies L-CP($\mathfrak{N}_I$), where

$$\mathfrak{N}_I(x) = \begin{cases} -0.5x+1, & \text{if } x \in [0, 0.5) , \\ -0.5x+0.5, & \text{if } x \in [0.5, 1] . \end{cases}$$

Moreover, $N \circ N' = \mathrm{id}_{[0,1]}$, where N' is defined by

$$N'(x) = \begin{cases} -0.5x+1, & \text{if } x \in [0, 0.5] , \\ -0.5x+0.5, & \text{if } x \in (0.5, 1] . \end{cases}$$

Hence, by Lemma 1.5.16, we see that I also satisfies L-CP(N').

Finally, we analyze the law of right contraposition (R-CP). The proof of the following lemma is similar to the proof of Lemma 1.5.13.

Lemma 1.5.20. *Let a function $I\colon [0,1]^2 \to [0,1]$ satisfy* (R-CP) *with respect to a continuous fuzzy negation N. Then*

(i) (I1) *holds if and only if* (I2) *holds.*
(ii) (LB) *holds if and only if* (RB) *holds.*

Lemma 1.5.21. *If a function $I\colon [0,1]^2 \to [0,1]$ satisfies* (NP) *and* (R-CP) *with a fuzzy negation N, then I satisfies* (I3), (I4), (I5) *and $N = N_I$.*

Proof. It is obvious from (NP) and (R-CP) that I satisfies (I3), (I4) and (I5). Further, for any $x \in [0,1]$ we have

$$N_I(x) = I(x,0) = I(x, N(1)) = I(1, N(x)) = N(x) . \qquad \square$$

Lemma 1.5.22. *If a function $I\colon [0,1]^2 \to [0,1]$ satisfies* (EP) *and N_I is a fuzzy negation, then I satisfies R-CP(N_I).*

Proof. For any $x, y \in [0,1]$ we get

$$I(x, N_I(y)) = I(x, I(y,0)) = I(y, I(x,0)) = I(y, N_I(x)) . \qquad \square$$

Example 1.5.23. Observe that the converse need not be true, i.e., I can satisfy (R-CP)(N_I) without satisfying (EP). As an example consider the fuzzy implication I_{MM} given in Remark 1.3.5(iii).

Lemma 1.5.24. *Let $I\colon [0,1]^2 \to [0,1]$ be any function and N_I a continuous fuzzy negation.*

(i) If I satisfies R-CP(N_I), then I satisfies (NP)*.*
(ii) If I satisfies (EP)*, then I satisfies* (I3), (I4), (I5), (NP) *and* (R-CP) *only with respect to N_I.*

Proof. (i) Let us fix $y \in [0,1]$. Since N_I is a continuous negation, there exists $y' \in [0,1]$ such that $N_I(y') = y$. Thus

$$I(1,y) = I(1, N_I(y')) = I(y', N_I(1)) = I(y', 0) = N_I(y') = y \,.$$

(ii) The property R-CP(N_I) follows from Lemma 1.5.22. From the previous point we deduce that it satisfies (NP) and thus, by Lemma 1.5.21, it satisfies (I3), (I4), (I5) and I satisfies (R-CP) only with respect to N_I. □

Corollary 1.5.25. *If $I \in \mathcal{FI}$ satisfies* (NP) *and* (EP)*, then I satisfies* (R-CP) *only with N_I.*

Example 1.5.26. From the above discussion we see that all the three laws of contraposition may not be equivalent to one another, when N is not a strong negation. This can also be seen from the following Table 1.8.

Table 1.8. Fuzzy implications and the laws of contraposition

Fuzzy implication	(CP)	(L-CP)	(R-CP)
$I_{\mathbf{NI}}$	×	×	×
$I_{\mathbf{YG}}$	×	×	✓
Example 1.5.10(i)	✓	×	✓
Example 1.4.20(ii)	×	✓	✓
$I_{\mathbf{LK}}$	✓	✓	✓

Example 1.5.27. Table 1.9 lists the 9 basic fuzzy implications from Example 1.1.4 along with the different laws of contraposition. Note that $I_{\mathbf{RS}}$ satisfies all the three laws of contraposition with any strictly decreasing negation. For the other 8 basic fuzzy implications, if they satisfy any of the three laws of contraposition with a negation N, then N is uniquely determined.

1.6 Reciprocal Fuzzy Implications

Interestingly, if a fuzzy implication I does not satisfy the law of contraposition (CP) with respect to some particular negation N, then we can obtain new fuzzy implications as follows.

Table 1.9. Basic fuzzy implications and the laws of contraposition

Fuzzy implication	(CP)	(L-CP)	(R-CP)
$I_{\mathbf{LK}}$	$N_{\mathbf{C}}$	$N_{\mathbf{C}}$	$N_{\mathbf{C}}$
$I_{\mathbf{GD}}$	×	×	$N_{\mathbf{D1}}$
$I_{\mathbf{RC}}$	$N_{\mathbf{C}}$	$N_{\mathbf{C}}$	$N_{\mathbf{C}}$
$I_{\mathbf{KD}}$	$N_{\mathbf{C}}$	$N_{\mathbf{C}}$	$N_{\mathbf{C}}$
$I_{\mathbf{GG}}$	×	×	$N_{\mathbf{D1}}$
$I_{\mathbf{RS}}$	$N_{\mathbf{C}}$	$N_{\mathbf{C}}$	$N_{\mathbf{C}}$
$I_{\mathbf{YG}}$	×	×	$N_{\mathbf{D1}}$
$I_{\mathbf{WB}}$	×	×	$N_{\mathbf{D2}}$
$I_{\mathbf{FD}}$	$N_{\mathbf{C}}$	$N_{\mathbf{C}}$	$N_{\mathbf{C}}$

Definition 1.6.1. *Let N be a fuzzy negation and $I \in \mathcal{FI}$. A function $I_N : [0,1]^2 \to [0,1]$ defined by*

$$I_N(x,y) = I(N(y), N(x)) , \qquad x, y \in [0,1] , \tag{1.17}$$

is called the N-reciprocal *of I. When N is the classical negation $N_{\mathbf{C}}$, then I_N is called the* reciprocal *of I and is denoted by I'.*

Theorem 1.6.2. *If N is a fuzzy negation and $I \in \mathcal{FI}$, then $I_N \in \mathcal{FI}$. In addition, if N, I are continuous, then I_N is also continuous.*

Proof. Let N be a fuzzy negation and $I \in \mathcal{FI}$. Let $x, y, z \in [0,1]$ be arbitrarily fixed. If $x \leq z$, then $N(z) \leq N(x)$ and from (I2) we have

$$I_N(x,y) = I(N(y), N(x)) \geq I(N(y), N(z)) = I_N(z,y) .$$

If $y \leq z$, then $N(z) < N(y)$ and from (I1)

$$I_N(x,y) = I(N(y), N(x)) \leq I(N(z), N(x)) = I_N(x,z) .$$

Hence I_N satisfies (I1) and (I2). Finally,

$$I_N(0,0) = I(1,1) = 1 , \quad I_N(1,1) = I(0,0) = 1 , \quad I_N(1,0) = I(1,0) = 0 ,$$

which shows that $I_N \in \mathcal{FI}$. It is obvious that I_N is continuous when N, I are continuous. □

Example 1.6.3. Let us consider the 9 basic fuzzy implications from Example 1.1.4. Since $I_{\mathbf{LK}}$, $I_{\mathbf{RC}}$, $I_{\mathbf{KD}}$, $I_{\mathbf{RS}}$, $I_{\mathbf{FD}}$ satisfy (CP) with $N_{\mathbf{C}}$, their reciprocals do not generate new implications. For the other examples we get 4 new fuzzy implications:

$$I'_{\mathbf{GD}}(x,y) = \begin{cases} 1, & \text{if } x \le y\,, \\ 1-x, & \text{otherwise}\,, \end{cases}$$

$$I'_{\mathbf{GG}}(x,y) = \begin{cases} 1, & \text{if } y = 1 \\ \min(1, \frac{1-x}{1-y}), & \text{otherwise} \end{cases} = \begin{cases} 1, & \text{if } x \le y\,, \\ \frac{1-x}{1-y}, & \text{otherwise}\,, \end{cases}$$

$$I'_{\mathbf{YG}}(x,y) = \begin{cases} 1, & \text{if } x = 1 \text{ and } y = 1\,, \\ (1-x)^{1-y}, & \text{otherwise}\,, \end{cases}$$

$$I'_{\mathbf{WB}}(x,y) = \begin{cases} 1, & \text{if } y > 0\,, \\ 1-x, & \text{otherwise}\,, \end{cases}$$

for all $x, y \in [0,1]$. Fig. 1.4 gives the plot of all these four reciprocal fuzzy implications.

Now, we show which basic properties are preserved by the operation (1.17). Very interestingly, as the following result shows, even if an $I \in \mathcal{FI}$ does not satisfy (NP), its reciprocal can satisfy (NP), under the following condition.

Proposition 1.6.4. *Let N be a fuzzy negation and let $I \in \mathcal{FI}$. The following statements are equivalent:*

(i) I_N *satisfies* (NP).
(ii) $N_I \circ N = \mathrm{id}_{[0,1]}$.

Proof. Let N be a fuzzy negation and $I \in \mathcal{FI}$. Note that the following equality is true for all $y \in [0,1]$:

$$I_N(1,y) = I(N(y), N(1)) = I(N(y), 0) = N_I(N(y))\,.$$

The result is now immediate. □

Example 1.6.5. Consider the fuzzy implication

$$I(x,y) = \max(1-x^2, y^2)\,, \qquad x, y \in [0,1]\,,$$

which does not satisfy (NP), but whose natural negation $N_I(x) = 1 - x^2$ is continuous. Now, if we consider the negation $N(x) = \sqrt{1-x}$, then, of course, $N_I \circ N = \mathrm{id}_{[0,1]}$ and the reciprocal of I with N gives a fuzzy implication which does satisfy (NP). In fact, in this case, $I_N = I_{\mathbf{KD}}$.

Proposition 1.6.6. *Let N be a fuzzy negation. If $I\colon [0,1] \to [0,1]$ satisfies* (IP) *((OP)), then I_N also satisfies* (IP) *((OP)).*

Proof. Let N be a fuzzy negation and $I\colon [0,1]^2 \to [0,1]$ be any function. Assume first that I satisfies (IP). Then, for all $x \in [0,1]$, we have

$$I_N(x,x) = I(N(x), N(x)) = 1\,.$$

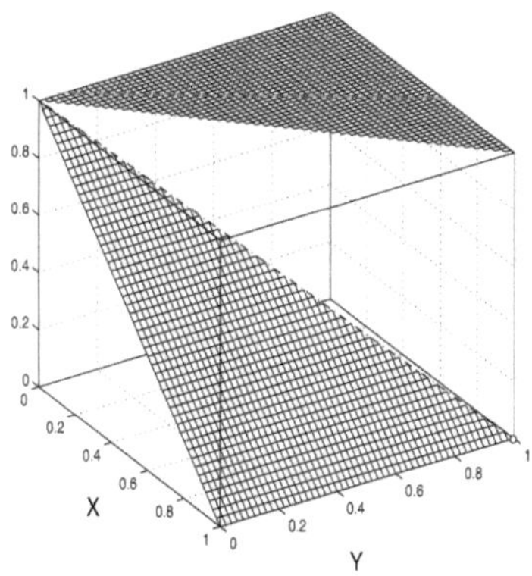

(a) Reciprocal Gödel implication $I'_{\mathbf{GD}}$

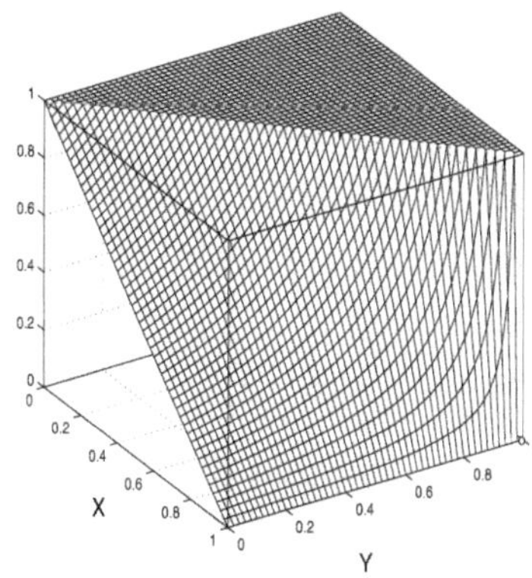

(b) Reciprocal Goguen implication $I'_{\mathbf{GG}}$

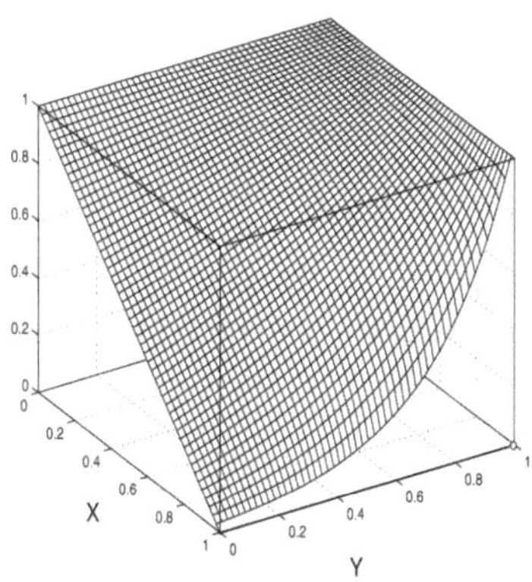

(c) Reciprocal Yager implication $I'_{\mathbf{YG}}$

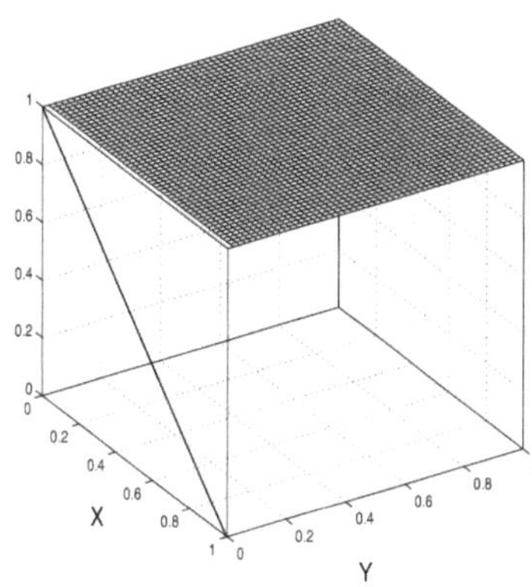

(d) Reciprocal Weber implication $I'_{\mathbf{WB}}$

Fig. 1.4. Plots of the reciprocals of Gödel, Goguen, Yager and Weber implications with respect to $N_{\mathbf{C}}$

Assume now that I satisfies (OP) and take any $x, y \in [0,1]$. We obtain

$$I_N(x,y) = 1 \iff I(N(y), N(x)) - 1 \iff N(y) \leq N(x) \iff x \leq y\,. \qquad \square$$

Remark 1.6.7. One can easily check that (EP) is not generally preserved by the above operation. Let us consider the Gödel implication $I_{\mathbf{GD}}$ and its reciprocal $I'_{\mathbf{GD}}$. We know that $I_{\mathbf{GD}}$ satisfies (EP), but

$$I'_{\mathbf{GD}}(0.7, I'_{\mathbf{GD}}(0.5, 0.2)) = \frac{3}{10} \neq \frac{1}{2} = I'_{\mathbf{GD}}(0.5, I'_{\mathbf{GD}}(0.7, 0.2))\,,$$

i.e., $I'_{\mathbf{GD}}$ does not satisfy (EP).

Observe also that if N is a strong negation, then the operation (1.17) is an involution, i.e.,

$$(I_N)_N = I \,. \tag{1.18}$$

Indeed, for any $x, y \in [0,1]$ we get

$$(I_N)_N(x,y) = I_N(N(y), N(x)) = I(N(N(x)), N(N(y))) = I(x,y) \,.$$

1.7 Bibliographical Remarks

It must be pleasantly surprising to know that JAN ŁUKASIEWICZ who is credited as being the father of what we now call multi-valued logic, in fact, began his exposition and exploration of his ideas in his seminal paper [157] by firstly defining a multi-valued implication operation thus:

> "The variables p, q stand for any real numbers in the interval $0 - 1$, including the limiting values of that interval.
> The formula '$p \supset q$' equals the number 1 if $p \leq q$, i.e.,
>
> $$p \supset q \quad .=. \quad 1 \qquad \text{for } p \leq q$$
>
> and
>
> $$p \supset q \quad .=. \quad 1 - p + q \qquad \text{for } p \geq q."$$

The above operation is what has come to be known as the Łukasiewicz implication in fuzzy logic. It is from this implication function he derived the other binary operations for his logic. Perhaps, in this sense, one can even claim that implication functions predate the existence of what we now consider as the fundamental binary operations of conjunction and disjunction.

In the scientific literature, especially in the early days of research on fuzzy logic, we can find different approaches to fuzzy implications. On the one hand a fuzzy implication was defined as a truth space fuzzy relation $I \subset [0,1]^2$, not necessarily as the extension of classical implication (see ZADEH [266], BALDWIN and PILSWORTH [27] for details) and used in many applications, like modeling of fuzzy IF-THEN rules with fuzzy predicates. On the other hand, a fuzzy implication was investigated as a connective, as in our context. Such investigations were done, for example, by BANDLER and KOHOUT [31] and WILLMOTT [257] in the theory of fuzzy power sets.

Once again, it might come as a surprise to the readers that despite the many important works on fuzzy implications that appeared during the 1980s there was no clear definition of a fuzzy implication, but only examples based on multi-valued implications. The first such definition can perhaps be found in the paper of ZADEH [266], wherein a fuzzy implication was defined as a truth space fuzzy relation, not necessarily as the extension of the classical implication. This allowed the use of such operations in many applications, like modeling of fuzzy IF-THEN rules with fuzzy predicates. In particular, Zadeh considered the function I_{-1} as an implication in his algorithm of fuzzy inference. Following this, MAMDANI [164], and later LARSEN [151], proposed to use the *minimum* and the *product* as the operations relating the antecedent to its consequent in a fuzzy IF-THEN rule. Although these operations were not fuzzy implications, since they do not satisfy the truth table for the classical implication (we will see that they are, in fact, examples of special and very important class of the well-known binary operations called t-norms), they found wide spread acceptance because of their success in practice, which follows from the fact that only positive cases of both the antecedent and consequent are important in applications.

Meanwhile, GAINES [110] and PAWELKA [198, 199, 200] proposed a class of operations obtained as the residuation of conjunctions. As we will see in Chap. 2 these are what are commonly known as R-implications and form one of the most important classes of fuzzy implications.

The next important work to appear on this topic was that of BALDWIN and PILSWORTH [27], wherein they take cognizance of the variation in the definition of a fuzzy implication, which were then predominantly employed in different approximate reasoning models. Still, there was no clear definition of fuzzy implications in their work. Instead, they proposed a few basic properties of fuzzy implications which are essentially connected with *modus ponens* and *modus tollens* rules of deduction (for more on these topics see Chap. 8). They only required that a fuzzy implication I was a fuzzy relation which is monotonic and satisfies the contrapositive symmetry (CP) with the standard negation $N_{\mathbf{C}}$.

In the same year, the work of BANDLER and KOHOUT [31] perhaps was the first paper to be published in a main stream journal where the term 'fuzzy implication' appeared in the title. Once again, the authors only enunciate the general nature of a fuzzy implication requiring it to be an extension of the Boolean implication to the multi-valued logic of the unit interval, i.e., a binary operator which agrees with the Boolean one at the corners $(0,0),(0,1),(1,0),(1,1)$, but do not propose any specific definition.

The works of TRILLAS and VALVERDE [238, 239] were the first to offer a concrete definition of a fuzzy implication. Their original definition was slightly more restrictive, in the sense that they required it to satisfy the left neutrality property (NP), the exchange principle (EP) and to be continuous. Soon many other definitions, not all of them equivalent to the above, started appearing in different works. For example, FODOR [101] echoes the sentiments of BANDLER and KOHOUT [31] and only expects that a fuzzy implication be an extension of the Boolean implication, while VILLAR et al. [250] ask for (I2), (I4) and (I5).

The definition given in Definition 1.1.1 is equivalent, by Remark 1.1.3, to the definition based on FODOR and ROUBENS [105]. In fact, in their book it is required for I to satisfy (I1), (I2), (LB), (RB) and (NC). The other equivalent definition of a fuzzy implication was used by BACZYŃSKI and DREWNIAK [11], where it is assumed that I satisfies the binary implication truth-table and is monotonic with respect to both variables. Such a definition, and consequently ours, is more or less well accepted now (see BACZYŃSKI and DREWNIAK [12], FODOR and YAGER [107], BUSTINCE et al. [45], IVANEK [124], DREWNIAK [81]). Still, it is possible to find works using older or different definitions, for example, CORDON et al. [59], SAINIO et al. [219].

Although the names used for the first 7 of the 9 basic fuzzy implications from Example 1.1.4 are the ones commonly employed in the literature, the last two, viz., $I_{\mathbf{WB}}$ and $I_{\mathbf{FD}}$, are known more commonly as the residuals of the drastic and nilpotent minimum t-norms. While $I_{\mathbf{FD}}$ was obtained by FODOR [102] during his work on contrapositive symmetry of fuzzy implications, it should be remarked that to the best of the authors' knowledge the first published work to contain the formula for $I_{\mathbf{WB}}$ was that of WEBER [255] and hence the nomenclature. It should be brought to the notice of the readers that the implication $I_{\mathbf{GD}}$ was firstly proposed by Heyting [122] and was later extended by Gödel to his system of many-valued logics. Hence, in this sense, the implication $I_{\mathbf{GD}}$ should be called Heyting-Gödel implication. However, in this monograph we only call this the Gödel implication keeping with the currently prevalent appellation in the literature. In early literature, see, for instance, RESCHER [210], GAINES [110], BANDLER and KOHOUT [31] the Rescher implication $I_{\mathbf{RS}}$ was sometimes referred to as the *standard-sequence* implication and the Gödel implication was usually denoted by S^* whose formula is given as

$$S^*(x,y) = \begin{cases} 1, & \text{if } x = y\,, \\ y, & \text{otherwise}\,, \end{cases} \qquad x, y \in [0,1]\,,$$

which is equivalent to $I_{\mathbf{GD}}$ by (I1).

It is not uncommon to derive negations from a given binary operator in fuzzy logic. In fact, in the realm of t-norms and t-conorms, considered the basic fuzzy logic connectives and which will be dealt with in considerable detail in the next chapter, the natural negations (also known as associated or the induced negations) associated with them play an important role. For example, for left-continuous t-norms - a major subclass of t-norms for which a complete representation is as yet unknown - many works dealing with their constructions have appeared, when their natural negations are strong. In fact, such t-norms have a very important role to play in the context of fuzzy implications, since (as we show in Chap. 4) the R-implications obtained from them are also (S,N)- and QL-implications.

In Sect. 1.3 we have listed only four basic properties of fuzzy implications. Of course, there are many additional properties of fuzzy implications, considered in different papers and also important in some applications. They are generally obtained as the translation of logical laws from propositional calculus into properties of fuzzy implications. We list here some of them:

$$I(x,y) \geq y\,,$$
$$I(y, I(x,y)) = y\,,$$
$$I(y, I(x,x)) = y\,,$$
$$I(I(x,y),y) = x \text{ if and only if } x \geq y\,,$$
$$I(x, I(I(x,y),y)) = 1\,,$$
$$I(x, I(y,z)) = I(I(x,y), I(x,z))\,.$$

In fact, some of these properties often play a significant role in the characterization of fuzzy implications, see, e.g., PEDRYCZ [203], MESIAR [181]. Probably the biggest list of such postulates for fuzzy implications was collected by DREWNIAK in [81].

The classical law of contraposition (CP) has been studied by many authors (cf. TRILLAS and VALVERDE [238], DUBOIS and PRADE [92], FODOR [103]). Note that, in general, it is required for N to be a strong negation and therefore, it is not necessary to consider three different laws of contraposition. However, when N is only a fuzzy negation with no additional assumptions, in the sense of Definition 1.4.1, then as we have shown the different laws of contraposition may not be equivalent. It can be observed that Table 1.8 is not fully complete and the following question naturally arises.

Problem 1.7.1. Give examples of fuzzy implications I such that

(i) I satisfies only (CP),
(ii) I satisfies only (L-CP),
(iii) I satisfies both (CP) and (L-CP) but not (R-CP),

with some fuzzy negation N.

Part I

Analytical Study of Fuzzy Implications

Mathematicians create by acts of insights
and intuition. Logic then sanctions
the conquest of intuition.
– Morris Kline (1908-1992)

As noted in Chap. 1, the implication in classical logic takes several distinct forms. While these are equivalent in classical logic, their natural generalizations in fuzzy logic are not and consequently give rise to different families of fuzzy implications, each satisfying different sets of properties. This is a distinguishing feature of fuzzy implications and perhaps the most singular motivation for research into these families.

While in Chap. 1 we have limited ourselves to the definition of a fuzzy implication, some examples, properties they usually satisfy and their interdependencies, in Part I of this treatise, which consists of 4 chapters, we deal with their systematic construction. In the literature, fuzzy implications have been constructed, so far, in the following ways:

- from binary operations on the unit interval $[0, 1]$, i.e., from basic fuzzy logic connectives, viz., t-norms, t-conorms and negations, uninorms, uninorms and negations (Chaps. 2 and 5);
- from unary functions on the unit interval $[0, 1]$, viz., generator functions from $[0, 1]$ to $[0, \infty]$ or from $[0, 1]$ to $[0, 1]$ (Chap. 3).

In Part I we discuss all of the above approaches to defining fuzzy implications which lead to many classes of fuzzy implications, including the following established classes, popularly known in the literature as (S,N)-, S-, R- and QL-implications in Chap. 2.

Yager's [261] novel methods of generating fuzzy implications from unary functions are presented in Chap. 3.

The investigation in this part is primarily focused on properties these classes possess, their characterizations and representations where available. The overlaps that exist among these families of fuzzy implications are investigated in Chap. 4.

In Chap. 5, we reproduce the generation process of Chap. 2 in the setting of uninorms, which include both the t-norms and t-conorms as special cases.

2 Fuzzy Implications from Fuzzy Logic Operations

The essence of mathematics is its freedom.
– Georg Cantor (1845-1918)

In this first chapter of Part I, we discuss families of fuzzy implications obtained from binary operations on the unit interval $[0, 1]$, i.e., from basic fuzzy logic connectives, viz., t-norms, t-conorms and negations. After giving the necessary background on the above connectives, we investigate three main ways of generating fuzzy implications from these connectives. We also study the properties possessed by the families of fuzzy implications thus generated, present their characterizations and their representations, in the cases where available.

2.1 Fuzzy Conjunctions: Triangular Norms

Triangular norms were originally introduced by MENGER [180], when generalizing the triangle inequality from the classical metric spaces to the probabilistic metric spaces. SCHWEIZER and SKLAR [221, 222] redefined axioms of triangular norms into the form used today. From the viewpoint of fuzzy logic, the triangular norms are suitable candidates for the generalization of the classical binary conjunction into a fuzzy intersection (see KLIR and YUAN [147] or GOTTWALD [113]). The following definitions and results, with proofs, can be found in the recent monograph by KLEMENT et al. [146].

Definition 2.1.1. *A function $T\colon [0,1]^2 \to [0,1]$ is called a* triangular norm *(shortly* t-norm*) if it satisfies, for all $x, y, z \in [0,1]$, the following conditions*

$$T(x,y) = T(y,x)\ , \tag{T1}$$

$$T(x, T(y,z)) = T(T(x,y), z)\ , \tag{T2}$$

$$\textit{if } y \leq z,\ \textit{then } T(x,y) \leq T(x,z)\ ,\ \textit{i.e., } T(x,\,\cdot\,) \textit{ is increasing}\ , \tag{T3}$$

$$T(x,1) = x\ . \tag{T4}$$

The commutativity (T1), the monotonicity (T3) and the boundary condition (T4) ensure that T on $\{0, 1\}$ behaves exactly like the classical conjunction. The associativity is self-explanatory. The class of t-norms is rather large and some subclasses of t-norms have been well investigated.

M. Baczyński and B. Jayaram: Fuzzy Implications, STUDFUZZ 231, pp. 41–108, 2008.
springerlink.com

Definition 2.1.2. *A t-norm T is called*

(i) continuous *if it is continuous in both the arguments;*
(ii) left-continuous *if it is left-continuous in each component;*
(iii) border continuous *if it is continuous on the boundary of the unit square $[0,1]^2$, i.e., on the set $[0,1]^2 \setminus (0,1)^2$;*
(iv) idempotent, *if $T(x,x) = x$ for all $x \in [0,1]$;*
(v) Archimedean, *if for all $x, y \in (0,1)$ there exists an $n \in \mathbb{N}$ such that $x_T^{[n]} < y$;*
(vi) strict, *if it is continuous and strictly monotone, i.e., $T(x,y) < T(x,z)$ whenever $x > 0$ and $y < z$;*
(vii) nilpotent, *if it is continuous and if each $x \in (0,1)$ is a nilpotent element of T, i.e., if there exists an $n \in \mathbb{N}$ such that $x_T^{[n]} = 0$;*
(viii) positive, *if $T(x,y) = 0$ then either $x = 0$ or $y = 0$.*

Example 2.1.3. Table 2.1 lists a few of the common t-norms along with their classification.

Table 2.1. Basic t-norms

Name	Formula	Properties
minimum	$T_{\mathbf{M}}(x,y) = \min(x,y)$	idempotent, continuous, positive
algebraic product	$T_{\mathbf{P}}(x,y) = xy$	strict, positive
Łukasiewicz	$T_{\mathbf{LK}}(x,y) = \max(x+y-1, 0)$	nilpotent
drastic product	$T_{\mathbf{D}}(x,y) = \begin{cases} 0, & \text{if } x,y \in [0,1) \\ \min(x,y), & \text{otherwise} \end{cases}$	Archimedean, non-continuous
nilpotent minimum	$T_{\mathbf{nM}}(x,y) = \begin{cases} 0, & \text{if } x+y \leq 1 \\ \min(x,y), & \text{otherwise} \end{cases}$	left-continuous

Remark 2.1.4. (i) By virtue of Theorem 1.2.1 and (T1), (T3) we can deduce that a continuity of any t-norm is equivalent to its continuity with respect to the first (or second) variable (see also [146], Proposition 1.19 and [113], Proposition 5.1.1).
(ii) The minimum t-norm $T_{\mathbf{M}}$ is the only idempotent t-norm (see [146], Proposition 1.9).
(iii) For a continuous t-norm T the Archimedean property is captured by the following simpler condition (see [113], Proposition 5.1.2):

$$T(x,x) < x\,, \qquad x \in (0,1)\,.$$

(iv) If a t-norm T is continuous and Archimedean, then T is nilpotent if and only if there exists some nilpotent element of T, which is equivalent to the existence of some zero divisor of T i.e., there exist $x, y \in (0,1)$ such that $T(x,y) = 0$ (see [146], Theorem 2.18).

(v) If a t-norm T is strict or nilpotent, then it is Archimedean (see [146], Proposition 2.15). Conversely, every continuous and Archimedean t-norm is strict or nilpotent (see [146], p. 16).
(vi) A continuous Archimedean t-norm is positive if and only if it is strict (see [105], p. 9).
(vii) If $\varphi \in \Phi$ and T is a continuous (Archimedean, strict, nilpotent) t-norm, then T_φ is also a continuous (Archimedean, strict, nilpotent) t-norm (see [146], Proposition 2.28 and Remark 2.30).
(viii) If T is a t-norm, then $T_\varphi = T$ for all $\varphi \in \Phi$ if and only if either $T = T_{\mathbf{M}}$ or $T = T_{\mathbf{D}}$ (see [146], Proposition 2.31).
(ix) On the family of all t-norms we can consider the partial order induced from the unit interval. In fact $T_{\mathbf{D}}$ is the least t-norm, while $T_{\mathbf{M}}$ is the greatest t-norm.

The following characterization theorem is based on properties of their underlying generators. In this form it was firstly proposed by Ling [154] (for the proof see also [146], Theorem 5.1).

Theorem 2.1.5. *For a function $T\colon [0,1]^2 \to [0,1]$ the following statements are equivalent:*

(i) T is a continuous Archimedean t-norm.
(ii) T has a continuous additive generator, i.e., there exists a continuous, strictly decreasing function $f\colon [0,1] \to [0,\infty]$ with $f(1) = 0$, which is uniquely determined up to a positive multiplicative constant, such that

$$T(x,y) = f^{(-1)}(f(x) + f(y)), \qquad x, y \in [0,1],$$

where $f^{(-1)}$ is the pseudo-inverse of f given by

$$f^{(-1)}(x) = \begin{cases} f^{-1}(x), & \text{if } x \in [0, f(0)], \\ 0, & \text{if } x \in (f(0), \infty]. \end{cases}$$

Example 2.1.6. Observe that $f(x) = -\ln x$ and $f(x) = 1 - x$ are additive generators for the product t-norm $T_{\mathbf{P}}$ and the Łukasiewicz t-norm $T_{\mathbf{LK}}$, respectively.

Remark 2.1.7. (i) The above representation of a t-norm T can also be written without using the pseudo-inverse in the following way

$$T(x,y) = f^{-1}(\min(f(x) + f(y), f(0))), \qquad x, y \in [0,1].$$

(ii) T is a strict t-norm if and only if each continuous additive generator f of T satisfies $f(0) = \infty$.
(iii) T is a nilpotent t-norm if and only if each continuous additive generator f of T satisfies $f(0) < \infty$.
(iv) If T is a continuous Archimedean t-norm with continuous additive generator f, then the function $\theta\colon [0,1] \to [0,1]$ defined by $\theta(x) = \exp(-f(x))$ is a

multiplicative generator of T, i.e., a continuous, strictly increasing function with $\theta(1) = 1$ such that

$$T(x,y) = \theta^{(-1)}(\theta(x) \cdot \theta(y)) , \qquad x,y \in [0,1] ,$$

where $\theta^{(-1)}$ is the pseudo-inverse of θ given by

$$\theta^{(-1)}(x) = \begin{cases} 0, & \text{if } x \in [0,\theta(0)) , \\ \theta^{-1}(x), & \text{if } x \in [\theta(0),1] . \end{cases}$$

We have also the following representation theorems for strict and nilpotent t-norms (see [146], Propositions 5.9 and 5.10).

Theorem 2.1.8. *For a function $T\colon [0,1]^2 \to [0,1]$ the following statements are equivalent:*

(i) T is a strict t-norm.
(ii) T is Φ-conjugate with the product t-norm $T_{\mathbf{P}}$, i.e., there exists $\varphi \in \Phi$, which is uniquely determined up to a positive constant exponent, such that

$$T(x,y) = (T_{\mathbf{P}})_{\varphi}(x,y) = \varphi^{-1}(\varphi(x) \cdot \varphi(y)) , \qquad x,y \in [0,1] .$$

Theorem 2.1.9. *For a function $T\colon [0,1]^2 \to [0,1]$ the following statements are equivalent:*

(i) T is a nilpotent t-norm.
(ii) T is Φ-conjugate with the Łukasiewicz t-norm $T_{\mathbf{LK}}$, i.e., there exists $\varphi \in \Phi$, which is uniquely determined, such that

$$T(x,y) = (T_{\mathbf{LK}})_{\varphi}(x,y) = \varphi^{-1}(\max(\varphi(x) + \varphi(y) - 1, 0)) , \qquad x,y \in [0,1] .$$

Finally, we have the following complete representation of continuous t-norms (cf. [146], Theorem 5.11).

Theorem 2.1.10. *For a function $T\colon [0,1]^2 \to [0,1]$ the following statements are equivalent:*

(i) T is a continuous t-norm.
(ii) T is uniquely representable as an ordinal sum of continuous Archimedean t-norms, i.e., there exist a uniquely determined (finite or countably infinite) index set A, a family of uniquely determined pairwise disjoint open sub-intervals $\{(a_\alpha, e_\alpha)\}_{\alpha \in A}$ of $[0,1]$ and a family of uniquely determined continuous Archimedean t-norms $(T_\alpha)_{\alpha \in A}$ such that

$$T(x,y) = \begin{cases} a_\alpha + (e_\alpha - a_\alpha) \cdot T_\alpha\left(\frac{x-a_\alpha}{e_\alpha - a_\alpha}, \frac{y-a_\alpha}{e_\alpha - a_\alpha}\right), & \textit{if } x,y \in [a_\alpha, e_\alpha] , \\ \min(x,y), & \textit{otherwise} . \end{cases}$$

In this case we will write $T = (\langle a_\alpha, e_\alpha, T_\alpha \rangle)_{\alpha \in A}$.

It should be noted that considering t-norms as real functions one can discuss their Lipschitzianity. Once again Lipschitzianity of a t-norm T immediately implies its continuity, though the converse is not true, in general (see [146], Example 1.26). One of the well-known families of t-norms that satisfies Lipschitzianity is the family of Frank t-norms $T_{\mathbf{F}}^{\lambda}$, where $\lambda \in [0, \infty]$, defined as follows

$$T_{\mathbf{F}}^{\lambda}(x,y) = \begin{cases} T_{\mathbf{M}}(x,y), & \text{if } \lambda = 0 \ , \\ T_{\mathbf{P}}(x,y), & \text{if } \lambda = 1 \ , \\ T_{\mathbf{LK}}(x,y), & \text{if } \lambda = \infty \ , \\ \log_{\lambda}\left(1 + \dfrac{(\lambda^x - 1)\cdot(\lambda^y - 1)}{\lambda - 1}\right), & \text{otherwise} \ , \end{cases}$$

for all $x, y \in [0,1]$. In fact, this family was obtained while characterizing the so called Frank functional equation (2.1) (see also Theorem 2.2.11). We will use them in the sequel.

2.2 Fuzzy Disjunctions: Triangular Conorms

The generalization of the classical binary disjunction is the fuzzy union, interpreted in many cases by the triangular conorms. The classical disjunction can be defined from the classical conjunction and negation as follows:

$$p \vee q = \neg(\neg p \wedge \neg q) \ .$$

We will see that this duality extends to fuzzy logic operations too and the development of the theory of triangular conorms largely mirrors this duality.

Definition 2.2.1. *A function $S\colon [0,1]^2 \to [0,1]$ is called a* triangular conorm *(shortly* t-conorm*) if it satisfies, for all $x, y, z \in [0,1]$, the following conditions*

$$S(x,y) = S(y,x) \ , \tag{S1}$$

$$S(x, S(y,z)) = S(S(x,y), z) \ , \tag{S2}$$

$$\textit{if } y \le z, \textit{ then } S(x,y) \le S(x,z) \ , \textit{ i.e., } S(x, \cdot\,) \textit{ is increasing} \ , \tag{S3}$$

$$S(x,0) = x \ . \tag{S4}$$

Again, the commutativity (S1), the monotonicity (S3) and the boundary condition (S4) ensure that S on $\{0,1\}$ behaves exactly like the classical disjunction.

Definition 2.2.2. *A t-conorm S is called*

(i) continuous *if it is continuous in both the arguments;*
(ii) right-continuous *if it is right-continuous in each component;*
(iii) idempotent, *if $S(x,x) = x$ for all $x \in [0,1]$;*
(iv) Archimedean, *if for all $x, y \in (0,1)$ there exists an $n \in \mathbb{N}$ such that $x_S^{[n]} > y$;*
(v) strict, *if it is continuous and strictly monotone, i.e., $S(x,y) < S(x,z)$ whenever $x < 1$ and $y < z$;*

(vi) nilpotent, *if it is continuous and if each* $x \in (0,1)$ *is a nilpotent element of* S, *i.e., if there exists an* $n \in \mathbb{N}$ *such that* $x_S^{[n]} = 1$;
(vii) positive, *if* $S(x,y) = 1$ *then either* $x = 1$ *or* $y = 1$.

As noted above we have the following duality between t-norms and t-conorms (see [222], [146]).

Proposition 2.2.3. *For a function* $S\colon [0,1]^2 \to [0,1]$ *the following statements are equivalent:*

(i) S *is a t-conorm.*
(ii) There exists a t-norm T *such that*

$$S(x,y) = 1 - T(1-x, 1-y)\,, \qquad x,y \in [0,1]\,.$$

Moreover, the t-norm T *is continuous (Archimedean, strict, nilpotent) if and only if the t-conorm* S *is continuous (Archimedean, strict, nilpotent).*

Example 2.2.4. Table 2.2 lists a few of the common t-conorms along with their classification. They are the counterparts of the t-norms in Example 2.1.3.

Table 2.2. Basic t-conorms

Name	Formula	Properties
maximum	$S_{\mathbf{M}}(x,y) = \max(x,y)$	idempotent, continuous, positive
probabilistic sum	$S_{\mathbf{P}}(x,y) = x + y - xy$	strict, positive
Łukasiewicz	$S_{\mathbf{LK}}(x,y) = \min(x+y, 1)$	nilpotent
drastic sum	$S_{\mathbf{D}}(x,y) = \begin{cases} 1, & \text{if } x,y \in (0,1] \\ \max(x,y), & \text{otherwise} \end{cases}$	Archimedean, non-continuous
nilpotent maximum	$S_{\mathbf{nM}}(x,y) = \begin{cases} 1, & \text{if } x+y \geq 1 \\ \max(x,y), & \text{otherwise} \end{cases}$	right-continuous

Remark 2.2.5. (i) Theorem 1.2.1 and (S1), (S3) imply that the continuity of any t-conorm is equivalent to its continuity with respect to the first (or second) variable.
(ii) The maximum t-conorm $S_{\mathbf{M}}$ is the only idempotent t-conorm.
(iii) For a continuous t-conorm S the Archimedean property is given by the simpler condition

$$S(x,x) > x\,, \qquad x \in (0,1)\,.$$

(iv) If a t-conorm S is continuous and Archimedean, then S is nilpotent if and only if there exists some nilpotent element of S, which is equivalent to the existence of some $x, y \in (0,1)$ such that $S(x,y) = 1$ (see [146], Theorem 2.18).

(v) If a t-conorm S is strict or nilpotent, then it is Archimedean. Conversely, every continuous and Archimedean t-conorm is either strict or nilpotent.
(vi) A continuous Archimedean t-conorm is positive if and only if it is strict.
(vii) If $\varphi \in \Phi$ and S is a continuous (Archimedean, strict, nilpotent) t-conorm, then S_φ is also a continuous (Archimedean, strict, nilpotent) t-conorm.
(viii) On the family of all t-conorms we can consider the partial order induced from the unit interval. Thus $S_{\mathbf{M}}$ is the least t-conorm, while $S_{\mathbf{D}}$ is the greatest t-conorm.

By the duality between t-conorms and t-norms the following representation theorems for t-conorms can be easily obtained from the characterizations of particular classes of t-norms (see [154], [146]).

Theorem 2.2.6. *For a function $S\colon [0,1]^2 \to [0,1]$ the following statements are equivalent:*

(i) S is a continuous Archimedean t-conorm.
(ii) S has a continuous additive generator, i.e., there exists a continuous, strictly increasing function $g\colon [0,1] \to [0,\infty]$ with $g(0) = 0$, which is uniquely determined up to a positive multiplicative constant, such that

$$S(x,y) = g^{(-1)}(g(x) + g(y)), \qquad x, y \in [0,1],$$

where $g^{(-1)}$ is the pseudo-inverse of g given by

$$g^{(-1)}(x) = \begin{cases} g^{-1}(x), & \text{if } x \in [0, g(1)], \\ 0, & \text{if } x \in (g(1), \infty]. \end{cases}$$

Remark 2.2.7. (i) The above representation of a t-conorm S can also be written without using the pseudo-inverse in the following way

$$S(x,y) = g^{-1}(\min(g(x) + g(y), g(1))), \qquad x, y \in [0,1].$$

(ii) S is a strict t-conorm if and only if each continuous additive generator g of S satisfies $g(1) = \infty$.
(iii) S is a nilpotent t-conorm if and only if each continuous additive generator g of S satisfies $g(1) < \infty$.
(iv) If S is a continuous Archimedean t-conorm with continuous additive generator g, then the function $\sigma\colon [0,1] \to [0,1]$ defined by $\sigma(x) = \exp(-g(x))$ is a multiplicative generator of S, i.e., a continuous, strictly decreasing function with $\sigma(0) = 1$ such that

$$S(x,y) = \sigma^{(-1)}(\sigma(x) \cdot \sigma(y)), \qquad x, y \in [0,1],$$

where $\sigma^{(-1)}$ is the pseudo-inverse of σ given by

$$\sigma^{(-1)}(x) = \begin{cases} 1, & \text{if } x \in [0, \sigma(1)), \\ \sigma^{-1}(x), & \text{if } x \in [\sigma(1), 1]. \end{cases}$$

Theorem 2.2.8. *For a function $S\colon [0,1]^2 \to [0,1]$ the following statements are equivalent:*

(i) S is a strict t-conorm.
(ii) S is Φ-conjugate with the probabilistic sum t-conorm $S_{\mathbf{P}}$, i.e., there exists $\varphi \in \Phi$ such that for all $x, y \in [0,1]$

$$S(x,y) = (S_{\mathbf{P}})_{\varphi}(x,y) = \varphi^{-1}(\varphi(x) + \varphi(y) - \varphi(x)\varphi(y)) .$$

Observe that in the above theorem the increasing bijection is not determined uniquely. It follows from the fact that φ can be defined as $\varphi(x) = 1 - \exp(-g(x))$, where g is a continuous additive generator of a strict t-conorm, which is uniquely determined up to a positive multiplicative constant.

Theorem 2.2.9. *For a function $S\colon [0,1]^2 \to [0,1]$ the following statements are equivalent:*

(i) S is a nilpotent t-conorm.
(ii) S is Φ-conjugate with the Łukasiewicz t-conorm $S_{\mathbf{LK}}$, i.e., there exists $\varphi \in \Phi$, which is uniquely determined, such that for all $x, y \in [0,1]$

$$S(x,y) = (S_{\mathbf{LK}})_{\varphi}(x,y) = \varphi^{-1}(\min(\varphi(x) + \varphi(y), 1)) .$$

For continuous t-conorms we have the following complete representation which is a dual result to Theorem 2.1.10 (see [146], Corollary 5.15).

Theorem 2.2.10. *For a function $S\colon [0,1]^2 \to [0,1]$ the following statements are equivalent:*

(i) S is a continuous t-conorm.
(ii) S is uniquely representable as an ordinal sum of continuous Archimedean t-conorms, i.e., there exists a uniquely determined (finite or countably infinite) index set A, a family of uniquely determined pairwise disjoint open sub-intervals $\{(a_\alpha, e_\alpha)\}_{\alpha \in A}$ of $[0,1]$ and a family of uniquely determined continuous Archimedean t-conorms $(S_\alpha)_{\alpha \in A}$ such that

$$S(x,y) = \begin{cases} a_\alpha + (e_\alpha - a_\alpha) \cdot S_\alpha\left(\frac{x-a_\alpha}{e_\alpha - a_\alpha}, \frac{y-a_\alpha}{e_\alpha - a_\alpha}\right), & \text{if } x, y \in [a_\alpha, e_\alpha] , \\ \max(x,y), & \text{otherwise} . \end{cases}$$

In this case we will write $S = (\langle a_\alpha, e_\alpha, S_\alpha \rangle)_{\alpha \in A}$.

From the Frank family of t-norms $T_{\mathbf{F}}^{\lambda}$, for every $\lambda \in [0, \infty]$, one can obtain their $N_{\mathbf{C}}$-dual t-conorms $S_{\mathbf{F}}^{\lambda}$ as follows:

$$S_{\mathbf{F}}^{\lambda}(x,y) = 1 - \left(T_{\mathbf{F}}^{\lambda}(1-x, 1-y)\right) , \qquad x, y \in [0,1] .$$

As noted earlier, these families of t-norms and t-conorms were obtained as the solutions of the Frank functional equation (2.1), as the following result shows (see [146], Theorem 5.14):

Theorem 2.2.11. *Let T be an Archimedean t-norm and S a t-conorm. Then the following statements are equivalent:*

(i) The pair (T, S) fulfills the following Frank functional equation

$$T(x, y) + S(x, y) = x + y , \qquad x, y \in [0, 1] . \tag{2.1}$$

(ii) There exists a $\lambda \in [0, \infty]$ such that $T = T_{\mathbf{F}}^{\lambda}$ and $S = S_{\mathbf{F}}^{\lambda}$.

2.3 Relationships between Negations, T-Norms and T-Conorms

By the duality of t-norms and t-conorms, a function T is a t-norm if and only if the function $1-T(1-x, 1-y)$ is a t-conorm. Moreover, if f and θ are the additive and multiplicative generators of the t-norm T, then $g(x) = f(1-x)$ and $\sigma(x) = \theta(1-x)$ are the additive and multiplicative generators of the t-conorm S (see KLEMENT et al. [146]). One can also consider the duality between t-norms and t-conorms for any strong (strict) negation other than the standard negation $N_{\mathbf{C}}$. Before dealing with De Morgan Triples, we discuss ways of obtaining negations from t-norms and t-conorms and the laws of excluded middle and contradiction.

2.3.1 Natural Negations of T-Norms and T-Conorms

As was noted at the end of previous chapter, one can associate a fuzzy negation to any t-norm or t-conorm (see [188], Definition 5.5.2; [146], p. 232).

Definition 2.3.1. *(i) Let T be a t-norm. A function $N_T \colon [0, 1] \to [0, 1]$ defined as*

$$N_T(x) = \sup\{y \in [0, 1] \mid T(x, y) = 0\} , \qquad x \in [0, 1] , \tag{2.2}$$

is called the natural negation *of T or the* negation induced *by T.*

(ii) Let S be a t-conorm. A function $N_S \colon [0, 1] \to [0, 1]$ defined as

$$N_S(x) = \inf\{y \in [0, 1] \mid S(x, y) = 1\} , \qquad x \in [0, 1] , \tag{2.3}$$

is called the natural negation *of S or the* negation induced *by S.*

Table 2.3 gives the natural negations of the basic t-norms and t-conorms.

Remark 2.3.2. (i) It is easy to prove that both N_T and N_S are fuzzy negations. In the literature N_T is also called the contour line C_0 of T, while N_S is called the contour line D_1 of S (see MAES and DE BAETS [158, 160]).

(ii) Since for any t-norm T and any t-conorm S we have $T(x, 0) = 0$ and $S(x, 1) = 1$ for all $x \in [0, 1]$, the appropriate sets in (2.2) and (2.3) are non-empty.

(iii) Notice that if $S(x, y) = 1$ for some $x, y \in [0, 1]$, then $y \geq N_S(x)$ and if $T(x, y) = 0$ for some $x, y \in [0, 1]$, then $y \leq N_T(x)$. Moreover, if $z < N_T(x)$, then $T(x, z) = 0$ and if $z > N_S(x)$, then $S(x, z) = 1$.

Table 2.3. Basic t-norms and t-conorms with their natural negations

t-norm T	N_T	t-conorm S	N_S
positive	$N_{\mathbf{D1}}$	positive	$N_{\mathbf{D2}}$
$T_{\mathbf{LK}}$	$N_{\mathbf{C}}$	$S_{\mathbf{LK}}$	$N_{\mathbf{C}}$
$T_{\mathbf{D}}$	$N_{\mathbf{D2}}$	$S_{\mathbf{D}}$	$N_{\mathbf{D1}}$
$T_{\mathbf{nM}}$	$N_{\mathbf{C}}$	$S_{\mathbf{nM}}$	$N_{\mathbf{C}}$

The next result, which was partially presented by MAES and DE BAETS [158], will be useful in the sequel.

Proposition 2.3.3. *If a t-conorm S is right-continuous, then*

(i) for every $x, y \in [0,1]$ the following equivalence holds:

$$S(x,y) = 1 \iff N_S(x) \leq y \, ; \tag{2.4}$$

(ii) the infimum in (2.3) is the minimum, i.e.,

$$N_S(x) = \min\{t \in [0,1] \mid S(x,t) = 1\} \, , \qquad x \in [0,1] \, , \tag{2.5}$$

where the right side exists for all $x \in [0,1]$;

(iii) N_S is right-continuous;

(iv) $S(N_S(x), x) = 1$ for all $x \in [0,1]$.

Proof. (i) Suppose that S is a right-continuous t-conorm and assume that $S(x,y) = 1$ for some $x, y \in [0,1]$, so $y \in \{t \in [0,1] \mid S(x,t) = 1\}$, and hence $N_S(x) \leq y$.
Now, let $N_S(x) \leq y$ for some $x, y \in [0,1]$. If $N_S(x) < y$, then there exists some $t' < y$ such that $S(x,t') = 1$, so the monotonicity of S implies that $S(x,y) = 1$. If $N_S(x) = y$, then either $y \in \{t \in [0,1] \mid S(x,t) = 1\}$ and therefore $S(x,y) = 1$, or $y \notin \{t \in [0,1] \mid S(x,t) = 1\}$. Thus, there exists a decreasing sequence $(t_i)_{i \in \mathbb{N}}$ such that $t_i > y$ and $S(x,t_i) = 1$ for all $i \in \mathbb{N}$ and $\lim_{i\to\infty} t_i = y$. By the right-continuity of S we get

$$S(x,y) = S(x, \lim_{i\to\infty} t_i) = \lim_{i\to\infty} S(x,t_i) = \lim_{i\to\infty} 1 = 1 \, ,$$

a contradiction.

(ii) From the previous point (i) we know that S and N_S satisfy (2.4). Since $N_S(x) \leq N_S(x)$ for all $x \in [0,1]$, one has $S(x, N_S(x)) = 1$, which means, by the definition of N_S, that the infimum in (2.3) is the minimum.

(iii) Let us assume that N_S is not right-continuous in some point $x_0 \in [0,1)$. Since N_S is decreasing, there exist $a, b \in [0,1]$ such that $a < b$ and

$$\begin{aligned} N_S(x) &\leq a \, , \qquad \text{for all } x > x_0 \, , \\ N_S(x_0) &= b \, . \end{aligned}$$

By (2.4) we get

$$S(x,a) = 1 , \qquad \text{for all } x > x_0 .$$

In the limit $x \to x_0$ we have $S(x_0, a) = 1$. Again from (2.4) we obtain $b = N_S(x_0) \leq a$, a contradiction to $a < b$. Therefore, N_S is a right-continuous function.

(iv) From point (ii) above we see that for every $x \in [0,1]$, $N_S(x)$ belongs to the set $\{t \in [0,1] \mid S(x,t) = 1\}$ in (2.5) and hence the claim. □

Similarly, we can prove the following:

Proposition 2.3.4. *If a t-norm T is left-continuous, then*

(i) for every $x, y \in [0,1]$ the following equivalence holds:

$$T(x,y) = 0 \iff N_T(x) \geq y , \tag{2.6}$$

(ii) the supremum in (2.2) is the maximum, i.e.,

$$N_T(x) = \max\{t \in [0,1] \mid T(x,t) = 0\} , \qquad x \in [0,1] ,$$

where the right side exists for all $x \in [0,1]$,

(iii) N_T is left-continuous,

(iv) $T(N_T(x), x) = 0$ for all $x \in [0,1]$.

The proof of the following result is given in Appendix A, Theorem A.0.7.

Theorem 2.3.5. *Let T be any t-norm and N_T its natural negation.*

(i) If N_T is continuous, then it is strong.

(ii) If N_T is discontinuous, then it is not strictly decreasing.

Similarly, we have

Theorem 2.3.6. *Let S be any t-conorm and N_S its natural negation.*

(i) If N_S is continuous, then it is strong.

(ii) If N_S is discontinuous then it is not strictly decreasing.

Corollary 2.3.7. *Let T be any t-norm and N_T its natural negation. Then the following statements are equivalent:*

(i) N_T is strictly decreasing.

(ii) N_T is continuous.

(iii) N_T is strict.

(iv) N_T is strong.

The problem of finding left-continuous t-norms with strong natural negation is deeply investigated in the literature. A few such families of left-continuous t-norms (up to a Φ-conjugation) with strong natural negations are known in the literature. The first of such families is the nilpotent class of t-norms, i.e., they are Φ-conjugate with the Łukasiewicz t-norm $T_{\mathbf{LK}}$. Another family consists of t-norms that are Φ-conjugate with the nilpotent minimum t-norm $T_{\mathbf{nM}}$. Yet another family is the class of t-norms that are Φ-conjugate with the Jenei t-norm family $T_{\mathbf{J}}^{\lambda}$, where

$$T_{\mathbf{J}}^{\lambda}(x,y) = \begin{cases} 0, & \text{if } x+y \le 1 \,, \\ x+y-1+\lambda, & \text{if } x+y > 1 \text{ and } x,y \in (\lambda, 1-\lambda] \,, \\ \min(x,y), & \text{otherwise} \,, \end{cases}$$

for $\lambda \in [0, 0.5]$, for $x, y \in [0,1]$. Note that $T_{\mathbf{J}}^{0} = T_{\mathbf{LK}}$ and $T_{\mathbf{J}}^{0.5} = T_{\mathbf{nM}}$. Recently, Maes and De Baets while studying fuzzified normal forms [159] obtained the following family of left-continuous t-norms $T_{\mathbf{MD}}^{\lambda}$, where

$$T_{\mathbf{MD}}^{\lambda}(x,y) = \begin{cases} 0, & \text{if } x+y \le 1 \,, \\ x+y-1, & \text{if } x+y > 1 \text{ and} \\ & \quad (\min(x,y) \in [0,\lambda] \text{ or } x+y \ge 2-\lambda) \,, \\ \min(x,y), & \text{if } x+y > 1 \text{ and } \min(x,y) \in (\lambda, 1-\lambda] \,, \\ 1-\lambda, & \text{otherwise} \,, \end{cases}$$

for $\lambda \in [0, 0.5)$, for $x, y \in [0,1]$. Note also that $T_{\mathbf{MD}}^{0} = T_{\mathbf{nM}}$.

For the methods to obtain these families see the works of Jenei [131, 132, 134, 135] connected with the rotation and the rotation-annihilation and the works of Maes and De Baets [159, 160, 161] connected with the triple rotation. In fact, it can be shown that every t-norm $T_{\mathbf{MD}}^{\lambda}$ can be obtained as a rotation-annihilation of a particular ordinal sum of the Łukasiewicz t-norm $T_{\mathbf{LK}}$ (see [159], p. 384).

2.3.2 Laws of Excluded Middle and Contradiction

In this section we will analyze the law of excluded middle and the law of contradiction, which in the classical case are tautologies and they have the following forms, respectively:

$$p \vee \neg p \qquad \text{and} \qquad p \wedge \neg p \,.$$

Definition 2.3.8. *Let S be a t-conorm and N a fuzzy negation. We say that the pair (S, N) satisfies the* law of excluded middle *if*

$$S(N(x), x) = 1 \,, \qquad x \in [0,1] \,. \tag{LEM}$$

Lemma 2.3.9. *If a t-conorm S and a fuzzy negation N satisfy* (LEM)*, then*

(i) $N \ge N_S$;
(ii) $N_S \circ N(x) \le x$, for all $x \in [0,1]$.

Proof. (i) On the contrary, if for some $x_0 \in [0,1]$ we have $N(x_0) < N_S(x_0)$, then $S(N(x_0), x_0) < 1$ by definition of N_S.
(ii) From Definition 2.3.1 we have

$$N_S(N(x)) = \inf\{y \in [0,1] \mid S(N(x), y) = 1\} \,, \qquad x \in [0,1] \,.$$

Now, since $S(N(x), x) = 1$ we have $x \ge N_S(N(x))$ for all $x \in [0,1]$. □

Remark 2.3.10. (i) Any t-conorm satisfies (LEM) with the greatest fuzzy negation $N_{\mathbf{D2}}$. Indeed, for any t-conorm S and $x \in [0,1]$ we have

$$S(N_{\mathbf{D2}}(x), x) = \begin{cases} S(1,x), & \text{if } x < 1 \\ S(0,x), & \text{if } x = 1 \end{cases} = \begin{cases} 1, & \text{if } x < 1 \\ x, & \text{if } x = 1 \end{cases} = 1 \,.$$

(ii) From Lemma 2.3.9 and Table 2.3 it follows that if S is a positive t-conorm (e.g., $S_{\mathbf{M}}$ or $S_{\mathbf{P}}$), then it satisfies (LEM) only with $N_{\mathbf{D2}}$.
(iii) Note that no t-conorm S satisfies (LEM) with the least negation $N_{\mathbf{D1}}$, since for any $x \in (0,1)$ we have $N_{\mathbf{D1}}(x) = 0$ and $S(N_{\mathbf{D1}}(x), x) = S(0,x) = x \neq 1$.
(iv) The fact that the conditions in Lemma 2.3.9 are only necessary and not sufficient follows from the following example. Consider the non-right-continuous nilpotent maximum t-conorm

$$S_{\mathbf{nM}^*}(x,y) = \begin{cases} 1, & \text{if } x + y > 1 \,, \\ \max(x,y), & \text{otherwise} \,, \end{cases} \qquad x, y \in [0,1] \,.$$

Then its natural negation is the classical negation, i.e., $N_{S_{\mathbf{nM}^*}} = N_{\mathbf{C}}$ and $N_{S_{\mathbf{nM}^*}} \circ N_{\mathbf{C}}(x) = x$ for all $x \in [0,1]$. However, the pair $(S_{\mathbf{nM}^*}, N_{\mathbf{C}})$ does not satisfy (LEM). Indeed, for $x = 0.5$ we get

$$S_{\mathbf{nM}^*}(N_S(0.5), 0.5) = S_{\mathbf{nM}^*}(0.5, 0.5) = 0.5 \,.$$

Interestingly, for the right-continuous t-conorms the condition (i) from Lemma 2.3.9 is both necessary and sufficient.

Proposition 2.3.11. *For a right-continuous t-conorm S and a fuzzy negation N the following statements are equivalent:*

(i) The pair (S, N) satisfies (LEM).
(ii) $N \geq N_S$.

Proof. From Lemma 2.3.9 it is enough to show that $(ii) \Longrightarrow (i)$. Assume that $N(x) \geq N_S(x)$ for all $x \in [0,1]$. By virtue of (2.4), we get that $S(x, N(x)) = 1$ for all $x \in [0,1]$, so the pair (S, N) satisfies (LEM). □

In the class of continuous functions we get the following important fact, which is a slight generalization of the result given by FODOR and ROUBENS [105] where the authors assume that N is a strict negation (see also [194] and [45]).

Proposition 2.3.12. *For a continuous t-conorm S and a continuous fuzzy negation N the following statements are equivalent:*

(i) The pair (S, N) satisfies (LEM).
(ii) S is a nilpotent t-conorm, i.e., S is Φ-conjugate with the Łukasiewicz t-conorm $S_{\mathbf{LK}}$, i.e., there exists $\varphi \in \Phi$, which is uniquely determined, such that S has the representation

$$S(x,y) = \varphi^{-1}(\min(\varphi(x) + \varphi(y), 1)) \,, \qquad x, y \in [0,1] \,,$$

and

$$N(x) \geq N_S(x) = \varphi^{-1}(1 - \varphi(x)) \,, \qquad x \in [0,1] \,.$$

Proof. $(i) \Longrightarrow (ii)$ Let S be a continuous t-conorm, N a continuous fuzzy negation and let the pair (S, N) satisfy (LEM). Since S is continuous, by Theorem 2.2.10, it is uniquely representable as an ordinal sum of continuous Archimedean t-conorms. The negation N is continuous, so Theorem 1.4.7 implies that it has exactly one fixed point $e \in (0,1)$ for which $S(e,e) = S(N(e),e) = 1$. Therefore, for all $x \geq e$ we have $S(x,x) = 1$. Thus S is neither idempotent nor strict. This fact follows also from Table 2.3, since they are both positive t-conorms.

Assume now that S is not Archimedean, so, in particular, it is not nilpotent. Since it is continuous, there exists $x_0 \in (0,1)$ such that $S(x_0, x_0) = x_0$ (cf. Remark 2.2.5(iii)). Hence there exists $a \in [x_0, e] \subset (0,1)$ such that the summand $\langle a, 1, S_1 \rangle$ of S is such that S_1 is a nilpotent t-conorm. Now, by Theorem 2.2.9, it is Φ-conjugate with the Łukasiewicz t-conorm $S_{\mathbf{LK}}$ for some unique $\phi \in \Phi$. One can calculate that in this case the natural negation of S has the following form:

$$N_S(x) = \begin{cases} 1, & \text{if } 0 \leq x \leq a\,, \\ a + (1-a)\phi^{-1}\left(1 - \phi\left(\frac{x-a}{1-a}\right)\right), & \text{if } a < x < 1\,, \\ 0, & \text{if } x = 1\,. \end{cases} \tag{2.7}$$

Indeed, firstly, we show that $N_S(x) = 1$ for $x \in [0,a]$. If we assume that $a \geq N_S(a)$, then by the monotonicity of S and Proposition 2.3.11 we get

$$a = S(a,a) \geq S(N_S(a), a) = 1\,,$$

contradictory to the assumption $a < 1$. Assume now that $a < N_S(a)$. The representation of an ordinal sum implies that $S(a,a) = a$ and $S(a,1) = 1$, so by the continuity of S there exists $y_0 > a$ such that $N_S(a) = S(a, y_0)$. Hence

$$N_S(a) = S(a, y_0) = S(S(a,a), y_0) = S(a, S(a, y_0)) = S(a, N_S(a)) = 1\,.$$

Since N_S is a negation, it is decreasing which implies $N_S(x) = 1$ for every $x \in [0,a]$. Next see that from the structure of the ordinal sum and the formula for $S_{\mathbf{LK}}$ we get, in particular, that

$$S(x,y) = a + (1-a)\min\left(\phi^{-1}\left(\phi\left(\frac{x-a}{1-a}\right) + \phi\left(\frac{y-a}{1-a}\right)\right), 1\right)\,,$$

for all $x, y \in [a,1]$ and $S(x,y) = \max(x,y)$ for all $x \in [a,1]$ and $y \in [0,a)$. Therefore, we obtain (2.7) for $x \in [a,1]$. Let us observe now that N_S is not continuous for $x = 1$, since $\lim_{x \to 1^-} N_S(x) = a < 1$. From Lemma 2.3.9(i) we get that $N \geq N_S$, but one can easily check that there does not exist any continuous fuzzy negation which is greater than or equal to N_S, a contradiction.

It shows that S is Archimedean, so it has to be nilpotent, and in this case $a = 0$. Therefore, S is Φ-conjugate with the Łukasiewicz t-conorm $S_{\mathbf{LK}}$ for some unique $\varphi \in \Phi$. Now, from (2.3) and (2.7), for $\phi = \varphi$ and $a = 0$ we have that $N_S(x) = \varphi^{-1}(1 - \varphi(x))$ for all $x \in [0,1]$ and because of Lemma 2.3.9(i) we get $N \geq N_S$.

$(ii) \Longrightarrow (i)$ The proof in this direction is immediate. □

Remark 2.3.13. The assumption that N is continuous is crucial above. As a counter-example consider the probabilistic sum t-conorm $S_{\mathbf{P}}$, which is continuous and satisfies (LEM) with the non-continuous negation $N_{\mathbf{D2}}$.

Definition 2.3.14. *Let T be a t-norm and N a fuzzy negation. We say that the pair (T, N) satisfies the* law of contradiction *if*

$$T(N(x), x) = 0 , \qquad x \in [0, 1] . \tag{LC}$$

By duality, all the results regarding the law of excluded middle LEM can be carried over to the law of contradiction (LC), of course, with the appropriate changes, viz., right-continuity substituted by left-continuity, reversing the direction of inequalities, changing 1 to 0, etc. We only list one result which is explicitly required in the sequel (cf. [105], Proposition 1.7).

Proposition 2.3.15. *For a continuous t-norm T and a continuous fuzzy negation N the following statements are equivalent:*

(i) The pair (T, N) satisfies (LC).
(ii) T is a nilpotent t-norm, i.e., T is Φ-conjugate with the Łukasiewicz t-norm $T_{\mathbf{LK}}$, i.e., there exists $\varphi \in \Phi$, which is uniquely determined, such that T has the representation

$$T(x, y) = \varphi^{-1}(\max(\varphi(x) + \varphi(y) - 1, 0)) , \qquad x, y \in [0, 1] ,$$

and

$$N(x) \le N_T(x) = \varphi^{-1}(1 - \varphi(x)) , \qquad x \in [0, 1] .$$

2.3.3 De Morgan Triples

It is well known that in classical set theory the following two De Morgan's laws are true: $(A \cup B)' = A' \cap B'$ and $(A \cap B)' = A' \cup B'$. Similarly, in classical logic, we have the following tautologies:

$$\neg(p \vee q) \equiv \neg p \wedge \neg q ,$$
$$\neg(p \wedge q) \equiv \neg p \vee \neg q .$$

In our context, they can be expressed, respectively, by

$$N(S(x, y)) = T(N(x), N(y)) , \qquad x, y \in [0, 1] , \tag{2.8}$$
$$N(T(x, y)) = S(N(x), N(y)) , \qquad x, y \in [0, 1] , \tag{2.9}$$

where T is a t-norm, S is a t-conorm and N is a fuzzy negation. In fuzzy logic, however, with myriad generalizations of the classical operations, the above two laws do not hold for all combinations of t-norms, t-conorms and fuzzy negations. Therefore, the following definition was introduced in the literature (see [105], Definition 1.12; [146], p. 232).

Definition 2.3.16. *Let T be a t-norm, S a t-conorm and N a strict negation. (T, S, N) is called a* De Morgan triple *if they satisfy* (2.8) *and* (2.9).

The following result can be easily obtained (see [146], p. 232).

Theorem 2.3.17. *For a t-norm T, t-conorm S and a strict negation N the following statements are equivalent:*

(i) The triple (T, S, N) is a De Morgan triple.
(ii) N is a strong negation and S is the N-dual of T, i.e.,

$$S(x, y) = N(T(N(x), N(y))) , \qquad x, y \in [0, 1] .$$

Using the above theorem it can be shown that the following relation exists between N_T and N_S.

Theorem 2.3.18. *Let T be a left-continuous t-norm and S a t-conorm. If (T, N_T, S) is a De Morgan triple, then*

(i) $N_S = N_T$ is a strong negation;
(ii) S is right-continuous.

Proof. (i) Firstly, observe that from Theorem 2.3.17 we get that N_T is a strong negation. Let us assume that $N_T \neq N_S$, i.e., there exists $x_0 \in (0, 1)$ such that $N_T(x_0) \neq N_S(x_0)$. Recall also from Remark 2.3.2 that if $T(x, y) = 0$, then $y \leq N_T(x)$. Now we consider the following two cases:

a) If $N_T(x_0) < N_S(x_0)$, then there exists $y \in (0, 1)$ such that $N_T(x_0) < y < N_S(x_0)$. Since N_T is a bijection, there exists $y_0 \in (0, 1)$ such that $N_T(y_0) = y$, i.e., $N_T(x_0) < N_T(y_0) < N_S(x_0)$. Now, by the monotonicity of S and T, S being the N_T-dual of T and the definitions of N_T, N_S we have

$$\begin{aligned} S(x_0, N_T(y_0)) \neq 1 &\Longrightarrow N_T(T(N_T(x_0), N_T \circ N_T(y_0))) \neq 1 \\ &\Longrightarrow T(N_T(x_0), y_0) \neq 0 \\ &\Longrightarrow N_T(y_0) < N_T(x_0) , \end{aligned}$$

a contradiction to our assumption.

b) If $N_T(x_0) > N_S(x_0)$, then there exists $y \in (0, 1)$ such that $N_T(x_0) > y > N_S(x_0)$. Since N_T is a bijection, there exists $y_0 \in (0, 1)$ such that $N_T(y_0) = y$, i.e., $N_T(x_0) > N_T(y_0) > N_S(x_0)$. Similarly as above we have

$$\begin{aligned} S(x_0, N_T(y_0)) = 1 &\Longrightarrow N_T(T(N_T(x_0), N_T \circ N_T(y_0))) = 1 \\ &\Longrightarrow T(N_T(x_0), y_0) = 0 \\ &\Longrightarrow N_T(y_0) \geq N_T(x_0) , \end{aligned}$$

a contradiction to our assumption.

(ii) If T is a left-continuous t-norm and N_T is a strong negation, then it is straight forward to see that S, as an N_T-dual of T, is right-continuous. □

2.4 (S,N)-Implications and S-Implications

2.4.1 Motivation, Definition and Examples

It is well-known in classical logic that the unary negation operator $\neg$ can combine with any other binary operator to generate rest of the binary operators. This distinction of the unary $\neg$ is also shared by the Boolean implication $\rightarrow$, if defined in the following usual way:

$$p \rightarrow q \equiv \neg p \vee q \ .$$

The above formula was the first to catch the attention of the researchers leading to the following class of fuzzy implications.

Definition 2.4.1. *A function $I\colon [0,1]^2 \rightarrow [0,1]$ is called an* (S,N)-implication *if there exist a t-conorm S and a fuzzy negation N such that*

$$I(x,y) = S(N(x),y) \ , \qquad x,y \in [0,1] \ . \tag{2.10}$$

If N is a strong fuzzy negation, then I is called a strong implication *or* S-implication. *Moreover, if I is an (S,N)-implication generated from S and N, then we will often denote it by $I_{S,N}$.*

Example 2.4.2. Table 2.4 lists few of the well-known (S,N)-implications along with their t-conorms and fuzzy negations from which they have been obtained. Fig. 2.7(a) gives the plot of the Dubois-Prade implication $I_{\mathbf{DP}}$ (see DUBOIS and PRADE [85]).

It is easy to see that for a fixed fuzzy negation N, if S_1, S_2 are two comparable t-conorms such that $S_1 \leq S_2$, then $I_{S_1,N} \leq I_{S_2,N}$. Similarly, if S is a fixed t-conorm, then for two comparable fuzzy negations N_1, N_2 such that $N_1 \leq N_2$ we get $I_{S,N_1} \leq I_{S,N_2}$. Thus, from Example 1.4.4 and Table 2.4 we have that

Table 2.4. Examples of basic (S,N)-implications

S	N	(S,N)-implication $I_{S,N}$
$S_{\mathbf{M}}$	$N_{\mathbf{C}}$	$I_{\mathbf{KD}}$
$S_{\mathbf{P}}$	$N_{\mathbf{C}}$	$I_{\mathbf{RC}}$
$S_{\mathbf{LK}}$	$N_{\mathbf{C}}$	$I_{\mathbf{LK}}$
$S_{\mathbf{D}}$	$N_{\mathbf{C}}$	$I_{\mathbf{DP}}(x,y) = \begin{cases} y, & \text{if } x = 1 \\ 1-x, & \text{if } y = 0 \\ 1, & \text{if } x < 1 \text{ and } y > 0 \end{cases}$
$S_{\mathbf{nM}}$	$N_{\mathbf{C}}$	$I_{\mathbf{FD}}$
any S	$N_{\mathbf{D1}}$	$I_{\mathbf{D}}(x,y) = \begin{cases} 1, & \text{if } x = 0 \\ y, & \text{if } x > 0 \end{cases}$
any S	$N_{\mathbf{D2}}$	$I_{\mathbf{WB}}$

$I_{\mathbf{D}}$ and $I_{\mathbf{WB}}$ are, respectively, the least and the greatest (S,N)-implications. Further, $I_{\mathbf{KD}}$ and $I_{\mathbf{DP}}$ are, respectively, the least and the greatest S-implications generated from the classical negation $N_{\mathbf{C}}$ (cf. [147], Theorem 11.1). Since there does not exist any least or greatest strong negation, there does not exist any least or greatest S-implication.

Firstly, we investigate some properties of (S,N)-implications. It is important to note that all (S,N)-implications are fuzzy implications in the sense of Definition 1.1.1.

Proposition 2.4.3. *If $I_{S,N}$ is an (S,N)-implication, then*

(i) $I_{S,N} \in \mathcal{FI}$ and $I_{S,N}$ satisfies (NP), (EP),
(ii) $N_{I_{S,N}} = N$,
(iii) $I_{S,N}$ satisfies R-CP(N),
(iv) If N is strict, then $I_{S,N}$ satisfies L-CP(N^{-1}),
(v) If N is strong, then $I_{S,N}$ satisfies CP(N).

Proof. (i) The monotonicity of $I_{S,N}$, i.e., (I1) and (I2), is the consequence of the monotonicity of the t-conorm S and the negation N. Additionally,

$$I_{S,N}(0,0) = S(N(0),0) = S(1,0) = 1 \ ,$$
$$I_{S,N}(1,1) = S(N(1),1) = S(0,1) = 1 \ ,$$
$$I_{S,N}(1,0) = S(N(1),0) = S(0,0) = 0 \ ,$$

which shows that $I_{S,N} \in \mathcal{FI}$. Further, $I_{S,N}$ satisfies (NP), since

$$I_{S,N}(1,y) = S(N(1),y) = S(0,y) = y \ , \qquad y \in [0,1] \ .$$

From the associativity and the commutativity of S we also have (EP), i.e., for any $x, y, z \in [0,1]$

$$\begin{aligned} I_{S,N}(x, I_{S,N}(y,z)) &= S(N(x), S(N(y),z)) = S(N(y), S(N(x),z)) \\ &= I_{S,N}(y, I_{S,N}(x,z)) \ . \end{aligned}$$

(ii) For any $x \in [0,1]$ we have

$$N_{I_{S,N}}(x) = I_{S,N}(x,0) = S(N(x),0) = N(x) \ .$$

(iii) Follows from Corollary 1.5.25.
(iv) If N is a strict negation, then because of Remark 1.5.19(i) we can deduce that $I_{S,N}$ satisfies L-CP(N^{-1}).
(v) If N is a strong negation, then because of Lemma 1.5.6(ii) we can deduce that $I_{S,N}$ satisfies CP(N). □

Remark 2.4.4. (i) The conditions (I1) - (I5), (NP) and (EP) are not enough to obtain an (S,N)-implication, i.e., they are necessary but not sufficient. Let us consider the Yager implication $I_{\mathbf{YG}}$. From Table 1.7 we know that it satisfies (NP) and (EP). Let us assume that it is an (S,N)-implication for a t-conorm S and a fuzzy negation N. By previous proposition $N_{I_{\mathbf{YG}}} = N$, but from Table 1.7 we have $N_{I_{\mathbf{YG}}} = N_{\mathbf{D1}}$. Hence from Table 2.4 we get that $I_{\mathbf{YG}} = I_{\mathbf{D}}$, a contradiction. Thus $I_{\mathbf{YG}}$ is not an (S,N)-implication. Similarly, one can show that the Gödel implication $I_{\mathbf{GD}}$, the Goguen implication $I_{\mathbf{GG}}$ and the Rescher implication $I_{\mathbf{RS}}$ are not (S,N)-implications.

(ii) While every (S,N)-implication in the Table 2.4 satisfies both (NP) and (EP), it can be easily verified that $I_{\mathbf{DP}}, I_{\mathbf{WB}}$ also satisfy (IP), while $I_{\mathbf{LK}}, I_{\mathbf{FD}}$ satisfy (OP) and hence (IP).

(iii) If an (S,N)-implication is generated from a non-strict negation, then it may satisfy neither the law of contraposition (CP) nor the law of left contraposition (L-CP). As an example consider the Weber implication $I_{\mathbf{WB}}$.

(iv) From Proposition 2.4.3(ii) it follows that an (S,N)-implication is generated by a unique negation.

The last result in this subsection shows some relationship between (S,N)-implications and their Φ-conjugations.

Theorem 2.4.5. *If $I_{S,N}$ is an (S,N)-implication, then the Φ-conjugate of $I_{S,N}$ is also an (S,N)-implication generated from the Φ-conjugate t-conorm of S and the Φ-conjugate fuzzy negation of N, i.e., if $\varphi \in \Phi$, then*

$$(I_{S,N})_\varphi = I_{S_\varphi, N_\varphi} .$$

Proof. Let $\varphi \in \Phi$ and let $I_{S,N}$ be an (S,N)-implication obtained from a t-conorm S and a fuzzy negation N. By Remark 2.2.5(vii) and Proposition 1.4.8, functions S_φ and N_φ are a t-conorm and a fuzzy negation, respectively. Hence I_{S_φ, N_φ} is an (S,N)-implication. Finally

$$\begin{aligned}(I_{S,N})_\varphi(x,y) &= \varphi^{-1}(I_{S,N}(\varphi(x), \varphi(y))) = \varphi^{-1}(S(N(\varphi(x)), \varphi(y))) \\ &= \varphi^{-1}(S(\varphi \circ \varphi^{-1}(N(\varphi(x))), \varphi(y))) \\ &= \varphi^{-1}(S(\varphi(N_\varphi(x)), \varphi(y))) = S_\varphi(N_\varphi(x), y) \\ &= I_{S_\varphi, N_\varphi}(x,y) ,\end{aligned}$$

for every $x, y \in [0,1]$. ⊔

2.4.2 Characterizations of (S,N)-Implications

Towards presenting characterizations of some classes of (S,N)-implications we consider the dual situation now, i.e., the method of obtaining t-conorms from fuzzy implications and fuzzy negations.

Proposition 2.4.6. *Let $I \in \mathcal{FI}$ and N be a fuzzy negation. Let us define a function $S_{I,N} \colon [0,1]^2 \to [0,1]$ as follows:*

$$S_{I,N}(x,y) = I(N(x), y) , \qquad x, y \in [0,1] . \tag{2.11}$$

Then

(i) $S_{I,N}(1,x) = S_{I,N}(x,1) = 1$, *for all* $x \in [0,1]$;
(ii) $S_{I,N}$ *is increasing in both variables, i.e.,* $S_{I,N}$ *satisfies* (S3);
(iii) $S_{I,N}$ *is commutative, i.e., it satisfies* (S1) *if and only if* I *satisfies L-CP(*N*).*

In addition, if I *satisfies L-CP(*N*), then*

(iv) $S_{I,N}$ *satisfies* (S4) *if and only if* I *satisfies* (NP);
(v) $S_{I,N}$ *is associative i.e., it satisfies* (S2) *if and only if* I *satisfies* (EP).

Proof. (i) By the boundary conditions (LB) and (RB), we have

$$S_{I,N}(1,x) = I(N(1),x) = I(0,x) = 1 \ ,$$
$$S_{I,N}(x,1) = I(N(x),1) = 1 \ ,$$

for all $x \in [0,1]$.

(ii) The fact that $S_{I,N}$ satisfies (S3) is a direct consequence of the monotonicity of I in the first and second variables and the monotonicity of N.

(iii) If I satisfies L-CP(N), then

$$S_{I,N}(x,y) = I(N(x),y) = I(N(y),x) = S_{I,N}(y,x) \ , \qquad x \in [0,1] \ .$$

Conversely, by the commutativity of $S_{I,N}$, we have

$$I(N(x),y) = S_{I,N}(x,y) = S_{I,N}(y,x) = I(N(y),x) \ , \qquad x,y \in [0,1] \ ,$$

i.e., I satisfies L-CP(N).

(iv) If I satisfies (NP), then by the commutativity of $S_{I,N}$, we have

$$S_{I,N}(x,0) = S_{I,N}(0,x) = I(N(0),x) = I(1,x) = x \ , \qquad x \in [0,1] \ .$$

Conversely,

$$I(1,x) = I(N(0),x) = S_{I,N}(0,x) = x \ , \qquad x \in [0,1] \ .$$

(v) If I satisfies (EP), then because of L-CP(N) we have

$$\begin{aligned} S_{I,N}(x, S_{I,N}(y,z)) &= I(N(x), I(N(y),z)) = I(N(x), I(N(z),y)) \\ &= I(N(z), I(N(x),y)) = S_{I,N}(z, S_{I,N}(x,y)) \\ &= S_{I,N}(S_{I,N}(x,y),z) \ , \end{aligned}$$

for any $x,y,z \in [0,1]$.

Conversely, let $S_{I,N}$ be associative. Since I satisfies L-CP(N), $S_{I,N}$ is also commutative and hence

$$\begin{aligned} I(x, I(y,z)) &= S_{I,N}(N(x), S_{I,N}(N(y),z)) \\ &= S_{I,N}(N(y), S_{I,N}(N(x),z)) = I(y, I(x,z)) \ , \end{aligned}$$

for all $x,y,z \in [0,1]$. □

Remark 2.4.7. From the above fact and Lemma 1.5.14 it follows that if we want to obtain a t-conorm by (2.11), we should consider a fuzzy implication I for which N_I is a continuous fuzzy negation. Again, (NP) and (EP) are necessary but not sufficient to define a t-conorm S by (2.11). As an interesting example we take the fuzzy implication $I_{\mathbf{D}}$. By Example 2.4.2, it is an (S,N)-implication and hence it satisfies (NP) and (EP). Since $N_{I_{\mathbf{D}}} = N_{\mathbf{D1}}$ is not continuous, we cannot obtain a t-conorm by (2.11) for any fuzzy negation.

If $I \in \mathcal{FI}$ satisfies (NP), (EP) and N_I is a continuous fuzzy negation, then by virtue of Corollary 1.5.18 and Proposition 2.4.6, we obtain that (2.11) can be considered for the modified pseudo-inverse of the natural negation of I.

Corollary 2.4.8. *If $I \in \mathcal{FI}$ satisfies* (NP), (EP) *and N_I is a continuous fuzzy negation, then the function S_I defined by*

$$S_I(x, y) = I(\mathfrak{N}_I(x), y) , \qquad x, y \in [0, 1] , \tag{2.12}$$

is a t-conorm, where $\mathfrak{N}_I$ is as defined in (1.16).

Remark 2.4.9. (i) If an $I \in \mathcal{FI}$ is such that it satisfies both (NP), (EP) and N_I is a strict negation, then $\mathfrak{N}_I = N_I^{-1}$. Now, because of Corollary 1.5.18, the fuzzy implication I satisfies (L-CP) with N_I^{-1}. By Lemma 1.5.14(v), (2.11) can be considered only for the inverse of the natural negation of I. In this case the formula for a t-conorm is the following:

$$S_I(x, y) = I(N_I^{-1}(x), y) , \qquad x, y \in [0, 1] .$$

(ii) In the special case, when N_I is a strong negation, we have that $N_I^{-1} = N_I$, i.e., the formula for a t-conorm is also unique and is the following:

$$S_I(x, y) = I(N_I(x), y) , \qquad x, y \in [0, 1] .$$

From the above discussion, we can state the following results which characterize some subclasses of (S,N)-implications.

Theorem 2.4.10. *For a function $I\colon [0, 1]^2 \to [0, 1]$ the following statements are equivalent:*

(i) I is an (S,N)-implication with a continuous fuzzy negation N.
(ii) I satisfies (I1), (EP) *and N_I is a continuous fuzzy negation.*

Moreover, the representation of (S,N)-implication (2.10) *is unique in this case.*

Proof. (i) $\Longrightarrow$ (ii) Assume that I is an (S,N)-implication based on a t-conorm S and a continuous negation N. By Proposition 2.4.3, it is a fuzzy implication which satisfies (NP) and (EP). In particular, I satisfies (I1). Moreover, by the same result and our assumptions $N_I = N$ is continuous.

(ii) $\Longrightarrow$ (i) Firstly, note that from Lemma 1.5.24 it follows that I satisfies (I3), (I4), (I5), (NP) and (R-CP) only with respect to N_I. Lemma 1.5.20(i) implies that I satisfies also (I2), so $I \in \mathcal{FI}$. By virtue of Corollary 1.5.18, the implication I satisfies L-CP($\mathfrak{N}_I$). Further, by Corollary 2.4.8, the function S_I defined by (2.12) is a t-conorm. We will show that $I_{S_I,N_I} = I$. Fix arbitrarily $x, y \in [0,1]$. If $x \in \mathrm{Ran}(\mathfrak{N}_I)$, then by (1.11) we have

$$I_{S_I,N_I}(x,y) = S_I(N_I(x), y) = I(\mathfrak{N}_I \circ N_I(x), y) = I(x,y) .$$

If $x \notin \mathrm{Ran}(\mathfrak{N}_I)$, then from the continuity of N_I there exists $x_0 \in \mathrm{Ran}(\mathfrak{N}_I)$ such that $N_I(x) = N_I(x_0)$. Firstly, see that $I(x,y) = I(x_0,y)$ for all $y \in [0,1]$. Indeed, let us fix arbitrarily $y \in [0,1]$. From the continuity of N_I there exists $y' \in [0,1]$ such that $N_I(y') = y$, so

$$\begin{aligned} I(x,y) &= I(x, N_I(y')) = I(y', N_I(x)) = I(y', N_I(x_0)) = I(x_0, N_I(y')) \\ &= I(x_0, y) . \end{aligned}$$

From the above fact we get

$$I_{S_I,N_I}(x,y) = S_I(N_I(x), y) = S_I(N_I(x_0), y) = I(x_0, y) = I(x,y) ,$$

and hence I is an (S,N)-implication.

Finally, assume that there exist two continuous fuzzy negations N_1, N_2 and two t-conorms S_1, S_2 such that $I(x,y) = S_1(N_1(x), y) = S_2(N_2(x), y)$ for all $x, y \in [0,1]$. Fix arbitrarily $x_0, y_0 \in [0,1]$. Observe that $N_1 = N_2 = N_I$. Now, since N_I is continuous, there exists $x_1 \in [0,1]$ such that $N_I(x_1) = x_0$. Thus $S_1(x_0, y_0) = S_1(N_I(x_1), y_0) = S_2(N_I(x_1), y_0) = S_2(x_0, y_0)$, i.e., $S_1 = S_2$. In fact, $S = S_I$ defined by (2.12). □

Note that, by virtue of Lemma 1.5.20(i), we can substitute in the above theorem the requirement (I1) by (I2). The next two results are special cases of Theorem 2.4.10.

Theorem 2.4.11. *For a function $I\colon [0,1]^2 \to [0,1]$ the following statements are equivalent:*

(i) I is an (S,N)-implication with a strict negation N.
(ii) I satisfies (I1), (EP) *and N_I is a strict negation.*

Theorem 2.4.12. *For a function $I\colon [0,1]^2 \to [0,1]$ the following statements are equivalent:*

(i) I is an S-implication with a strong negation N.
(ii) I satisfies (I1), (EP) *and N_I is a strong negation.*

Remark 2.4.13. (i) The representation of an (S,N)-implication in the above two theorems is unique.
(ii) In both the above theorems we can substitute the axiom (I1) by (I2).
(iii) In Table 2.5 we show that the properties in Theorems 2.4.10, 2.4.11 and 2.4.12 are independent from each other. Here N_F means, of course, the function defined by (1.13), i.e., $N_F(x) = F(x,0)$ for $x \in [0,1]$.

As an important consequence of Theorem 2.4.10 we get the following result.

Table 2.5. The mutual independence of properties in Theorem 2.4.10

Function F	(I1)	(EP)	N_F is a continuous negation
$F = I_{\mathbf{RS}}$	✓	×	×
$F = T_{\mathbf{P}}$	×	✓	×
$F(x,y) = \begin{cases} 0, & \text{if } x < 0.5 \text{ and } y > 0 \\ 1, & \text{if } x \geq 0.5 \text{ and } y > 0 \\ 1-x, & \text{if } y = 0 \end{cases}$	×	×	✓
$F = I_{\mathbf{YG}}$	✓	✓	×
$F(x,y) = \begin{cases} 1-x, & \text{if } y = 0 \\ y^2, & \text{if } x = 1 \\ 1, & \text{otherwise} \end{cases}$	✓	×	✓
$F(x,y) = \begin{cases} 1-x, & \text{if } y = 0 \\ y, & \text{if } x = 1 \\ 0.5, & \text{otherwise} \end{cases}$	×	✓	✓

Proposition 2.4.14. *For a function $I\colon [0,1]^2 \to [0,1]$ the following statements are equivalent:*

(i) I is a continuous (S,N)-implication.
(ii) I is an (S,N)-implication with continuous S and N.

Proof. *(i)* $\Longrightarrow$ *(ii)* Let I be a continuous (S,N)-implication generated from some t-conorm S and some fuzzy negation N. The negation $N = N_I$ is continuous since $I(x,0) = S(N(x),0) = N(x)$ is continuous. Now, because of Theorem 2.4.10, we see that $S = S_I$, where S_I is given by (2.12). We show that S is continuous. From Remark 2.2.5(i) it is enough to show the continuity of S with respect to the second variable. Let us assume that S is not continuous with respect to the second variable in some point $(x_0, y_0) \in [0,1]^2$. Let $x_1 = \mathfrak{N}_I(x_0)$. By (2.12), we get that I is not continuous with respect to the second variable in the point (x_1, y_0), a contradiction. Therefore, S is continuous.

(ii) $\Longrightarrow$ *(i)* This is obvious, since a composition of continuous functions is continuous. □

Remark 2.4.15. If we assume that $I \in \mathcal{FI}$ satisfies (NP), (EP) and N_I is a strict negation, then using the De Morgan laws we obtain the first method of generating t-norms from fuzzy implications (see also Theorem 2.5.30):

$$\begin{aligned} T_I(x,y) &= N_I(S_I(N_I(x), N_I(y))) = N_I(I(N_I^{-1}(N_I(x)), N_I(y))) \\ &= N_I(I(x, N_I(y))) \,, \qquad x,y \in [0,1] \,. \end{aligned} \tag{2.13}$$

In the rest of this section we discuss (S,N)-implications with respect to the other desirable properties from Definition 1.3.1 that they satisfy.

2.4.3 (S,N)-Implications and the Identity Principle

Since not all (S,N)-implications, even S-implications, satisfy the identity principle (IP) (e.g., $I_{\mathbf{RC}}$ and $I_{\mathbf{KD}}$), in this section we analyze this property for this family.

Lemma 2.4.16. *For a t-conorm S and a fuzzy negation N the following statements are equivalent:*

(i) The (S,N)-implication $I_{S,N}$ satisfies (IP).
(ii) The pair (S, N) satisfies (LEM).

Proof. If $I_{S,N}$ satisfies (IP), then $S(N(x), x) = I(x, x) = 1$, for all $x \in [0, 1]$. Conversely, if the pair (S, N) satisfies (LEM), then $I(x, x) = S(N(x), x) = 1$ for all $x \in [0, 1]$. □

Now, using Propositions 2.3.12 and 2.4.14, we get (cf. TRILLAS and VALVERDE [239], Theorem 3.3)

Theorem 2.4.17. *For a t-conorm S and a fuzzy negation N the following statements are equivalent:*

(i) $I_{S,N}$ is a continuous (S,N)-implication that satisfies (IP).
(ii) S is a nilpotent t-conorm and $N(x) \geq N_S(x)$ for all $x \in [0, 1]$.

By virtue of Theorems 2.4.12, 1.4.13 and 2.2.9, we have also the following characterization (cf. [45], Theorem 7).

Corollary 2.4.18. *For a function $I\colon [0, 1]^2 \to [0, 1]$ the following statements are equivalent:*

(i) I is continuous and satisfies (I1) *(or* (I2)*),* (NP), (EP), (IP) *and N_I is a strong negation.*
(ii) I is an S-implication obtained from a continuous t-conorm S and satisfies (IP).
(iii) There exist unique $\varphi \in \Phi$ and $\psi \in \Phi$ such that

$$I(x, y) = \varphi^{-1}(\min(\varphi(\psi^{-1}(1 - \psi(x))) + \varphi(y), 1)) , \qquad x, y \in [0, 1] ,$$

and

$$\psi^{-1}(1 - \psi(x)) \geq \varphi^{-1}(1 - \varphi(x)) , \qquad x \in [0, 1] .$$

2.4.4 (S,N)-Implications and the Ordering Property

As noted earlier, not all natural generalizations of the classical implication to multi-valued logic satisfy (OP). In the following section we discuss results on (S,N)-implications with respect to their ordering property.

Theorem 2.4.19. *For a t-conorm S and a fuzzy negation N the following statements are equivalent:*

(i) The (S,N)-implication $I_{S,N}$ satisfies (OP).
(ii) $N = N_S$ is a strong negation and the pair (S, N_S) satisfies (LEM).

Proof. (i) $\Longrightarrow$ (ii) If an (S,N)-implication I satisfies (OP), then it satisfies (IP). By Lemma 2.4.16, the pair (S, N) satisfies (LEM). Therefore, from Lemma 2.3.9 we know that $N(x) \geq N_S(x)$ and $N_S \circ N(x) \leq x$ for all $x \in [0,1]$. Assume that for some $x_0 \in [0,1]$ we have $N_S \circ N(x_0) < x_0$. Then there exists $y \in [0,1]$ such that $N_S \circ N(x_0) < y < x_0$. Now, $I(x_0, y) = S(N(x_0), y) = 1$, i.e., $x_0 \leq y$ from (OP), which is a contradiction. Hence $x = N_S \circ N(x)$ for all $x \in [0,1]$. By virtue of Lemma 1.4.9, we get that N_S is a continuous fuzzy negation and N is a strictly decreasing fuzzy negation. Further, since N_S is a negation, we get

$$N_S(N_S(x)) \geq N_S(N(x)) = x\,, \qquad x \in [0,1]\,. \tag{2.14}$$

Let us fix arbitrary $x_0 \in [0,1]$. Thus, again from the monotonicity of N_S and (2.14), we have

$$N_S(N_S(N_S(x_0))) \leq N_S(x_0)\,.$$

Now, since (2.14) is true for all $x \in [0,1]$, it is also true for $x = N_S(x_0)$, i.e.,

$$N_S(N_S(N_S(x_0))) \geq N_S(x_0)\,.$$

Since x_0 was arbitrarily fixed, as a result we obtain

$$N_S(N_S(N_S(x))) = N_S(x)\,, \qquad x \in [0,1]\,.$$

However, N_S is a continuous negation and by the Darboux property (see Appendix A), for every $y \in [0,1]$ there exists $x \in [0,1]$ such that $y = N_S(x)$ and

$$N_S(N_S(y)) = N_S(N_S(N_S(x))) = N_S(x) = y\,, \qquad y \in [0,1]\,.$$

Thus N_S is a strong negation and it is obvious now that $N = N_S$ from (2.14).

(ii) $\Longrightarrow$ (i) Assume that S is any t-conorm such that N_S is strong and the pair (S, N_S) satisfies (LEM). We will show that the (S,N)-implication generated from S and N_S satisfies (OP). To this end, fix arbitrarily $x, y \in [0,1]$ such that $x \leq y$. By the monotonicity of S we have

$$I_{S,N_S}(x,y) = S(N_S(x), y) \geq S(N_S(x), x) = 1\,.$$

If $I_{S,N_S}(x,y) = S(N_S(x), y) = 1$ for some $x, y \in [0,1]$, then $y \geq N_S \circ N_S(x) = x$, by Remark 2.3.2(iii). Thus I_{S,N_S} satisfies (OP). □

In the class of continuous functions we have the following very important theorem (cf. Trillas and Valverde [239], Smets and Magrez [226], Fodor and Roubens [105] and Baczyński [9]).

Theorem 2.4.20. *For a function* $I\colon [0,1]^2 \to [0,1]$ *the following statements are equivalent:*

(i) I *is continuous and satisfies* (EP), (OP).
(ii) I *is an (S,N)-implication obtained from a continuous t-conorm* S *and a continuous fuzzy negation* N*, which satisfies* (OP).
(iii) I *is a continuous (S,N)-implication, which satisfies* (OP).
(iv) I *is an (S,N)-implication obtained from a nilpotent t-conorm and its natural negation.*
(v) I *is* Φ*-conjugate with the Łukasiewicz implication* $I_{\mathbf{LK}}$*, i.e., there exists* $\varphi \in \Phi$*, which is uniquely determined, such that for all* $x, y \in [0,1]$*,*

$$I(x,y) = (I_{\mathbf{LK}})_{\varphi}(x,y) = \varphi^{-1}(\min(1 - \varphi(x) + \varphi(y), 1)) . \tag{2.15}$$

Proof. *(i)* $\Longrightarrow$ *(ii)* Assume that I is a continuous function which satisfies (EP) and (OP). This implies that N_I is a continuous function. By virtue of Corollary 1.5.12, we obtain that $I \in \mathcal{FI}$ and I satisfies (NP), (IP). Moreover, N_I is a strong negation. Now, from Theorem 2.4.10, it follows that I is an (S,N)-implication generated from some t-conorm S and some strong negation N. Since I is continuous, Proposition 2.4.14 implies that S is also continuous.

(ii) $\Longleftrightarrow$ *(iii)* This equivalence is a consequence of Proposition 2.4.14.

(iv) $\Longleftrightarrow$ *(v)* This equivalence is obvious and follows from the representation of nilpotent t-conorms presented in Theorem 2.2.9.

(ii) $\Longrightarrow$ *(v)* Let us assume that I is an (S,N)-implication obtained from a continuous t-conorm S and a continuous fuzzy negation N, which satisfies (OP). From Proposition 2.4.3(i) and (ii) we see that I also satisfies (NP), (EP) and $N = N_I$. Again, by virtue of Corollary 1.5.12, we get that N_I is a strong negation and I satisfies also (IP). From Corollary 2.4.18 there exist unique $\varphi \in \Phi$ and $\psi \in \Phi$ such that

$$I(x,y) = \varphi^{-1}(\min(\varphi(\psi^{-1}(1 - \psi(x))) + \varphi(y), 1)) , \qquad x, y \in [0,1] ,$$

and

$$N_I(x) = \psi^{-1}(1 - \psi(x)) \geq \varphi^{-1}(1 - \varphi(x)) = N_S(x) , \qquad x \in [0,1] .$$

Since I satisfies (OP), by virtue of Theorem 2.4.19, we know that $N_I = N_S$. In particular, I has the form (2.15).

(v) $\Longrightarrow$ *(i)* Observe that $I_{\mathbf{LK}} \in \mathcal{CFI}$ satisfies (EP) and (OP). By Proposition 1.3.6, the function I given by (2.15) is also continuous and satisfies (EP), (OP). □

Remark 2.4.21. (i) The continuity of I is important in Theorem 2.4.20. Consider the Fodor implication $I_{\mathbf{FD}}$. It is a non-continuous (S,N)-implication obtained from the right-continuous t-conorm $S_{\mathbf{nM}}$, so it is not Φ-conjugate with the Łukasiewicz implication $I_{\mathbf{LK}}$. However, it satisfies both (EP) and (OP).

(ii) Note that all the properties in Theorem 2.4.20(i) are mutually independent. For example, the Goguen implication $I_{\mathbf{GG}}$ satisfies both (EP) and (OP) but is not continuous at the point $(0,0)$, while the Reichenbach implication $I_{\mathbf{RC}}$ is continuous and satisfies (EP) but does not satisfy (OP), whereas the Baczyński implication $I_{\mathbf{BZ}}$ is continuous and satisfies (OP) but does not satisfy (EP).

2.4.5 Intersections between Subfamilies of (S,N)-Implications

In this subsection we summarize the known intersections among some subfamilies of (S,N)-implications based on the results cited and obtained so far. Let us denote the different families of (S,N)-implications as follows:

- $\mathbb{I}_{\mathbb{S},\mathbb{N}}$ – the family of all (S,N)-implications;
- $^{\mathbb{C}}\mathbb{I}_{\mathbb{S},\mathbb{N}}$ – the family of all continuous (S,N)-implications;
- $\mathbb{I}_{\mathbb{S}_{\mathbb{C}},\mathbb{N}_{\mathbb{C}}}$ – the family of all (S,N)-implications obtained from continuous t-conorms and continuous negations;
- $\mathbb{I}_{\mathbb{S}}$ – the family of all S-implications;
- $\mathbb{I}_{\mathbb{S},\mathbb{N}_{\mathbb{S}}}$ – the family of all (S,N)-implications obtained from t-conorms and their natural negations;
- $\mathbb{I}_{\mathbb{S}^*,\mathbb{N}_{\mathbb{S}}^*}$ – the family of all (S,N)-implications obtained from right-continuous t-conorms and their natural negations which are strong;
- $\mathbb{I}_{\mathbb{LK}}$ – the family of all implications Φ-conjugate with the Łukasiewicz implication $I_{\mathbf{LK}}$.

By Proposition 2.4.14, we get

$$^{\mathbb{C}}\mathbb{I}_{\mathbb{S},\mathbb{N}} = \mathbb{I}_{\mathbb{S}_{\mathbb{C}},\mathbb{N}_{\mathbb{C}}} \,.$$

By Propositions 2.4.3(ii), 2.3.3 and 2.3.12, We also have the following equalities:

$$\mathbb{I}_{\mathbb{LK}} = \mathbb{I}_{\mathbb{S},\mathbb{N}_{\mathbb{S}}} \cap \mathbb{I}_{\mathbb{S}_{\mathbb{C}},\mathbb{N}_{\mathbb{C}}} = \mathbb{I}_{\mathbb{S}_{\mathbb{C}},\mathbb{N}_{\mathbb{C}}} \cap \mathbb{I}_{\mathbb{S}^*,\mathbb{N}_{\mathbb{S}}^*} \,.$$

We will only prove the first equality, since the second has a similar reasoning. It is obvious that any I that is Φ-conjugate with the Łukasiewicz implication $I_{\mathbf{LK}}$ belongs to the above intersection. Conversely, let $I \in \mathbb{I}_{\mathbb{S},\mathbb{N}_{\mathbb{S}}} \cap \mathbb{I}_{\mathbb{S}_{\mathbb{C}},\mathbb{N}_{\mathbb{C}}}$. From Proposition 2.4.3(ii), we see that $N_I = N_S$ is continuous. Since N_S is the natural negation of the t-conorm S, from Theorem 2.3.6 we have that N_S is strong. Moreover, by the uniqueness of the representation of the (S,N)-implication I and hence of the t-conorm S, see Remark 2.4.13(i), we have that S is continuous. Now, by Proposition 2.3.3(iv), we have that (S, N_S) satisfies the law of excluded middle (LEM). Hence, by Theorem 2.4.19, I satisfies (OP). Thus I is continuous and satisfies both (EP) and (OP). Now, the result follows from Theorem 2.4.20.

One can also easily show that

$$\mathbb{I}_{\mathbb{LK}} \subsetneq \mathbb{I}_{\mathbb{S}^*,\mathbb{N}_{\mathbb{S}}^*} \subsetneq \mathbb{I}_{\mathbb{S},\mathbb{N}_{\mathbb{S}}} \cap \mathbb{I}_{\mathbb{S}} \,.$$

Final result is also diagrammatically represented in Fig. 2.1. The fuzzy implications $I_{S_{\mathbf{nM}^*},N_{\mathbf{C}}}$ and $I_{\mathbf{MK}}$ are given in Example 2.4.22.

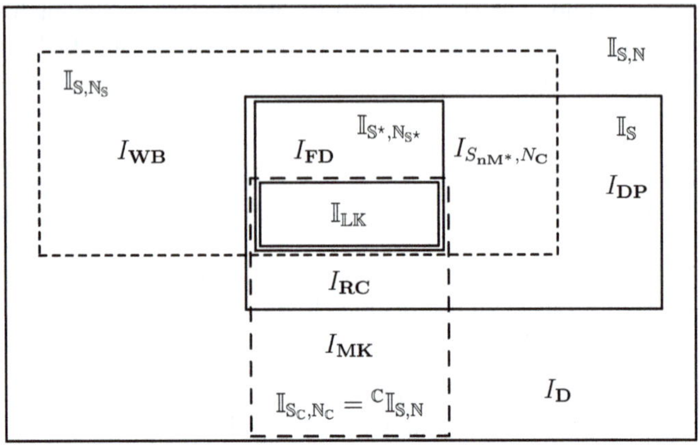

Fig. 2.1. Intersections among the subfamilies of (S,N)-implications

Example 2.4.22. (i) Let us consider the non-right-continuous nilpotent maximum t-conorm $S_{\mathbf{nM}^*}$ given in Remark 2.3.10(iv). Then the (S,N)-implication generated from $S_{\mathbf{nM}^*}$ and its natural negation, which is the classical negation $N_{\mathbf{C}}$, is given by

$$I_{S_{\mathbf{nM}^*},N_{\mathbf{C}}}(x,y) = \begin{cases} 1, & \text{if } x < y\,, \\ \max(1-x,y), & \text{if } x \geq y\,, \end{cases} \qquad x,y \in [0,1]\,.$$

In particular, $I_{S_{\mathbf{nM}^*},N_{\mathbf{C}}}$ is not continuous.

(ii) Let us denote by $I_{\mathbf{MK}}$ the (S,N)-implication obtained from $S_{\mathbf{M}}$ and the strict but non-involutive negation $N_{\mathbf{K}}(x) = 1 - x^2$, i.e.,

$$I_{\mathbf{MK}}(x,y) = \max(1-x^2,y)\,, \qquad x,y \in [0,1]\,.$$

2.5 R-Implications

2.5.1 Motivation, Definition and Examples

From Sect. 2.4 we see that (S,N)-implications are a generalization of the material implication of the classical two-valued logic to fuzzy logic. From the isomorphism that exists between classical two-valued logic and classical set theory one can immediately note the following set theoretic identity:

$$A' \cup B = (A \setminus B)' = \bigcup \{C \subseteq X \mid A \cap C \subseteq B\}\,,$$

where A, B are subsets of some universal set X. The above identity gives another way of defining Boolean implications and is employed in intuitionistic logic. Fuzzy implications obtained as the generalization of the above identity form the family of residuated implications, usually called as R-implications in the literature. In this section, we investigate this family of fuzzy implications, analogous to our treatment of (S,N)-implications in Sect. 2.4.

Definition 2.5.1. *A function* $I\colon [0,1]^2 \to [0,1]$ *is called an* R-implication *if there exists a t-norm* T *such that*

$$I(x,y) = \sup\{t \in [0,1] \mid T(x,t) \le y\}\,, \qquad x, y \in [0,1]\,. \tag{2.16}$$

If I *is an R-implication generated from a t-norm* T*, then we will often denote it by* I_T*.*

Since for every t-norm T and all $x \in [0,1]$ we have $T(x,0) = 0$, one can easily observe that the appropriate set in (2.16) is non-empty. The name 'R-implication' is a short version of 'residual implication', and I_T is also called as 'the residuum of T'. This class of implications is related to a residuation concept from the intuitionistic logic. Although the function I_T obtained as in Definition 2.5.1 is a fuzzy implication for any t-norm T, see Theorem 2.5.4, the following result shows that only in the case the t-norm T is left-continuous, the pair (T, I_T) has a very important property, known in the literature under various names, namely, *the Galois connection, adjunction property, residuation property* (cf. Gottwald [113], Proposition 5.4.2 and Corollary 5.4.1).

Proposition 2.5.2. *For a t-norm* T *the following statements are equivalent:*

(i) T *is left-continuous.*

(ii) T *and* I_T *form an adjoint pair, i.e., they satisfy the following residual principle*

$$T(x,z) \le y \iff I_T(x,y) \ge z\,, \qquad x, y, z \in [0,1]\,. \tag{RP}$$

(iii) The supremum in (2.16) *is the maximum, i.e.,*

$$I_T(x,y) = \max\{t \in [0,1] \mid T(x,t) \le y\}\,, \tag{2.17}$$

where the right side exists for all $x, y \in [0,1]$.

Proof. *(i)* $\Longrightarrow$ *(ii)* Suppose firstly that T is a left-continuous t-norm and assume that $T(x,z) \le y$ for some $x, y, z \in [0,1]$. This implies, in particular, that $z \in \{t \in [0,1] \mid T(x,t) \le y\}$, and hence $I_T(x,y) \ge z$.

Conversely, assume that $z \le I_T(x,y)$ for some $x, y, z \in [0,1]$. We consider two cases now. If $z < I_T(x,y)$, then there exists some $t' > z$ such that $T(x,t') \le y$, so (T3) implies $T(x,z) \le y$. If $z = I_T(x,y)$, then either $z \in \{t \in [0,1] \mid T(x,t) \le y\}$ and therefore $T(x,z) \le y$, or $z \notin \{t \in [0,1] \mid T(x,t) \le y\}$. Thus, there exists an increasing sequence $(t_i)_{i \in \mathbb{N}}$ such that $t_i < z$, $T(x,t_i) \le y$ for all $i \in \mathbb{N}$ and $\lim_{i\to\infty} t_i = z$. By the left-continuity of T we get

$$T(x,z) = T(x, \lim_{i\to\infty} t_i) = \lim_{i\to\infty} T(x,t_i) \le y\,.$$

(ii) $\Longrightarrow$ *(iii)* Assume that T and I_T form an adjoint pair, i.e. they satisfy (RP). Since $I_T(x,y) \le I_T(x,y)$, we have $T(x, I_T(x,y)) \le y$, which means, by the definition of I_T, that the supremum in (2.16) is the maximum.

(iii) $\Longrightarrow$ *(i)* Since any t-norm T is increasing and commutative, it is enough to show that T is infinitely sup-distributive, i.e., $T(x, \sup_{s\in S} y_s) = \sup_{s\in S} T(x, y_s)$,

where $x, y_s \in [0,1]$ for every $s \in S$. Observe firstly that from the monotonicity of T we always have one inequality

$$T(x, \sup_{s\in S} y_s) \geq \sup_{s\in S} T(x, y_s) .$$

Let $y = \sup_{s\in S} T(x, y_s)$. This implies that $T(x, y_s) \leq y$ for every $s \in S$. Therefore, $y_s \in \{t \in [0,1] | T(x,t) \leq y\}$ for every $s \in S$ and, consequently, $y_s \leq I_T(x, y)$ for every $s \in S$. Thus $\sup_{s\in S} y_s \leq I_T(x, y)$. By the monotonicity of T, we have

$$T(x, \sup_{s\in S} y_s) \leq T(x, I_T(x, y)) \leq y = \sup_{s\in S} T(x, y_s) .$$

From the above inequalities we get that T is infinitely sup-distributive, and therefore, left-continuous. □

Example 2.5.3. Table 2.6 lists few of the well-known R-implications along with their t-norms from which they have been obtained. Note that $T_{\mathbf{D}}$ is not left-continuous t-norm, but still $I_{\mathbf{WB}}$ is a fuzzy implication.

Table 2.6. Examples of basic R-implications

t-norm T	R-implication I_T
$T_{\mathbf{M}}$	$I_{\mathbf{GD}}$
$T_{\mathbf{P}}$	$I_{\mathbf{GG}}$
$T_{\mathbf{LK}}$	$I_{\mathbf{LK}}$
$T_{\mathbf{D}}$	$I_{\mathbf{WB}}$
$T_{\mathbf{nM}}$	$I_{\mathbf{FD}}$

2.5.2 Properties of R-Implications

Now we examine the basic properties of R-implications. Firstly, it is easy to see that if T_1 and T_2 are comparable t-norms such that $T_1 \leq T_2$, then $I_{T_1} \geq I_{T_2}$. Thus, from Table 2.6 we have that $I_{\mathbf{GD}}$ and $I_{\mathbf{WB}}$ are, respectively, the least and the greatest R-implications. Further, since there does not exist any least left-continuous t-norm, there does not exist any greatest R-implication generated from left-continuous t-norm. $I_{\mathbf{GD}}$ is again the least R-implication generated from a left-continuous t-norm.

Similarly as for (S,N)-implications, it is important to note that all R-implications are fuzzy implications in the sense of Definition 1.1.1.

Theorem 2.5.4. *If T is a t-norm, then $I_T \in \mathcal{FI}$. Moreover I_T satisfies* (NP), (IP) *and the natural negation of I_T is also the natural negation of the t-norm T.*

Proof. Let T be a t-norm and let I_T be a function defined by (2.16). Firstly, we show that I_T satisfies conditions from Definition 1.1.1. Let $x_1, x_2, y \in [0,1]$ and

$x_1 \leq x_2$. We have to show that $I_T(x_1, y) \geq I_T(x_2, y)$, which is equivalent to the inequality

$$\sup\{t \in [0,1] \mid T(x_1, t) \leq y\} \geq \sup\{t \in [0,1] \mid T(x_2, t) \leq y\} ,$$

and so it is enough to show the inclusion

$$\{t \in [0,1] \mid T(x_1, t) \leq y\} \supset \{t \in [0,1] \mid T(x_2, t) \leq y\} .$$

Take any $t \in [0, 1]$ such that $T(x_2, t) \leq y$. Since $x_1 \leq x_2$, from the monotonicity of a t-norm T we get $T(x_1, t) \leq T(x_2, t)$, thus $T(x_1, t) \leq y$. Therefore, I_T satisfies (I1). Now, let $x, y_1, y_2 \in [0, 1]$ and $y_1 \leq y_2$. By similar deduction, we get that (I2) follows from

$$\{t \in [0,1] \mid T(x, t) \leq y_1\} \subset \{t \in [0,1] \mid T(x, t) \leq y_2\} ,$$

which is the consequence of the assumption that $y_1 \leq y_2$. Moreover,

$$\begin{aligned}
I_T(0,0) &= \sup\{t \in [0,1] \mid T(0,t) \leq 0\} = \sup\{t \in [0,1] \mid 0 \leq 0\} = 1 , \\
I_T(1,1) &= \sup\{t \in [0,1] \mid T(1,t) \leq 1\} = \sup\{t \in [0,1] \mid t \leq 1\} = 1 , \\
I_T(1,0) &= \sup\{t \in [0,1] \mid T(1,t) \leq 0\} = \sup\{t \in [0,1] \mid t \leq 0\} = 0 ,
\end{aligned}$$

which shows that $I_T \in \mathcal{FI}$. Further, for any $y \in [0, 1]$ we get

$$I_T(1, y) = \sup\{t \in [0,1] \mid T(1,t) \leq y\} = \sup\{t \in [0,1] \mid t \leq y\} = y ,$$

so I_T satisfies (NP). Finally, since $T(x, 1) = x$ for all $x \in [0, 1]$, we have

$$I_T(x, x) = \sup\{t \in [0,1] \mid T(x,t) \leq x\} = 1 .$$

The fact that $N_{I_T} = N_T$ can be seen from Definition 2.3.1(i). □

Remark 2.5.5. From the above result and Table 1.7 we get that the Reichenbach implication $I_{\mathbf{RC}}$, the Kleene-Dienes implication $I_{\mathbf{KD}}$, the Rescher implication $I_{\mathbf{RS}}$ and the Yager implication $I_{\mathbf{YG}}$ are not R-implications.

Without any additional assumptions on a t-norm T, the R-implication I_T may not satisfy other basic properties.

Example 2.5.6. (i) The Weber implication $I_{\mathbf{WB}}$ is an R-implication (see Table 2.6) which satisfies (EP), but does not satisfy (OP). Moreover, since the natural negation of $I_{\mathbf{WB}}$ is discontinuous, $I_{\mathbf{WB}}$ does not satisfy either (CP) or (L-CP) with any fuzzy negation (cf. Table 1.9).

(ii) Consider the border continuous but non-left-continuous t-norm given in [146], Example 1.24(i) as follows:

$$T_{\mathbf{B}^*}(x, y) = \begin{cases} 0, & \text{if } (x, y) \in (0, 0.5)^2 , \\ \min(x, y), & \text{otherwise} . \end{cases}$$

Then the R-implication generated from T is given by

$$I_{\mathbf{TB}^*}(x,y) = \begin{cases} 1, & \text{if } x \le y\,, \\ 0.5, & \text{if } x > y \text{ and } x \in [0,0.5)\,, \\ y, & \text{otherwise}\,. \end{cases}$$

See Fig. 2.2(a) for the plot of $I_{\mathbf{TB}^*}$. Obviously, $I_{\mathbf{TB}^*}$ satisfies (OP), but not (EP). To see this, for $x = 0.4, y = 0.5$ and $z = 0.3$ we have

$$\begin{aligned} I_{\mathbf{TB}^*}(x, I_{\mathbf{TB}^*}(y,z)) &= I_{\mathbf{TB}^*}(0.4, I_{\mathbf{TB}^*}(0.5,0.3)) \\ &= I_{\mathbf{TB}^*}(0.4,0.3) = 0.5\,, \end{aligned}$$

while

$$\begin{aligned} I_{\mathbf{TB}^*}(y, I_{\mathbf{TB}^*}(x,z)) &= I_{\mathbf{TB}^*}(0.5, I_{\mathbf{TB}^*}(0.4,0.3)) = I_{\mathbf{TB}^*}(0.5,0.5) \\ &= 1 \neq 0.5\,. \end{aligned}$$

(iii) Consider now the t-norm $T_{\mathbf{B}}$ given in [146], Example 1.24(ii) as follows:

$$T_{\mathbf{B}}(x,y) = \begin{cases} 0, & \text{if } (x,y) \in (0,1)^2 \setminus [0.5,1)^2\,, \\ \min(x,y), & \text{otherwise}\,, \end{cases}$$

which is neither border continuous nor left-continuous. Then the R-implication generated from $T_{\mathbf{B}}$ is given by

$$I_{\mathbf{TB}}(x,y) = \begin{cases} 1, & \text{if } x \le y \text{ or } x,y \in [0,0.5)\,, \\ 0.5, & \text{if } x \in [0.5,1) \text{ and } y \in [0,0.5)\,, \\ y, & \text{otherwise}\,. \end{cases}$$

See Fig. 2.2(b) for the plot of $I_{\mathbf{TB}}$. It is obvious that $I_{\mathbf{TB}}$ does not satisfy (OP). Now putting $x = 0.8, y = 0.5$ and $z = 0.3$ we have

$$I_{\mathbf{TB}}(x, I_{\mathbf{TB}}(y,z)) = I_{\mathbf{TB}}(0.8, I_{\mathbf{TB}}(0.5,0.3)) = I_{\mathbf{TB}}(0.8,0.5) = 0.5\,,$$

while

$$I_{\mathbf{TB}}(y, I_{\mathbf{TB}}(x,z)) = I_{\mathbf{TB}}(0.5, I_{\mathbf{TB}}(0.8,0.3)) = I_{\mathbf{TB}}(0.5,0.5) = 1\,.$$

Hence $I_{\mathbf{TB}}$ does not satisfy (EP), too.

(iv) Interestingly, if we consider the following non-left-continuous nilpotent minimum t-norm (see [160], p. 851)

$$T_{\mathbf{nM}^*}(x,y) = \begin{cases} 0, & \text{if } x + y < 1\,, \\ \min(x,y), & \text{otherwise}\,, \end{cases}$$

then the R-implication generated from $T_{\mathbf{nM}^*}$ is the Fodor implication $I_{\mathbf{FD}}$, which satisfies both (EP) and (OP).

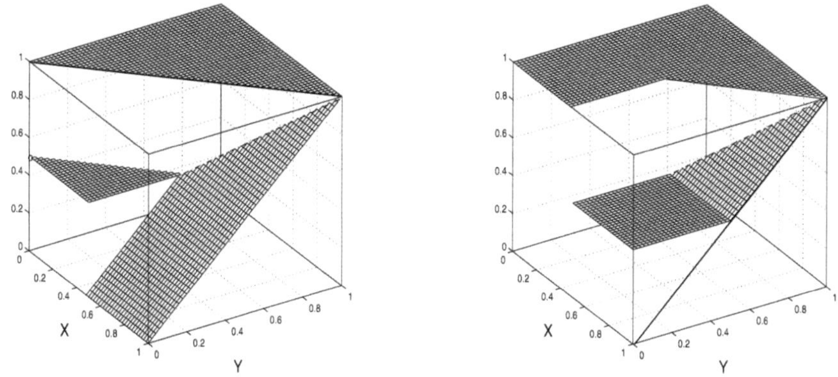

(a) $I_{\mathbf{TB}^*}$ (b) $I_{\mathbf{TB}}$

Fig. 2.2. Plots of the R-implications $I_{\mathbf{TB}^*}$ and $I_{\mathbf{TB}}$ (see Example 2.5.6(ii) and (iii))

When a t-norm T is left-continuous, then the R-implication generated from it satisfies all basic properties.

Theorem 2.5.7. *If T is a left-continuous t-norm, then $I_T \in \mathcal{FI}$ and it satisfies* (NP), (EP), (IP), (OP). *Moreover, I_T is left-continuous with respect to the first variable and right-continuous with respect to the second variable.*

Proof. From Theorem 2.5.4 we know that $I_T \in \mathcal{FI}$ and it satisfies (NP), (IP). To prove (EP) it is enough to show

$$I_T(x, I_T(y, z)) = I_T(T(x, y), z) , \qquad z, y, z \in [0, 1] ,$$

and use the commutativity of T. By Proposition 2.5.2, the R-implication I_T is given by (2.17) and the pair (T, I_T) satisfies (RP), so

$$\begin{aligned} I_T(x, I_T(y, z)) &= \max\{t \in [0, 1] \mid T(x, t) \leq I_T(y, z)\} \\ &= \max\{t \in [0, 1] \mid T(y, T(x, t)) \leq z\} \\ &= \max\{t \in [0, 1] \mid T(T(y, x), t) \leq z\} \\ &= \max\{t \in [0, 1] \mid T(T(x, y), t) \leq z\} \\ &= I_T(T(x, y), z) , \end{aligned}$$

for all $x, y, z \in [0, 1]$.

Now we show that I_T satisfies (OP). Let $x, y \in [0, 1]$. If $x \leq y$, then $T(x, 1) = x \leq y$, so $I_T(x, y) = 1$. Conversely, if $I_T(x, y) = 1$, then because of (RP) we get $T(x, 1) \leq y$, i.e., $x \leq y$.

Let us assume now, that I_T is not left-continuous with respect to the first variable in some point $(x_0, y_0) \in (0, 1] \times [0, 1]$. Since I_T satisfies (I1), there exist $a, b \in [0, 1]$, such that $a > b$ and

$$\begin{aligned} I_T(x, y_0) &\geq a , \qquad \text{for all } x < x_0 , \\ I_T(x_0, y_0) &= b . \end{aligned}$$

By (RP), we get

$$T(x, a) \leq y_0 , \qquad \text{for all } x > x_0 .$$

Since T is left-continuous, in the limit $x \to x_0$ we have $T(x_0, a) \leq y_0$. Once again, from (RP) we obtain $b = I_T(x_0, y_0) \geq a$, a contradiction to $a > b$. Therefore, I_T is a left-continuous function with respect to the first variable.

Finally, let us assume that I_T is not right-continuous with respect to the second variable in some point $(x_0, y_0) \in [0,1] \times [0,1)$. Since I_T satisfies (I2), there exist $a, b \in [0,1]$, such that $a > b$ and

$$I_T(x_0, y) \geq a , \qquad \text{for all } y > y_0 ,$$
$$I_T(x_0, y_0) = b .$$

By (RP) we get

$$T(x_0, a) \leq y , \qquad \text{for all } y > y_0 .$$

In the limit $y \to y_0$ we have $T(x_0, a) \leq y_0$. Again from (RP) we obtain $b = I_T(x_0, y_0) \geq a$, a contradiction to $a > b$. Therefore, I_T is a right-continuous function with respect to the second variable. □

Remark 2.5.8. Although both the t-norms $T_{\mathbf{B}^*}$ and $T_{\mathbf{B}}$ are not left-continuous, $T_{\mathbf{B}^*}$ is a border continuous t-norm which, as we show below, is an equivalent condition for the corresponding I_T to satisfy (OP).

Proposition 2.5.9. *For a t-norm T the following statements are equivalent:*

(i) T is border continuous.
(ii) I_T satisfies (OP).

Proof. Let T be any t-norm and for any fixed $x \in [0,1]$, consider the vertical segment $T_x(\cdot) = T(x, \cdot)$. Obviously, T_x is a one-variable function from $[0,1]$ to $[0,x]$. Note that, if T is a border continuous t-norm, then for every $x \in (0,1)$ there exists a neighborhood $U_x = (a_x, 1]$, where $a_x \in (0,1)$ is dependent on the chosen x, such that T_x is continuous on U_x.

(i) $\Longrightarrow$ *(ii)* Let T be a border continuous t-norm. On the contrary, let I_T not satisfy (OP). Since for any T we have that $I_T(x, y) = 1$ whenever $x \leq y$, there exist $x_0, y_0 \in (0,1)$ such that $x_0 > y_0$ and $I_T(x_0, y_0) = 1$. Let

$$x' = T_{x_0}(a_{x_0}) = T(x_0, a_{x_0}) \leq x_0 .$$

On the one hand, if $y_0 > x'$, then there exists $1 \neq t \in U_{x_0}$ such that $T_{x_0}(t) = y_0$, contradicting our assumption that $I_T(x_0, y_0) = 1$. On the other hand, if $y_0 \leq x'$, then by definition, $I_T(x_0, y_0) \leq a_x < 1$. Hence I_T satisfies (OP).

(ii) $\Longrightarrow$ *(i)* Let I_T satisfy (OP). On the contrary, if T is not border continuous, then there exists an $x_0 \in (0,1)$ such that $\lim_{y \to 1^-} T(x_0, y) = z < x_0$. Now, by definition $I_T(x_0, z) = \sup\{t \in [0,1] \mid T(x_0, t) \leq z\} = 1$, a contradiction to the fact that I_T satisfies (OP). □

From Theorem 2.5.7 and Proposition 2.5.9 we see that any left-continuous t-norm is border continuous, but the converse is not true (see $T_{\mathbf{B}^*}$ in Example 2.5.6(ii)).

The following result shows that the Φ-conjugate of an R-implication generated from any t-norm is also an R-implication.

Proposition 2.5.10. *If I_T is an R-implication, then the Φ-conjugate of I_T is also an R-implication generated from the Φ-conjugate t-norm of T, i.e., if $\varphi \in \Phi$, then*

$$(I_T)_\varphi = I_{T_\varphi} . \tag{2.18}$$

Proof. Let $\varphi \in \Phi$. If T is any t-norm, then by Remark 2.1.4(vii) the function T_φ is also a t-norm. Hence I_{T_φ} is an R-implication. Now, from the continuity of the bijection φ we have

$$\begin{aligned}
(I_T)_\varphi(x,y) &= \varphi^{-1}(I_T(\varphi(x),\varphi(y))) \\
&= \varphi^{-1}\left(\sup\{t \in [0,1] \mid T(\varphi(x),t) \le \varphi(y)\}\right) \\
&= \sup\{\varphi^{-1}(t) \in [0,1] \mid \varphi^{-1}(T(\varphi(x),t) \le y)\} \\
&= \sup\{t \in [0,1] \mid \varphi^{-1}(T(\varphi(x),\varphi(t)) \le y)\} \\
&= \sup\{t \in [0,1] \mid T_\varphi(x,t) \le y\} \\
&= I_{T_\varphi}(x,y) ,
\end{aligned}$$

for any $x, y \in [0,1]$. □

From Remark 2.1.4(viii) and the above Proposition 2.5.10 we have the following corollary (see also [81]).

Proposition 2.5.11. *For a t-norm T the following statements are equivalent:*

(i) $(I_T)_\varphi = I_T$, *for all* $\varphi \in \Phi$.
(ii) $I_T = I_{\mathbf{WB}}$ *or* $I_T = I_{\mathbf{GD}}$.

2.5.3 Characterizations and Representations of R-Implications

Our main goal in this subsection is to present the characterization of R-implications. In fact, currently, such a characterization is available only for R-implications obtained from left-continuous t-norms. We also discuss the representations of R-implications for some special classes of left-continuous t-norms. To do this we consider the dual situation now, i.e., the method of obtaining t-norms from fuzzy implications by a residuation principle. Since for every $I \in \mathcal{FI}$ we have (RB), the function $T_I \colon [0,1]^2 \to [0,1]$ defined by

$$T_I(x,y) = \inf\{t \in [0,1] \mid I(x,t) \ge y\} , \qquad x,y \in [0,1] , \tag{2.19}$$

is a well defined function of two variables, i.e., the appropriate set in (2.19) is non-empty.

Remark 2.5.12. It is important to note that (2.19) does not always generate a t-norm. For example, if I is the Reichenbach implication, $I_{\mathbf{RC}}$, then for $x > 0$ we obtain $T_I(x,1) = 1$, so T_I does not satisfy (T4).

Similar to Proposition 2.5.2 we have the following result.

Proposition 2.5.13. *For $I \in \mathcal{FI}$ the following statements are equivalent:*

(i) I is right-continuous with respect to the second variable.
(ii) I and T_I form an adjoint pair, i.e., they satisfy

$$T_I(x,y) \le z \iff I(x,z) \ge y\,, \qquad x,y,z \in [0,1]\,. \tag{RP*}$$

(iii) The infimum in (2.19) is the minimum, i.e.,

$$T_I(x,y) = \min\{t \in [0,1] \mid I(x,t) \ge y\}\,, \tag{2.20}$$

where the right side exists for all $x,y \in [0,1]$.

Proof. *(i)* $\Longrightarrow$ *(ii)* Suppose firstly that I is a fuzzy implication which is right-continuous with respect to the second variable and also that $I(x,z) \ge y$ for some $x,y,z \in [0,1]$. This implies that $z \in \{t \in [0,1] \mid I(x,t) \ge y\}$ and hence $T_I(x,y) \le z$. Conversely, let us assume that $T_I(x,y) \le z$ for some $x,y,z \in [0,1]$. We consider two cases. If $T_I(x,y) < z$, then there exists some $t' < z$ such that $I(x,t') \ge y$, so from (I2) we have $I(x,z) \ge y$. If $z = T_I(x,y)$, then either $z \in \{t \in [0,1] \mid I(x,t) \ge y\}$ and thus $I(x,z) \ge y$, or $z \notin \{t \in [0,1] \mid I(x,t) \ge y\}$. Therefore, there exists a decreasing sequence $(t_i)_{i\in\mathbb{N}}$ such that $t_i > z$, $I(x,t_i) \ge y$ for all $i \in \mathbb{N}$ and $\lim_{i\to\infty} t_i = z$. By the right-continuity of I with respect to the second variable, we get $I(x,z) = I(x, \lim_{i\to\infty} t_i) = \lim_{i\to\infty} I(x,t_i) \ge y$.

(ii) $\Longrightarrow$ *(iii)* Assume that $I \in \mathcal{FI}$ and T_I form an adjoint pair. Since $T_I(x,y) \le T_I(x,y)$, one has $I(x, T_I(x,y)) \ge y$, which means, by the definition of T_I that the infimum in (2.19) is the minimum.

(iii) $\Longrightarrow$ *(i)* Since any fuzzy implication is increasing with respect to the second variable, it is enough to show that I is infinitely inf-distributive with respect to the second variable, i.e., $I(x, \inf_{s\in S} y_s) = \inf_{s\in S} I(x,y_s)$, where $x, y_s \in [0,1]$ for every $s \in S$. Observe firstly that from the monotonicity of I for the second variable (I2) we always have the following inequality

$$I(x, \inf_{s\in S} y_s) \le \inf_{s\in S} I(x,y_s)\,.$$

Let $y = \inf_{s\in S} I(x,y_s)$. This implies that $I(x,y_s) \ge y$ for every $s \in S$. Therefore, $y_s \in \{t \in [0,1] \mid I(x,t) \ge y\}$ for every $s \in S$ and, consequently, $y_s \ge I_T(x,y)$ for every $s \in S$. Thus $\inf_{s\in S} y_s \ge T_I(x,y)$. By (I2), we have

$$I(x, \inf_{s\in S} y_s) \ge I(x, T_I(x,y)) \ge y = \inf_{s\in S} I(x,y_s)\,.$$

From the above inequalities we get that I is infinitely inf-distributive in the second variable, and therefore, right-continuous with respect to the second variable. □

The next theorem is very important in this section. We will show that with some additional assumptions we can obtain a left-continuous t-norm (cf. FODOR and ROUBENS [105], Theorem 1.14; GOTTWALD [113], Theorem 5.4.1).

Theorem 2.5.14. *If a function* $I\colon [0,1]^2 \to [0,1]$ *satisfies* (I2), (EP), (OP) *and is right-continuous with respect to the second variable, then* T_I *defined by* (2.20) *is a left-continuous t-norm. Moreover* $I = I_{T_I}$, *i.e.,*

$$I(x,y) = \max\{t \in [0,1] \mid T_I(x,t) \le y\}\,, \qquad x,y \in [0,1]\,.$$

Proof. Since I satisfies (EP) and (OP), by Lemma 1.3.4, it satisfies (I1), (I3), (I4), (I5), (NP) and (IP). In particular, since I satisfies (I2), it is a fuzzy implication. Now we show that T_I is a t-norm.

(T1) (*Commutativity*). Let $x,y \in [0,1]$ be fixed. Then

$$T_I(x,y) = \min\{t \in [0,1] \mid I(x,t) \ge y\}\,,$$
$$T_I(y,x) = \min\{t \in [0,1] \mid I(y,t) \ge x\}\,.$$

Therefore, it is enough to show that

$$I(x,t) \ge y \iff I(y,t) \ge x\,, \qquad \text{for every } t \in [0,1]\,.$$

Take any $t \in [0,1]$. From (EP) and (OP) we have

$$I(x,t) \ge y \iff I(y,I(x,t)) = 1 \iff I(x,I(y,t)) = 1 \iff I(y,t) \ge x\,.$$

(T3) (*Monotonicity*). Assume that $x,y_1,y_2 \in [0,1]$ and $y_1 \le y_2$. By easy calculations we see that (T3) follows from

$$\{t \in [0,1] \mid I(x,t) \ge y_1\} \supset \{t \in [0,1] \mid I(x,t) \ge y_2\}\,,$$

which is the consequence of the assumption that $y_1 \le y_2$.

(T2) (*Associativity*). Let $x,y,z \in [0,1]$ be arbitrarily fixed. From the commutativity of T_I proven above we have

$$T_I(T_I(x,y),z) = T_I(z,T_I(x,y))\,,$$

thus, it is enough to show

$$T_I(z,T_I(x,y)) = T_I(x,T_I(z,y))\,,$$

and hence we should show

$$\min\{t \in [0,1] \mid I(z,t) \ge T_I(x,y)\} = \min\{t \in [0,1] \mid I(x,t) \ge T_I(z,y)\}\,.$$

We will prove more, i.e., we show that

$$\{t \in [0,1] \mid I(z,t) \ge T_I(x,y)\} = \{t \in [0,1] \mid I(x,t) \ge T_I(z,y)\}\,,$$

which is equivalent to showing that, for any $t \in [0,1]$

$$I(z,t) \ge T_I(x,y) \iff I(x,t) \ge T_I(z,y)\,. \tag{2.21}$$

Observe that, by the symmetry of x and z, it is enough to show only one implication. To do this, let us fix arbitrarily $t \in [0,1]$ and assume that

$$I(z,t) \geq T_I(x,y) . \tag{2.22}$$

Since $T_I(x,y) \leq T_I(x,y)$, by (RP*) we get

$$I(x, T_I(x,y)) \geq y . \tag{2.23}$$

By virtue of (I2), (2.22) and (2.23) we get $I(x, I(z,t)) \geq I(x, T_I(x,y)) \geq y$, so from proven (T3) and (EP) we have

$$T_I(z,y) \leq T_I(z, I(x, I(z,t))) = T_I(z, I(z, I(x,t))) .$$

Now, from the definition of T_I

$$T_I(x, I(x,y)) = \min\{t \in [0,1] \mid I(x,t) \geq I(x,y)\} \leq y , \tag{2.24}$$

so $T_I(z, I(z, I(x,t))) \leq I(x,t)$ and hence, $T_I(z,y) \leq I(x,t)$, which ends the proof of (2.21) and (T2).

(T4) (*Neutral element*). Fix arbitrarily $x \in [0,1]$. By (NP), we get

$$\begin{aligned} T_I(x,1) = T_I(1,x) &= \min\{t \in [0,1] \mid I(1,t) \geq x\} \\ &= \min\{t \in [0,1] \mid t \geq x\} = x . \end{aligned}$$

We have shown that T is a t-norm. Now we will prove that it is left-continuous. Since T is commutative, it is enough to show the left-continuity of T with respect to the second variable. On the contrary, let us assume that T is not left-continuous with respect to the second variable at some point $(x_0, y_0) \in [0,1] \times (0,1]$. Since as a t-norm T is increasing, there exist $a, b \in [0,1]$ such that $a < b$ and

$$\begin{aligned} T_I(x_0, y) &\leq a , && \text{for all } y < y_0 , \\ T_I(x_0, y_0) &= b . \end{aligned}$$

By virtue of (RP*), we have

$$I(x_0, a) \geq y , \qquad \text{for all } y < y_0 .$$

In the limit $y \to y_0$ we get $I(x_0, a) \geq y_0$. Using (RP*) again, we obtain $b = T(x_0, y_0) \leq a$, a contradiction to $a < b$. Therefore, T_I is a left-continuous t-norm.

Consider now the R-implication I_{T_I}. We show that

$$I_{T_I}(x,y) = I(x,y) , \qquad x, y \in [0,1] . \tag{2.25}$$

Let $x, y \in [0,1]$ be arbitrarily fixed. From (2.24) we have

$$I(x,y) \in \{t \in [0,1] \mid T_I(x,t) \leq y\} ,$$

so

$$I(x,y) \le I_{T_I}(x,y) . \tag{2.26}$$

From (2.23) we have $I(x, T_I(x,z)) \ge z$ for any $z \in [0,1]$. Let $z = I_{T_I}(x,y)$. Then

$$I_{T_I}(x,y) \le I(x, T_I(x, I_{T_I}(x,y))) . \tag{2.27}$$

In addition, since T_I is left-continuous, from Proposition 2.5.2 we know that the pair (T_I, I_{T_I}) satisfies (RP). Now, noting that $I_{T_I}(x,y) \ge I_{T_I}(x,y)$, we get $T_I(x, I_{T_I}(x,y)) \le y$. From the inequality (2.27) and using (I2), we get

$$I_{T_I}(x,y) \le I(x,y) . \tag{2.28}$$

By (2.26) and (2.28) we obtain (2.25). □

We also have the following connection between left-continuous t-norms and R-implications generated from them.

Lemma 2.5.15. *If T is a left-continuous t-norm, then $T = T_{I_T}$.*

Proof. From Theorem 2.5.7 we know that the function $I_T \in \mathcal{FI}$ satisfies (EP), (OP) and is right-continuous with respect to the second variable. Therefore, by Theorem 2.5.14, the function T_{I_T} is a left-continuous t-norm. We show that

$$T_{I_T}(x,y) = T(x,y) , \qquad x,y \in [0,1] . \tag{2.29}$$

Fix arbitrarily $x, y \in [0,1]$. Since T is left-continuous, from Proposition 2.5.2 we know that the pair (T, I) satisfies (RP), and hence noting that $T(x,y) \le T(x,y)$, we have

$$I_T(x, T(x,y)) = \max\{t \in [0,1] \mid T(x,t) \le T(x,y)\} \ge y ,$$

i.e., $T(x,y) \in \{t \in [0,1] \mid I_T(x,t) \ge y\}$, and hence

$$T(x,y) \ge T_{I_T}(x,y) . \tag{2.30}$$

Conversely, since T is left-continuous, from Proposition 2.5.2 it follows that $T(x, I_T(x,z)) \le z$, for any $z \in [0,1]$. If $z = T_{I_T}(x,y)$, then

$$T_{I_T}(x,y) \ge T(x, I_T(x, T_{I_T}(x,y))) . \tag{2.31}$$

Further, since I_T is right-continuous, from Proposition 2.5.13 we get also $I_T(x, T_{I_T}(x,y)) \ge y$. Using this inequality in (2.31) and by (T3) we have

$$T_{I_T}(x,y) \ge T(x,y) . \tag{2.32}$$

From (2.30) and (2.32) we get (2.29). □

Remark 2.5.16. Once again, it should be noted that in the case when T is a non-left-continuous t-norm, then T_{I_T} can be different from T. For example, consider the t-norm $T_{\mathbf{nM}^*}$ from Example 2.5.6(iv). Now, the R-implication is equal to the Fodor implication, i.e., $I_{T_{\mathbf{nM}^*}} = I_{\mathbf{FD}}$, while $T_{I_{\mathbf{FD}}} = T_{\mathbf{nM}} \neq T_{\mathbf{nM}^*}$.

From the above discussion we get one of the main results of this section - the characterization of R-implications generated from left-continuous t-norms.

Theorem 2.5.17. *For a function* $I\colon [0,1]^2 \to [0,1]$ *the following statements are equivalent:*

(i) *I is an R-implication generated from a left-continuous t-norm.*
(ii) *I satisfies* (I2), (EP), (OP) *and it is right-continuous with respect to the second variable.*

Moreover, the representation of R-implication, up to a left-continuous t-norm, is unique in this case.

Proof. (*i*) $\Longrightarrow$ (*ii*) Assume that a function $I\colon [0,1]^2 \to [0,1]$ is an R-implication generated from a left-continuous t-norm. From Theorem 2.5.7 it satisfies (I2), (EP), (OP) and it is right-continuous with respect to the second variable.

(*ii*) $\Longrightarrow$ (*i*) Assume now that a function $I\colon [0,1]^2 \to [0,1]$ satisfies (I2), (EP), (OP) and it is right-continuous with respect to the second variable. By Theorem (2.5.14) we get that $I = I_{T_I}$, where T_I defined by (2.20) is a left-continuous t-norm. Hence I is an R-implication generated from a left-continuous t-norm T_I.

Observe that the uniqueness of the representation of an R-implication up to a left-continuous t-norm follows from Lemma 2.5.15. □

Remark 2.5.18. It should be noted that in contrast to the characterization of (S,N)-implications the problem of mutual-independence of all the above properties is still an open problem. Therefore, in Table 2.7 we only show that some properties in Theorem 2.5.17 are independent from each other.

As a consequence of the above results we obtain also the following characterization of left-continuous t-norms (see BACZYŃSKI [9]).

Corollary 2.5.19. *For a function* $T\colon [0,1]^2 \to [0,1]$ *the following statements are equivalent:*

(i) *T is a left-continuous t-norm.*
(ii) *There exists* $I \in \mathcal{FI}$, *which satisfies* (EP), (OP) *and is right-continuous with respect to the second variable, such that T is given by* (2.20).

Lemma 2.5.20. *If a function* $I\colon [0,1]^2 \to [0,1]$ *is continuous except at the point* $(0,0)$, *satisfies* (EP), (OP), (I2) *and is such that* $I(\cdot,0) = N_{\mathbf{D1}}$, *then the function* T_I *defined by* (2.19) *is a continuous, positive Archimedean t-norm.*

Proof. Firstly, notice that I is still right continuous with respect to the second variable. Hence, from Theorem 2.5.14, we see that T_I is a left continuous t-norm. Moreover, (2.19) reduces to (2.20) and I and T_I form an adjoint pair, i.e., they satisfy (RP*). Thus, it suffices to prove that T_I is continuous, Archimedean and positive. In fact, by Theorem 1.2.1 and commutativity of T_I, we must show only the right continuity of T_I.

Table 2.7. The mutual independence of some properties in Theorem 2.5.17

Function F	(I2)	(EP)	(OP)	Right–continuity
$F(x,y) = \begin{cases} 1, & \text{if } x < y \\ 0, & \text{if } x \geq y \end{cases}$	✓	×	×	×
$F(x,y) = \begin{cases} 1, & \text{if } x,y \in [0,0.5] \text{ or } x,y \in (0.5,1] \\ 0, & \text{otherwise} \end{cases}$	×	✓	×	×
$F(x,y) = \begin{cases} 1, & \text{if } x \leq y \\ 0, & \text{if } x > y \text{ and } ((x,y) \in [0,0.5)^2 \\ & \qquad \text{or } (x,y) \in (0.5,1]^2) \\ 0.5, & \text{otherwise} \end{cases}$	×	×	✓	×
$F = I_{-2}$	×	×	×	✓
$F(x,y) = \begin{cases} \min(1-x,y), & \text{if } \max(1-x,y) \leq 0.5 \\ \max(1-x,y), & \text{otherwise} \end{cases}$	✓	✓	×	×
$F(x,y) = \begin{cases} 1, & \text{if } x \leq y \\ 0.5, & \text{if } x > y \text{ and } y > 0 \\ 0, & \text{otherwise} \end{cases}$	✓	×	✓	×
$F = I_{-1}$	✓	×	×	✓
$F(x,y) = \begin{cases} 1, & \text{if } x \leq y \\ 0.75, & \text{if } x \in (0.25,0.75] \text{ and } y = 0.25 \\ y, & \text{otherwise} \end{cases}$	×	✓	✓	×
$F(x,y) = 1 - y$	×	✓	×	✓
$F(x,y) = \begin{cases} 1, & \text{if } x \leq y \\ 0, & \text{if } x > y \text{ and } ((x,y) \in [0,0.5)^2 \\ & \qquad \text{or } (x,y) \in [0.5,1]^2) \\ 0.5, & \text{otherwise} \end{cases}$	×	×	✓	✓
$F = I_{\mathbf{KD}}$	✓	✓	×	✓
$F = I_{\mathbf{RS}}$	✓	×	✓	✓

On the contrary, if T_I is not right continuous, then there exist $x, y \in [0,1]$ and $z, z' \in (0,1)$ such that

$$z = T_I(x,y) < z' < \lim_{h \to 0^+} T_I(x, y+h)\,.$$

From (RP*) we have $I(x,z) \geq y$ and by (I2), we have $I(x,z') \geq y$. Applying (RP*) again, we have that $z' \leq T_I(x,y)$, a contradiction to our assumption above. Thus, T_I is right-continuous and hence, is continuous too.

Now, we show that T_I satisfies the Archimedean property. For an $x \in (0,1)$, we have $T_I(x,x) = \min\{t \in [0,1] : I(x,t) \geq x\}$. Also $I(x,x) = 1 \geq x$, since I

satisfies (OP). By (RP*) we get $T_I(x,x) \leq x$, i.e., $x \in \{t \in [0,1] : I(x,t) \geq x\}$. Let us define, for some arbitrary but fixed $x \in (0,1)$, a function $I_x\colon [0,x] \to [0,1]$ by the formula $I_x(t) := I(x,t)$. It is obvious that this function is continuous. Observe that from (OP) we have $I_x(0) < 1$ and $I_x(x) = 1 > x$. If $I_x(0) \geq x$, then $T_I(x,x) = 0 < x$. If $I_x(0) < x$, then from the Darboux property (see Appendix A), there exists $z \in (0,x)$ such that $I_x(z) = x$, i.e., $I(x,z) = x$. Therefore, $T_I(x,x) \leq z < x$. Now, from Remark 2.1.4(iii) we see that T_I is Archimedean.

Finally, let us assume that T_I is not positive, i.e., there exist $x, y \in (0,1)$, such that $T_I(x,y) = 0$. From (RP*) it follows that $I(x,0) \geq y$. However, by our assumption, it implies that $x = 0$, a contradiction. Hence, T_I is a positive, continuous Archimedean t-norm. □

For some classes of t-norms we have the following representations of R-implications, whose proofs follow from the representations of these classes of t-norms (see [105], Theorem 1.16, [72] and [113], Proposition 5.4.3).

Theorem 2.5.21. *If T is a continuous Archimedean t-norm with the additive generator f as given in Theorem 2.1.5, then*

$$I_T(x,y) = f^{-1}(\max(f(y) - f(x), 0)) \,, \qquad x, y \in [0,1] \,. \tag{2.33}$$

Proof. Since f is continuous and strictly decreasing, we get

$$\begin{aligned} I_T(x,y) &= \max\{t \in [0,1] \mid T(x,t) \leq y\} \\ &= \max\{t \in [0,1] \mid f^{-1}(\min(f(x) + f(t), f(0))) \leq y\} \\ &= \max\{t \in [0,1] \mid \min(f(x) + f(t), f(0)) \geq f(y)\} \\ &= \max\{t \in [0,1] \mid f(x) + f(t) \geq f(y)\} \\ &= \max\{t \in [0,1] \mid f(t) \geq f(y) - f(x)\} \\ &= f^{-1}(\max(f(y) - f(x), 0)) \,, \end{aligned}$$

for all $x, y \in [0,1]$. □

Lemma 2.5.22. *If T is a strict t-norm, then I_T is Φ-conjugate with the Goguen implication $I_{\mathbf{GG}}$, i.e., there exists $\varphi \in \Phi$, which is uniquely determined up to a positive constant exponent, such that for all $x, y \in [0,1]$,*

$$I_T(x,y) = (I_{\mathbf{GG}})_\varphi(x,y) = \begin{cases} 1, & \text{if } x \leq y \,, \\ \min\left(1, \varphi^{-1}\left(\frac{\varphi(y)}{\varphi(x)}\right)\right), & \text{if } x > y \,. \end{cases} \tag{2.34}$$

Proof. If T is a strict t-norm, then from Theorem 2.1.8 we know that T is Φ-conjugate with the product t-norm $T_{\mathbf{P}}$, i.e., $T(x,y) = \varphi^{-1}(\varphi(x)\varphi(y))$ for some $\varphi \in \Phi$. Since $T_{\mathbf{P}}$ is continuous, so is T and hence by Proposition 2.5.2, I_T is given by (2.17), whence we obtain

$$\begin{aligned} I_T(x,y) &= \max\{t \in [0,1] \mid T(x,t) \leq y\} \\ &= \max\{t \in [0,1] \mid \varphi^{-1}(\varphi(x)\varphi(t)) \leq y\} \\ &= \max\{t \in [0,1] \mid \varphi(x)\varphi(t) \leq \varphi(y)\} \,, \end{aligned}$$

for any $x, y \in [0,1]$. Now, fix arbitrarily $x, y \in [0,1]$. If $x \leq y$, then of course $I_T(x,y) = 1$. If $x > y$, then we will show that

$$\max\{t \in [0,1] \mid \varphi(x)\varphi(t) \leq \varphi(y)\} = \varphi^{-1}\left(\frac{\varphi(y)}{\varphi(x)}\right) .$$

Let $B(x,y) := \{t \in [0,1] \mid \varphi(x)\varphi(t) \leq \varphi(y)\}$ and $t_0 = \varphi^{-1}\left(\frac{\varphi(y)}{\varphi(x)}\right)$. Firstly, we show that $t_0 \in B(x,y)$. Indeed,

$$\varphi(x)\varphi(t_0) = \varphi(x)\varphi\left(\varphi^{-1}\left(\frac{\varphi(y)}{\varphi(x)}\right)\right) = \varphi(x) \cdot \frac{\varphi(y)}{\varphi(x)} = \varphi(y) . \tag{2.35}$$

Finally, suppose that there exists $t_1 \in B(x,y)$, such that $t_0 < t_1$. It means that $\varphi(x)\varphi(t_1) \leq \varphi(y)$, but from (2.35) we have $\varphi(x)\varphi(t_1) \leq \varphi(x)\varphi(t_0)$, hence $\varphi(t_1) \leq \varphi(t_0)$ and $t_1 \leq t_0$, a contradiction. It proves that $\max B(x,y) = t_0$. Therefore, I_T has the form (2.34). □

The proof of the following result can be obtained along the above lines, by using the representation result given in Theorem 2.1.9.

Lemma 2.5.23. *If T is a nilpotent t-norm, then I_T is Φ-conjugate with the Łukasiewicz implication $I_{\mathbf{LK}}$, i.e., there exists unique $\varphi \in \Phi$, such that I has the form* (2.15).

Theorem 2.5.24. *If T is a continuous t-norm with the ordinal sum structure as given in Theorem 2.1.10, then*

$$I_T(x,y) = \begin{cases} 1, & \text{if } x \leq y\ , \\ a_\alpha + (e_\alpha - a_\alpha) \cdot I_{T_\alpha}\left(\frac{x-a_\alpha}{e_\alpha - a_\alpha}, \frac{y-a_\alpha}{e_\alpha - a_\alpha}\right), & \text{if } x, y \in [a_\alpha, e_\alpha]\ , \\ y, & \text{otherwise}\ . \end{cases}$$

Proof. Firstly, it is obvious that if $x \leq y$, then $I_T(x,y) = 1$.

Let $e_\alpha \geq x > y > a_\alpha$, for some $\alpha \in A$. Then, for any $t > e_\alpha$, by the definition of T, we have $T(x,t) = \min(x,t) = x > y$, while if $t \leq a_\alpha$, then $T(x,t) \leq a_\alpha < y$. Hence, we have

$$\begin{aligned} I_T(x,y) &= \sup\{t \in (a_\alpha, e_\alpha) \mid T(x,t) \leq y\} \\ &= \sup\left\{t \in (a_\alpha, e_\alpha) \mid a_\alpha + (e_\alpha - a_\alpha) \cdot T_\alpha\left(\frac{x-a_\alpha}{e_\alpha - a_\alpha}, \frac{t-a_\alpha}{e_\alpha - a_\alpha}\right) \leq y\right\} \\ &= \sup\left\{t \in (a_\alpha, e_\alpha) \mid T_\alpha\left(\frac{x-a_\alpha}{e_\alpha - a_\alpha}, \frac{t-a_\alpha}{e_\alpha - a_\alpha}\right) \leq \frac{y-a_\alpha}{e_\alpha - a_\alpha}\right\} \\ &= a_\alpha + (e_\alpha - a_\alpha) \cdot I_{T_\alpha}\left(\frac{x-a_\alpha}{e_\alpha - a_\alpha}, \frac{y-a_\alpha}{e_\alpha - a_\alpha}\right) . \end{aligned}$$

Finally, if $y < x$, then $x \in (a_\alpha, e_\alpha)$, for some $\alpha \in A$ and $y \leq a_\alpha$. On the one hand, if $t > x$ then $T(x,t) = x > y$. On the other hand, if $t \leq a_\alpha$, we have $T(x,t) = \min(x,t) = t \leq y$, while if $a_\alpha < t \leq x$, once again, $T(x,t) = t$. Hence, we have $I_T(x,y) = y$. □

Example 2.5.25. Consider the continuous t-norm

$$T = (\langle 0.1, 0.5, T_{\mathbf{P}} \rangle, \langle 0.7, 0.9, T_{\mathbf{LK}} \rangle)$$

given as the ordinal sum of $T_{\mathbf{P}}$ and $T_{\mathbf{LK}}$ as follows:

$$T(x,y) = \begin{cases} 0.1 + 2.5(x-0.1)(y-0.1), & \text{if } x,y \in [0.1, 0.5]\,, \\ 0.7 + \max(x+y-1.6, 0), & \text{if } x,y \in [0.7, 0.9]\,, \\ \min(x,y), & \text{otherwise}\,. \end{cases}$$

From Theorem 2.5.24, the R-implication obtained from T is given as

$$I_T(x,y) = \begin{cases} 1, & \text{if } x \le y\,, \\ 0.1 + 0.4\left(\dfrac{y-0.1}{x-0.1}\right), & \text{if } x,y \in [0.1, 0.5]\,, \\ 0.9 - x + y, & \text{if } x,y \in [0.7, 0.9]\,, \\ y, & \text{otherwise}\,. \end{cases}$$

The 2D- and 3D-plots of the above I_T are given in Fig.2.3.

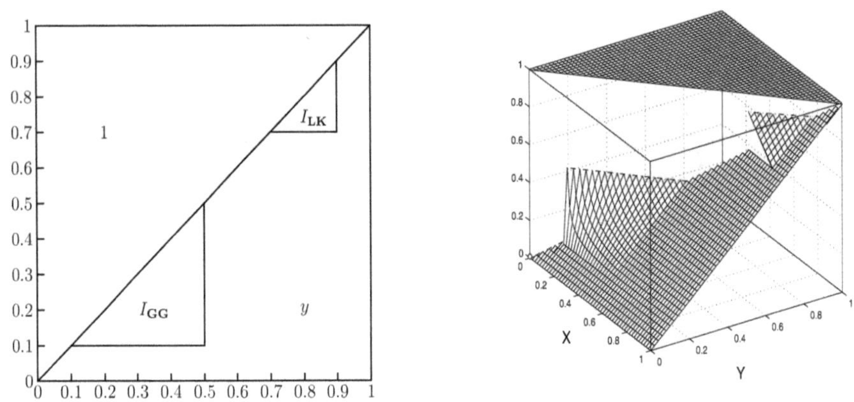

(a) 2D-plot of I_T (b) 3D-plot of I_T

Fig. 2.3. 2D- and 3D-plots of the residual implication of the continuous ordinal sum t-norm $T = (\langle 0.1, 0.5, T_{\mathbf{P}} \rangle, \langle 0.7, 0.9, T_{\mathbf{LK}} \rangle)$ (see Example 2.5.25)

Note that, in the case of a t-norm T which has an ordinal sum representation, it is the squares along the main diagonal that play an important role, whereas, in the corresponding residual it is the triangles along the main diagonal that play a key role. A detailed algorithm to draw the R-implication generated from a given left-continuous t-norm is presented in Miyakoshi and Shimbo [183].

In the case when a left-continuous t-norm T can be represented as an ordinal sum of left-continuous t-subnorms, a similar representation for I_T can be obtained (see Mesiar and Mesiarová [182]).

2.5.4 R-Implications and Laws of Contraposition

Since every R-implication satisfies (NP) (see Theorem 2.5.4) we get the following result concerning the laws of contraposition.

Proposition 2.5.26. *Let T be a t-norm.*

(i) If I_T satisfies (R-CP) *with some fuzzy negation N, then $N = N_{I_T}$.*
(ii) If I_T satisfies (L-CP) *with some fuzzy negation N, then $N = N_{I_T}$ is a strong negation.*
(iii) If I_T satisfies (CP) *with some fuzzy negation N, then $N = N_{I_T}$ is a strong negation.*

Proof. (i) This fact follows from Lemma 1.5.21.
(ii) From Lemma 1.5.14 we see that I_T can satisfy (L-CP) only with a strictly decreasing negation N, such that $N_{I_T} \circ N = \mathrm{id}_{[0,1]}$ and N_{I_T} is a continuous fuzzy negation. Since $N_{I_T} = N_T$, the natural negation of the t-norm T, from Theorem 2.3.5(i) it follows that $N = N_{I_T}$ is a strong negation.
(iii) This fact follows from Corollary 1.5.8. □

Remark 2.5.27. It should be emphasized that without any additional assumptions on the t-norm T the converse of Proposition 2.5.26 may not hold. Consider the R-implication $I_{\mathbf{TB}^*}$ from Example 2.5.6(ii) whose natural negation is given by

$$N_{I_{\mathbf{TB}^*}} = N_{\mathbf{TB}^*} = \begin{cases} 1, & x = 0\,, \\ 0.5, & x \in [0, 0.5)\,, \\ 0, & x \in [0.5, 1]\,. \end{cases}$$

Letting $N = N_{I_{\mathbf{TB}^*}}$, we have $I_{\mathbf{TB}^*}(0.4, N(0.5)) = 0.5 \neq 1 = I_{\mathbf{TB}^*}(0.5, N(0.4))$, thus $I_{\mathbf{TB}^*}$ does not satisfy R-CP(N), even if $N = N_{I_{\mathbf{TB}^*}}$.

Observe that an R-implication based on a left-continuous t-norm satisfies all basic properties, in particular (EP), and hence, the converse of Proposition 2.5.26 is also true, as can be seen below.

Proposition 2.5.28. *Let T be a left-continuous t-norm.*

(i) I_T satisfies (R-CP) *with some fuzzy negation N if and only if $N = N_{I_T}$.*
(ii) I_T satisfies (L-CP) *with some fuzzy negation N if and only if $N = N_{I_T}$ is a strong negation.*
(iii) I_T satisfies (CP) *with some fuzzy negation N if and only if $N = N_{I_T}$ is a strong negation.*

Proof. For the proofs in the forward direction, see Proposition 2.5.26.

(i) The converse follows from properties of I_T generated from left-continuous t-norm (see Theorem 2.5.7) and Corollary 1.5.25.
(ii) Conversely, if $N = N_{I_T}$ is a strong negation, then from Remark 1.5.19(i) implication I_T satisfies (L-CP) with N^{-1}. Since N is strong, $N^{-1} = N_{I_T}$.
(iii) This equivalence follows immediately from Corollary 1.5.9. □

Remark 2.5.29. From Remark 2.5.27, we see that the left-continuity of T is important for the corresponding I_T to satisfy R-CP(N_{I_T}). However, left-continuity of the t-norm is not necessary for an I_T to satisfy CP(N_{I_T}). Consider, for example, the remarkable t-norm of VICENÍK [249] and its residual:

$$T_{\mathbf{VC}}(x,y) = \begin{cases} 0.5, & \text{if } 0.5 \leq \min(x,y) \text{ and } x+y \leq 1.5 \,, \\ \max(0, x+y-1), & \text{otherwise} \,, \end{cases}$$

$$I_{\mathbf{VC}}(x,y) = \begin{cases} 0.5, & \text{if } x > 0.5 \;\&\; y < 0.5 \text{ and } y \geq x - 0.5 \,, \\ \min(1, 1-x+y), & \text{otherwise} \,. \end{cases}$$

(i) $T_{\mathbf{VC}}$ is not a left-continuous t-norm, but $I_{\mathbf{VC}}$ satisfies CP($N_{I_{\mathbf{VC}}}$), where $N_{I_{\mathbf{VC}}} = N_{\mathbf{C}}$.

(ii) $I_{\mathbf{VC}}$ satisfies (I2), (OP) and is right-continuous with respect to the second variable. However, it does not satisfy (EP), since

$$\begin{aligned} I_{\mathbf{VC}}(0.55, I_{\mathbf{VC}}(0.95, 0.45)) &= 0.95 \\ &\neq 0.55 = I_{\mathbf{VC}}(0.95, I_{\mathbf{VC}}(0.55, 0.45)) \,, \end{aligned}$$

which implies that we cannot obtain it as a residual of any left-continuous t-norm, and conversely, we cannot obtain any left-continuous t-norm from $I_{\mathbf{VC}}$ by using (2.20). Note that the non-left-continuous version of the nilpotent minimum $T_{\mathbf{nM}^*}$ gives rise to the Fodor implication $I_{\mathbf{FD}}$ as its residual, which can alternately be obtained as the residual of the left-continuous $T_{\mathbf{nM}}$.

(iii) It should also be highlighted that $T_{\mathbf{VC}}$ is a border continuous t-norm (with a strong associated negation) that cannot be suitably re-defined to make it a left-continuous t-norm. Compare this with $T_{\mathbf{nM}^*}$ (with a strong associated negation) or the border continuous t-norm $T_{\mathbf{B}^*}$ (whose associated negation is discontinuous) from Example 2.5.6(ii).

Theorem 2.5.30. *If a function* $I\colon [0,1]^2 \to [0,1]$ *satisfies* (OP), (EP) *and* N_I *is strong, then a function* $T\colon [0,1]^2 \to [0,1]$ *defined as*

$$T(x,y) = N_I(I(x, N_I(y))) \,, \qquad x, y \in [0,1] \,, \tag{2.36}$$

is a t-norm such that T *and* I *satisfy* (RP). *In particular,* $T = T_I$ *given by* (2.20).

Proof. Let us assume that I satisfies the required conditions. Firstly, observe that Corollary 1.5.12 implies that I satisfies (NP), (IP) and (CP) only with respect to N_I. Now, we prove that T given by (2.36) is commutative. By (CP), we obtain

$$T(x,y) = N_I(I(x, N_I(y))) = N_I(I(y, N_I(x))) = T(y,x) \,,$$

for all $x, y \in [0,1]$. Since T is commutative, to show that T is associative, it suffices to prove that

$$T(x, T(z,y)) = T(z, T(x,y)) \,, \qquad x, y, z \in [0,1] \,.$$

Using (EP), we have, for all $x, y, z \in [0,1]$,

$$\begin{aligned} T(x, T(z, y)) &= N_I(I(x, N_I(T(z, y)))) = N_I(I(x, N_I(N_I(I(z, N_I(y)))))) \\ &= N_I(I(x, I(z, N_I(y)))) = N_I(I(z, I(x, N_I(y)))) \\ &= T(z, T(x, y)) . \end{aligned}$$

By (NP), T has 1 as its neutral element, since

$$T(x,1) = T(1,x) = N_I(I(1, N_I(x))) = N_I(N_I(x)) = x , \qquad x \in [0,1] .$$

The monotonicity of T is the consequence of the monotonicity of I and N_I.

Now we will show that I and T form an adjoint pair. Let $x, y, z \in [0,1]$. By virtue of our assumptions, we obtain

$$\begin{aligned} T(x,y) \le z &\Longleftrightarrow N_I(I(x, N_I(y))) \le z \Longleftrightarrow I(x, N_I(y)) \ge N_I(z) \\ &\Longleftrightarrow I(N_I(z), I(x, N_I(y))) = 1 \Longleftrightarrow I(x, I(N_I(z), N_I(y))) = 1 \\ &\Longleftrightarrow I(x, I(y, z)) = 1 \Longleftrightarrow I(y, I(x, z)) = 1 \Longleftrightarrow y \le I(x, z) , \end{aligned}$$

which proves that I and T satisfy (RP).

Finally, we show that $T = T_I$. Observe that, since I and T satisfy (RP), we get $I(x, T(x,y)) \ge y$ for $x, y \in [0,1]$, i.e., $T(x,y) \in \{t \in [0,1] \mid I(x,t) \ge y\}$. Therefore, $T_I \le T$. Let us now assume that there exist $x_0, y_0 \in [0,1]$ such that $T_I(x_0, y_0) < T(x_0, y_0)$. This implies that there exists a $t_0 \in (0,1)$ such that $T_I(x_0, y_0) < t_0 < T(x_0, y_0)$ and by the formula (2.20) of T_I we obtain $I(x_0, t_0) \ge y_0$. Finally, since the pair (T, I) satisfies (RP), we have that $T(x_0, y_0) \le t_0$, a contradiction. □

As a consequence we get the following useful characterization result (cf. FODOR [103]).

Corollary 2.5.31. *For a left-continuous t-norm T the following statements are equivalent:*

(i) I_T satisfies (CP) *with some fuzzy negation N.*
(ii) $N = N_{I_T}$ is a strong negation and

$$I_T(x,y) = N_I(T(x, N_I(y))) , \qquad x, y \in [0,1] .$$

Observe now that the problem of finding R-implications with strong N_{I_T} is equivalent to finding t-norms with strong induced negation.

Example 2.5.32. As we have noted in Sect. 2.3 only a few families of left-continuous t-norms (up to a Φ-conjugation) with strong natural negations are known in the literature. Now, using Theorem 2.5.30 we can easily obtain the formulas for R-implications generated from such t-norms.

(i) If we consider the Jenei t-norm family $T_{\mathbf{J}}^{\lambda}$, for $\lambda \in [0, 0.5]$, then we will obtain the following Jenei family $I_{\mathbf{J}}^{\lambda}$ of R-implications with strong natural negation:

$$I_{\mathbf{J}}^{\lambda}(x,y) = \begin{cases} 1, & \text{if } x \leq y\,, \\ 1 - x + y - \lambda, & \text{if } x > y \text{ and } x, 1-y \in (\lambda, 1-\lambda]\,, \\ \max(1-x, y), & \text{otherwise}\,, \end{cases}$$

for $x, y \in [0,1]$. Note that $I_{\mathbf{J}}^{0} = I_{\mathbf{LK}}$ and $I_{\mathbf{J}}^{0.5} = I_{\mathbf{FD}}$.

(ii) If we consider the Maes and De Baets t-norm family $T_{\mathbf{MD}}^{\lambda}$, where $\lambda \in [0, 0.5)$, then we will obtain the following Maes and De Baets family $I_{\mathbf{MD}}^{\lambda}$ of R-implications with strong natural negation:

$$I_{\mathbf{MD}}^{\lambda}(x,y) = \begin{cases} 1, & \text{if } x \leq y\,, \\ 1 - x + y, & \text{if } x > y \text{ and } (\min(x, 1-y) \in [0, \lambda] \\ & \qquad\qquad \text{or } x - y \geq 1 - \lambda)\,, \\ \max(1-x, y), & \text{if } x > y \text{ and } \min(x, 1-y) \in (\lambda, 1-\lambda]\,, \\ \lambda, & \text{otherwise}\,, \end{cases}$$

for $x, y \in [0,1]$. Note also that $I_{\mathbf{MD}}^{0} = I_{\mathbf{FD}}$.

Plots of these implications for $\lambda = \frac{1}{3}$ are given in Fig. 2.4.

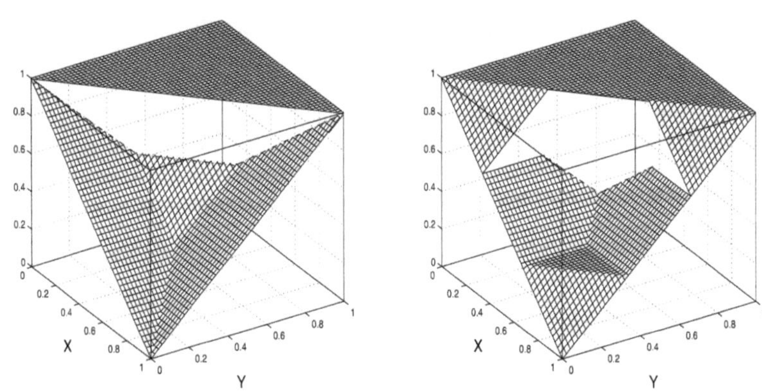

(a) $I_{\mathbf{J}}^{\lambda}$ for $\lambda = \frac{1}{3}$ (b) $I_{\mathbf{MD}}^{\lambda}$ for $\lambda = \frac{1}{3}$

Fig. 2.4. Plots of the R-implications $I_{\mathbf{J}}^{\lambda}$ and $I_{\mathbf{MD}}^{\lambda}$ for $\lambda = \frac{1}{3}$ (see Example 2.5.32)

It is interesting and important to note that for continuous R-implications generated from left-continuous t-norms we have the following result, which is an other version of Theorem 2.4.20 (cf. Miyakoshi and Shimbo [183], Fodor [103]).

Theorem 2.5.33. *For a function $I\colon [0,1]^2 \to [0,1]$ the following statements are equivalent:*

(i) I is continuous and satisfies both (EP) *and* (OP).
(ii) I is a continuous R-implication based on some left-continuous t-norm.
(iii) I is an R-implication based on some continuous t-norm, with a strong natural negation N_I.
(iv) I is an R-implication based on some nilpotent t-norm.
(v) I is Φ-conjugate with the Łukasiewicz implication $I_{\mathbf{LK}}$, i.e., there exists unique $\varphi \in \Phi$, such that I has the form (2.15).

Proof. *(i)* $\Longrightarrow$ *(ii)* Assume that I is a continuous function which satisfies (EP) and (OP). By virtue of Corollary 1.5.12 we obtain, in particular, that I satisfies (I2). Now, from Theorem 2.5.17, it follows that I is a continuous R-implication generated from some left-continuous t-norm T.

(ii) $\Longrightarrow$ *(iii)* Let I be a continuous R-implication based on some left-continuous t-norm. From Theorem 2.5.17 it satisfies (I2), (EP) and (OP). Moreover, the natural negation N_I is continuous. By Corollary 1.5.12, we get that N_I is a strong negation. Therefore, from Theorem 2.5.30, a t-norm T from which I is generated has the form (2.36), i.e., $T(x,y) = N_I(I(x, N_I(y)))$. Since I is continuous, T is also continuous.

(iii) $\Longrightarrow$ *(iv)* Assume that I is an R-implication based on some continuous t-norm T with a strong natural negation N_I. By Theorem 2.5.17, I satisfies (I2), (EP) and (OP). Therefore, from Theorem 2.5.30, a t-norm T from which I is generated has the form (2.36), i.e., $T(x,y) = N_I(I(x, N_I(y)))$, for all $x, y \in [0,1]$. Since N_I is strong we obtain the following formula for I:

$$I(x,y) = N_I(T(x, N_I(y))) \, , \qquad x, y \in [0,1] \, .$$

By Corollary 1.5.12, we see that I satisfies also (IP), and hence we get that T satisfies the law of contradiction (LC) with the strong negation $N_I = N_T$, the natural negation of T. By virtue of Proposition 2.3.15, we get the thesis here.

(iv) $\Longrightarrow$ *(v)* This implication is obvious and follows from the representation of nilpotent t-norms.

(v) $\Longrightarrow$ *(i)* This implication has been shown in the proof of Theorem 2.4.20. □

2.5.5 Intersections between Subfamilies of R-Implications

In this subsection we summarize the known intersections among some subfamilies of R-implications based on the results cited and obtained earlier. The discussion in this section is also diagrammatically represented in Fig. 2.5.

Let us denote the different families of R-implications as follows:

- $\mathbb{I}_{\mathbb{T}}$ – the family of all R-implications;
- $^{\mathbb{C}}\mathbb{I}_{\mathbb{T}}$ – the family of all continuous R-implications;
- $\mathbb{I}_{\mathbb{T}_{\mathbb{LC}}}$ – the family of all R-implications obtained from left-continuous t-norms;
- $\mathbb{I}_{\mathbb{T}_{\mathbb{C}}}$ – the family of all R-implications obtained from continuous t-norms;

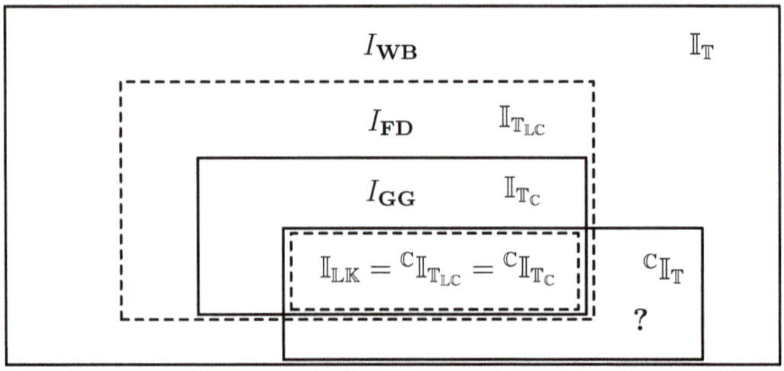

Fig. 2.5. Intersections among the subfamilies of R-implications

- ${}^{\mathbb{C}}\mathbb{I}_{\mathbb{T}}$ – the family of all continuous R-implications;
- ${}^{\mathbb{C}}\mathbb{I}_{\mathbb{T}_{\mathbb{LC}}}$ – the family of all continuous R-implications obtained from left-continuous t-norms.

By Theorems 2.5.7 and 2.5.33, we have

$$ {}^{\mathbb{C}}\mathbb{I}_{\mathbb{T}_{\mathbb{LC}}} = \mathbb{I}_{\mathbb{LK}} . $$

Quite obviously, we have the following containments (see Fig. 2.5):

$$ {}^{\mathbb{C}}\mathbb{I}_{\mathbb{T}_{\mathbb{LC}}} = \mathbb{I}_{\mathbb{LK}} \subsetneq \mathbb{I}_{\mathbb{T}_{\mathbb{C}}} \subsetneq \mathbb{I}_{\mathbb{T}_{\mathbb{LC}}} \subsetneq \mathbb{I}_{\mathbb{T}} . $$

Moreover, from Theorem 2.5.33, we get the following equalities:

$$ {}^{\mathbb{C}}\mathbb{I}_{\mathbb{T}} \cap \mathbb{I}_{\mathbb{T}_{\mathbb{C}}} = {}^{\mathbb{C}}\mathbb{I}_{\mathbb{T}} \cap \mathbb{I}_{\mathbb{T}_{\mathbb{LC}}} = {}^{\mathbb{C}}\mathbb{I}_{\mathbb{T}_{\mathbb{C}}} = {}^{\mathbb{C}}\mathbb{I}_{\mathbb{T}_{\mathbb{LC}}} = \mathbb{I}_{\mathbb{LK}} . $$

However, it is still an open problem to find, if there exists at all, a continuous R-implication generated from a non-continuous t-norm.

2.6 QL-Implications

While the (S,N)- and R-implications, dealt with in the earlier sections, are the generalizations of material and intuitionistic-logic implications, in this section we deal with yet another popular way of obtaining fuzzy implications - as the generalization of the following implication defined in quantum logic:

$$ p \to q \equiv \neg p \vee (p \wedge q) . $$

Needless to state, when the truth values are restricted to $\{0, 1\}$ its truth table coincides with that of the material and intuitionistic-logic implications. In this section we deal with the generalization of the above implication usually called the QL-implication owing to its origins. Once again our exploration in this section parallels that of the earlier sections on (S,N)- and R-implications.

2.6.1 Definition, Examples and Basic Properties

Definition 2.6.1. *A function* $I\colon [0,1]^2 \to [0,1]$ *is called a* QL-operation *if there exist a t-norm* T*, a t-conorm* S *and a fuzzy negation* N *such that*

$$I(x,y) = S(N(x), T(x,y)) , \qquad x,y \in [0,1] . \tag{2.37}$$

If I *is a QL-operation generated from the triple* (T,S,N)*, then we will often denote it by* $I_{T,S,N}$.

Firstly, we investigate some properties of QL-operations. We will see that not all QL-operations are fuzzy implications in the sense of Definition 1.1.1.

Proposition 2.6.2. *If* $I_{T,S,N}$ *is a QL-operation, then*

(i) $I_{T,S,N}$ *satisfies* (I2), (I3), (I4), (I5), (NC), (LB) *and* (NP)*;*
(ii) $N_{I_{T,S,N}} = N$.

Proof. (i) The monotonicity with respect to the second variable (I2) of a QL-operation $I_{T,S,N}$ is the consequence of the monotonicity of the t-norm T and the t-conorm S. Indeed, let us assume that $x, y_1, y_2 \in [0,1]$ and $y_1 \le y_2$. Then, by (T3), we have $T(x,y_1) \le T(x,y_2)$ and, further, $S(N(x),T(x,y_1)) \le S(N(x),T(x,y_2))$ by (S3). Additionally,

$$\begin{aligned} I_{T,S,N}(1,1) &= S(N(1),T(1,1)) = S(0,1) = 1 , \\ I_{T,S,N}(1,0) &= S(N(1),T(1,0)) = S(0,0) = 0 , \\ I_{T,S,N}(0,y) &= S(N(0),T(0,y)) = S(1,0) = 1 , \end{aligned}$$

for any $y \in [0,1]$, which shows that $I_{T,S,N}$ satisfies (I3), (I4), (I5), (NC) and (LB). Finally, $I_{T,S,N}$ satisfies (NP), since

$$I_{T,S,N}(1,y) = S(N(1),T(1,y)) = S(0,y) = y , \qquad y \in [0,1] .$$

(ii) For any $x \in [0,1]$ we have

$$N_{I_{T,S,N}}(x) = I_{T,S,N}(x,0) = S(N(x),T(x,0)) = S(N(x),0) = N(x) . \qquad \square$$

From the above proposition, it follows that a QL-operation is generated by a unique negation.

Remark 2.6.3. A QL-operation does not always satisfy (I1). For example, consider the function I_{-1}. As can be seen in Table 2.8, it is the QL-operation obtained from the triple $(T_{\mathbf{M}}, S_{\mathbf{M}}, N_{\mathbf{C}})$, but it does not satisfy (I1) (cf. Table 1.2). However, the QL-operation obtained from the triple $(T_{\mathbf{LK}}, S_{\mathbf{LK}}, N_{\mathbf{C}})$ satisfies (I1). In fact, it is the Kleene-Dienes implication $I_{\mathbf{KD}}$ which is a fuzzy implication.

Example 2.6.4. Table 2.8 lists QL-operations obtained from the basic t-norms, t-conorms and fuzzy negations. In the last column we indicate whether the QL-operation is also a fuzzy implication.

Table 2.8. Examples of basic QL-operations

T	S	N	QL-operation $I_{T,S,N}$	$I_{T,S,N} \in \mathcal{FI}$
$T_{\mathbf{M}}$	$S_{\mathbf{M}}$	$N_{\mathbf{C}}$	I_{-1}	×
$T_{\mathbf{M}}$	$S_{\mathbf{P}}$	$N_{\mathbf{C}}$	$I(x,y) = \begin{cases} 1-x+x^2, & \text{if } x \leq y \\ 1-x+xy, & \text{otherwise} \end{cases}$	×
$T_{\mathbf{M}}$	$S_{\mathbf{LK}}$	$N_{\mathbf{C}}$	$I_{\mathbf{LK}}$	✓
$T_{\mathbf{M}}$	$S_{\mathbf{D}}$	$N_{\mathbf{C}}$	$I_{\mathbf{DP}}$	✓
$T_{\mathbf{M}}$	$S_{\mathbf{nM}}$	$N_{\mathbf{C}}$	$I_{\mathbf{FD}}$	✓
$T_{\mathbf{P}}$	$S_{\mathbf{M}}$	$N_{\mathbf{C}}$	$I(x,y) = \max(1-x, xy)$	×
$T_{\mathbf{P}}$	$S_{\mathbf{P}}$	$N_{\mathbf{C}}$	$I(x,y) = 1-x+x^2y$	×
$T_{\mathbf{P}}$	$S_{\mathbf{LK}}$	$N_{\mathbf{C}}$	$I_{\mathbf{RC}}$	✓
$T_{\mathbf{P}}$	$S_{\mathbf{D}}$	$N_{\mathbf{C}}$	$I_{\mathbf{DP}}$	✓
$T_{\mathbf{P}}$	$S_{\mathbf{nM}}$	$N_{\mathbf{C}}$	$I(x,y) = \begin{cases} 1, & \text{if } y = 1 \\ \max(1-x, xy), & \text{otherwise} \end{cases}$	×
$T_{\mathbf{LK}}$	$S_{\mathbf{M}}$	$N_{\mathbf{C}}$	$I(x,y) = \max(1-x, x+y-1)$	×
$T_{\mathbf{LK}}$	$S_{\mathbf{P}}$	$N_{\mathbf{C}}$	$I(x,y) = \begin{cases} 1-x, & \text{if } y \leq 1-x \\ 1+x^2+xy-2x, & \text{otherwise} \end{cases}$	×
$T_{\mathbf{LK}}$	$S_{\mathbf{LK}}$	$N_{\mathbf{C}}$	$I_{\mathbf{KD}}$	✓
$T_{\mathbf{LK}}$	$S_{\mathbf{D}}$	$N_{\mathbf{C}}$	$I(x,y) = \begin{cases} y, & \text{if } x = 1 \\ 1-x, & \text{if } y \leq 1-x \\ 1, & \text{otherwise} \end{cases}$	×
$T_{\mathbf{LK}}$	$S_{\mathbf{nM}}$	$N_{\mathbf{C}}$	$I(x,y) = \begin{cases} 1, & \text{if } x = 0 \text{ or } y = 1 \\ 1-x, & \text{if } y \leq 2-2x \\ y, & \text{otherwise} \end{cases}$	×
$T_{\mathbf{D}}$	any S	$N_{\mathbf{C}}$	$I(x,y) = \begin{cases} S(N(x),x), & \text{if } y = 1 \\ y, & \text{if } x = 1 \\ 1-x, & \text{otherwise} \end{cases}$	×
$T_{\mathbf{nM}}$	$S_{\mathbf{nM}}$	$N_{\mathbf{C}}$	$I(x,y) = \begin{cases} 1, & \text{if } x \leq y \text{ and } y > 1-x \\ y, & \text{if } x > y \text{ and } y > 1-x \\ 1-x, & \text{otherwise} \end{cases}$	×
any T	any S	$N_{\mathbf{D1}}$	$I(x,y) = \begin{cases} T(x,y), & \text{if } x > 0 \\ 1, & \text{if } x = 0 \end{cases}$	×
any T	any S	$N_{\mathbf{D2}}$	$I_{\mathbf{WB}}$	✓

Therefore, the first main problem is the characterization of those QL-operations which satisfy (I1). Unfortunately, only partial results are known in the literature. Following the terminology used in TRILLAS et al. [236] and MAS et al. [174], only if the QL-operation is a fuzzy implication we use the term *QL-implication*.

Lemma 2.6.5. *If a QL-operation $I_{T,S,N} \in \mathcal{FI}$, then the pair (S,N) satisfies* (LEM).

Proof. If $I_{T,S,N}$ is a fuzzy implication, then by Remark 1.1.3 it satisfies (RB). Thus $I_{T,S,N}(x,1) = 1$ if and only if $S(N(x),T(x,1)) = 1$, i.e., $S(N(x),x) = 1$, for every $x \in [0,1]$. □

Remark 2.6.6. (i) From Remark 2.3.10(iii) we know that there does not exist any t-conorm S such that the pair $(S, N_{\mathbf{D1}})$ satisfies (LEM). Therefore, by Lemma 2.6.5, we see that no QL-operation obtained from the triple (T,S,N), where $N = N_{\mathbf{D1}}$ is the least fuzzy negation, can be a fuzzy implication. Since the natural negations of $I_{\mathbf{GD}}, I_{\mathbf{GG}}, I_{\mathbf{RS}}$ and $I_{\mathbf{YG}}$ equal $N_{\mathbf{D1}}$, from Proposition 2.6.2(ii), we see that the above four basic fuzzy implications are not QL-implications. However, from Table 2.8 we see that all the other five basic implications, viz., $I_{\mathbf{KD}}, I_{\mathbf{LK}}, I_{\mathbf{FD}}, I_{\mathbf{RC}}$ and $I_{\mathbf{WB}}$ are QL-implications obtained from some triple (T,S,N).

(ii) The fact that the condition in Lemma 2.6.5 is only necessary and not sufficient can be seen from the QL-operation I obtained from the triple $(T_{\mathbf{P}}, S_{\mathbf{nM}}, N_{\mathbf{C}})$, which is given in Table 2.8. Although the pair $(S_{\mathbf{nM}}, N_{\mathbf{C}})$ satisfies (LEM), it can be easily verified, by letting $x_1 = 0.8$, $x_2 = 0.9$ and $y = 0.3$, that $x_1 < x_2$ but $I(0.8, 0.3) = 0.24 < 0.27 = I(0.9, 0.3)$, so this I does not satisfy (I1).

(iii) From Lemma 2.3.9, it is easy to see that if a negation N in the triple (T,S,N) is less than the natural negation of S, i.e., if $N(x) < N_S(x)$ for some $x \in [0,1]$, then the pair (S,N) does not satisfy (LEM) and hence the QL-operation $I_{T,S,N}$ is not a fuzzy implication.

(iv) Let S be any t-conorm and $N = N_{\mathbf{D2}}$, the greatest fuzzy negation. From Remark 2.3.10(i) we see that the pair $(S, N_{\mathbf{D2}})$ satisfies (LEM). Now, for any t-norm T we have that the QL-operation obtained from the triple $(T, S, N_{\mathbf{D2}})$ is a fuzzy implication and is, in fact, the Weber implication $I_{\mathbf{WB}}$ (see also Table 2.8).

However, from Remark 2.3.10(ii) we know that if S is a positive t-conorm, then S satisfies (LEM) only with the greatest fuzzy negation $N_{\mathbf{D2}}$ and hence we obtain the following fact.

Proposition 2.6.7. *A QL-operation $I_{T,S,N}$, where S is a positive t-conorm, is a fuzzy implication if and only if $N = N_{\mathbf{D2}}$. Moreover, $I_{T,S,N} = I_{\mathbf{WB}}$ in this case.*

From Lemmas 1.5.13(ii) and 1.5.14 we have the following sufficient conditions.

Lemma 2.6.8. *If a QL-operation $I_{T,S,N}$ satisfies* (L-CP) *with a continuous negation N^*, then $I_{T,S,N} \in \mathcal{FI}$ and N is such that $N \circ N^* = \mathrm{id}_{[0,1]}$.*

Corollary 2.6.9. *If a QL-operation $I_{T,S,N}$ satisfies* (L-CP) *with a strict (strong, respectively) negation N^*, then $I_{T,S,N} \in \mathcal{FI}$ and $N = (N^*)^{-1}$ ($N = N^*$, respectively).*

The following examples give some more sufficient conditions for the QL-operation $I_{T,S,N}$ to be a fuzzy implication.

Example 2.6.10. Let S be a t-conorm and N a fuzzy negation such that the pair (S, N) satisfies (LEM).

(i) If T is the minimum t-norm $T_{\mathbf{M}}$, then it can be easily seen that the QL-operation obtained from the triple $(T_{\mathbf{M}}, S, N)$ is always a fuzzy implication given by

$$I_{T_{\mathbf{M}},S,N}(x,y) = \begin{cases} 1, & \text{if } x \leq y , \\ S(N(x),y), & \text{if } x > y , \end{cases} \qquad x, y \in [0,1] . \tag{2.38}$$

(ii) If T is the drastic t-norm $T_{\mathbf{D}}$, then the QL-operation obtained from the triple $(T_{\mathbf{D}}, S, N)$ is given by

$$I_{T_{\mathbf{D}},S,N}(x,y) = \begin{cases} 1, & \text{if } y = 1 , \\ y, & \text{if } x = 1 , \\ N(x), & \text{otherwise} , \end{cases} \qquad x, y \in [0,1] .$$

This function is not always a fuzzy implication, even if S and N satisfy (LEM). Observe that it is a fuzzy implication if and only if $N(x) \geq y$ for all $x, y \in [0,1)$, which means that $N = N_{\mathbf{D2}}$. In this case, of course, the QL-operation reduces, once again, to the Weber implication $I_{\mathbf{WB}}$.

Before considering special examples of QL-implications, we show some relationship between the Φ-conjugates of QL-implications.

Theorem 2.6.11. *If $I_{T,S,N}$ is a QL-implication (QL-operation, respectively), then the Φ-conjugate of $I_{T,S,N}$ is also a QL-implication (QL-operation, respectively) generated from the Φ-conjugate t-norm of T, the Φ-conjugate t-conorm of S and the Φ-conjugate fuzzy negation of N, i.e., if $\varphi \in \Phi$, then*

$$(I_{T,S,N})_\varphi = I_{T_\varphi,S_\varphi,N_\varphi} .$$

Proof. Let $\varphi \in \Phi$ and let $I_{T,S,N}$ be a QL-implication based on the suitable functions. We now know that the functions T_φ, S_φ and N_φ are a t-norm, t-conorm and a fuzzy negation, respectively. Further, Proposition 1.1.8 implies that if $I_{T,S,N}$ is a fuzzy implication, then $(I_{T,S,N})_\varphi$ is also a fuzzy implication. Now, for every $x, y \in [0,1]$,

$$\begin{aligned}
(I_{T,S,N})_\varphi(x,y) &= \varphi^{-1}(I_{T,S,N}(\varphi(x),\varphi(y))) \\
&= \varphi^{-1}\left(S\left(N(\varphi(x)), T(\varphi(x),\varphi(y))\right)\right) \\
&= \varphi^{-1}(S(\varphi \circ \varphi^{-1}(N(\varphi(x))), \varphi \circ \varphi^{-1}(T(\varphi(x),\varphi(y))))) \\
&= \varphi^{-1}(S(\varphi(N_\varphi(x)), \varphi(T_\varphi(x,y)))) = S_\varphi(N_\varphi(x), T_\varphi(x,y)) \\
&= I_{T_\varphi,S_\varphi,N_\varphi}(x,y) .
\end{aligned}$$

□

Now, let us consider QL-implications obtained from triples (T, S, N), where S is some continuous t-conorm. Firstly, if S is a continuous but positive t-conorm, from Proposition 2.6.7 we know that for the QL-operation obtained from the triple (T, S, N) to be a fuzzy implication N has to be the greatest negation $N_{\mathbf{D2}}$ and that $I_{T,S,N} = I_{\mathbf{WB}}$ in this situation. Hence we consider only non-positive continuous t-conorms.

Let S be a continuous t-conorm and N a continuous fuzzy negation such that the pair (S, N) satisfies (LEM). Then, from Proposition 2.3.12, there exists a unique $\varphi \in \Phi$ such

$$S(x, y) = (S_{\mathbf{LK}})_\varphi(x, y) = \varphi^{-1}(\min(\varphi(x) + \varphi(y), 1)) ,$$
$$N(x) \geq (N_{\mathbf{C}})_\varphi(x) = \varphi^{-1}(1 - \varphi(x)) ,$$

for all $x, y \in [0, 1]$. Note that, in this case S is a nilpotent t-conorm, i.e., it is non-positive and continuous. Let us consider the extreme case when $N(x) = (N_{\mathbf{C}})_\varphi(x) = \varphi^{-1}(1 - \varphi(x))$ (with the same bijection φ), in which case, from Theorem 1.4.13, we have that N is a strong negation. Now, if we consider the QL-operation obtained from the triple $(T, (S_{\mathbf{LK}})_\varphi, (N_{\mathbf{C}})_\varphi)$, then since $T(x, y) \leq x$ for any t-norm T and $x \in [0, 1]$, we obtain the following function, denoted by $I_{\varphi,T}$ for ease of notation (see also [236], [174]):

$$\begin{aligned} I_{\varphi,T}(x, y) &= (S_{\mathbf{LK}})_\varphi \left((N_{\mathbf{C}})_\varphi(x), T(x, y)\right) \\ &= \varphi^{-1}(1 - \varphi(x) + \varphi(T(x, y))) , \end{aligned} \tag{2.39}$$

for $x, y \in [0, 1]$. The following result has been obtained by Mas et al. [174].

Theorem 2.6.12. *For a QL-operation $I_{\varphi,T}$ given by* (2.39)*, where T is any t-norm and $\varphi \in \Phi$, the following statements are equivalent:*

(i) $I_{\varphi,T} \in \mathcal{FI}$.
(ii) $T_{\varphi^{-1}}$ satisfies the Lipschitz condition (1.6) *with the constant $c = 1$.*

Proof. Let us assume that $a_1, a_2, b \in [0, 1]$ and $a_1 \leq a_2$. Since φ is an increasing bijection there exist $x_1, x_2, y \in [0, 1]$ such that $x_1 = \varphi^{-1}(a_1), x_2 = \varphi^{-1}(a_2)$ and $y = \varphi^{-1}(b)$. Since $I_{\varphi,T}$ is a fuzzy implication, we know that it satisfies (I1) and we have the following equivalences:

$$\begin{aligned} & I_{\varphi,T}(x_1, y) \geq I_{\varphi,T}(x_2, y) \\ & \quad \Longleftrightarrow \varphi^{-1}(1 - \varphi(x_2) + \varphi(T(x_2, y))) \leq \varphi^{-1}(1 - \varphi(x_1) + \varphi(T(x_1, y))) \\ & \quad \Longleftrightarrow \varphi(T(x_2, y)) - \varphi(T(x_1, y)) \leq \varphi(x_2) - \varphi(x_1) \\ & \quad \Longleftrightarrow \varphi(T(\varphi^{-1}(a_2), \varphi^{-1}(b))) - \varphi(T(\varphi^{-1}(a_1), \varphi^{-1}(b))) \leq a_2 - a_1 \\ & \quad \Longleftrightarrow T_{\varphi^{-1}}(a_2, b) - T_{\varphi^{-1}}(a_1, b) \leq a_2 - a_1 , \end{aligned}$$

i.e., if and only if $T_{\varphi^{-1}}$ satisfies the Lipschitz condition with $c = 1$. □

Remark 2.6.13. Since the class of t-norms satisfying the Lipschitz condition is contained in the class of continuous t-norms, we have that $T_{\varphi^{-1}}$, and hence T itself, is a continuous t-norm.

The case when T is an Archimedean or an idempotent t-norm has been investigated by FODOR [103]. In fact, it is shown there that an equivalence relation exists between the t-norms T employed below and the resulting QL-implications.

Example 2.6.14. All QL-operations $I_{\varphi,T}$ obtained using the following t-norms satisfy (I1) and hence are fuzzy implications (cf. Table 2.8).

(i) If the t-norm T in (2.39) is Φ-conjugate with the Łukasiewicz t-norm $T_{\mathbf{LK}}$ with the same $\varphi \in \Phi$, then $I_{\varphi,(T_{\mathbf{LK}})_\varphi}$ is Φ-conjugate with the Kleene-Dienes implication $I_{\mathbf{KD}}$, i.e., for all $x, y \in [0,1]$,

$$I_{\varphi,(T_{\mathbf{LK}})_\varphi}(x,y) = (I_{\mathbf{KD}})_\varphi(x,y) = \max(N_\varphi(x), y) .$$

In particular, when $\varphi = \mathrm{id}$, we obtain the Kleene-Dienes implication $I_{\mathbf{KD}}$.

(ii) If the t-norm T in (2.39) is Φ-conjugate with the product t-norm $T_{\mathbf{P}}$ with the same $\varphi \in \Phi$, then $I_{\varphi,(T_{\mathbf{P}})_\varphi}$ is Φ-conjugate with the Reichenbach implication $I_{\mathbf{RC}}$, i.e., for all $x, y \in [0,1]$,

$$I_{\varphi,(T_{\mathbf{P}})_\varphi}(x,y) = (I_{\mathbf{RC}})_\varphi(x,y) = \varphi^{-1}(1 - \varphi(x) + \varphi(x)\varphi(y)) .$$

In particular, when $\varphi = \mathrm{id}$, we obtain the Reichenbach implication $I_{\mathbf{RC}}$.

(iii) Firstly, note that from Remark 2.1.4(viii), we have that $(T_{\mathbf{M}})_\varphi = T_{\mathbf{M}}$ for any $\varphi \in \Phi$. Now, if the t-norm T in (2.39) is the minimum t-norm $T_{\mathbf{M}}$, then $I_{\varphi,T_{\mathbf{M}}}$ is Φ-conjugate with the Łukasiewicz implication $I_{\mathbf{LK}}$, i.e., for all $x, y \in [0,1]$,

$$I_{\varphi,T_{\mathbf{M}}}(x,y) = (I_{\mathbf{LK}})_\varphi(x,y) = \min(\varphi^{-1}(1 - \varphi(x) + \varphi(y), 1)) .$$

In particular, when $\varphi = \mathrm{id}$, we obtain the Łukasiewicz implication $I_{\mathbf{LK}}$.

In the following we show yet other examples of QL-implications generated from continuous functions.

Example 2.6.15. Let S be the Schweizer-Sklar t-conorm $S^\lambda_{\mathbf{SS}}$ for $\lambda = 2$ given by

$$S^2_{\mathbf{SS}}(x,y) = 1 - \left(\max\left((1-x)^2 + (1-y)^2 - 1, 0\right)\right)^{\frac{1}{2}} , \qquad x, y \in [0,1] ,$$

and T be the product t-norm $T_{\mathbf{P}}$. It can be easily verified that the pairs $(S^2_{\mathbf{SS}}, N_{\mathbf{C}})$ and $(S^2_{\mathbf{SS}}, N_{\mathbf{R}})$ satisfy (LEM).

(i) The QL-operation obtained from the triple $(T_{\mathbf{P}}, S^2_{\mathbf{SS}}, N_{\mathbf{C}})$ is given by

$$I_{\mathbf{PC}}(x,y) = 1 - \left(\max\left(x(x + xy^2 - 2y), 0\right)\right)^{\frac{1}{2}} , \qquad x, y \in [0,1] .$$

(ii) The QL-operation obtained from the triple $(T_{\mathbf{P}}, S^2_{\mathbf{SS}}, N_{\mathbf{R}})$ is given by

$$I_{\mathbf{PR}}(x,y) = 1 - \left(\max\left(x(1 + xy^2 - 2y), 0\right)\right)^{\frac{1}{2}} , \qquad x, y \in [0,1] .$$

It can be easily checked that both $I_{\mathbf{PC}}$ and $I_{\mathbf{PR}}$ satisfy (I1) and hence are QL-implications, whose plots are given in Fig. 2.6.

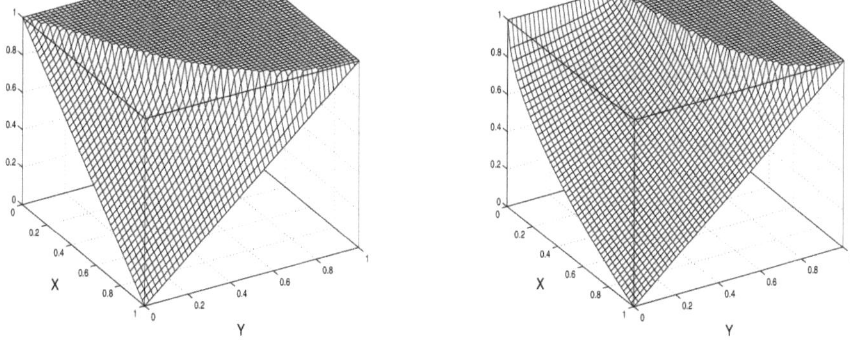

(a) QL-implication $I_{\mathbf{PC}}$ from the triple $(T_{\mathbf{P}}, S^2_{\mathbf{SS}}, N_{\mathbf{C}})$

(b) QL-implication $I_{\mathbf{PR}}$ from the triple $(T_{\mathbf{P}}, S^2_{\mathbf{SS}}, N_{\mathbf{R}})$

Fig. 2.6. Plots of the QL-implications $I_{\mathbf{PC}}$ and $I_{\mathbf{PR}}$ (see Example 2.6.15)

In the rest of this section we give examples of QL-implications obtained from triples (T, S, N), where S is a non-continuous t-conorm.

Example 2.6.16. Let S be the drastic t-conorm $S_{\mathbf{D}}$ and N any non-vanishing negation. Then the pair $(S_{\mathbf{D}}, N)$ satisfies (LEM). If the t-norm T is positive, then, as can be verified, the QL-operation obtained from the triple $(T, S_{\mathbf{D}}, N)$ is a fuzzy implication given by

$$I_{T,S_{\mathbf{D}},N}(x,y) = \begin{cases} y, & \text{if } x = 1\,, \\ N(x), & \text{if } y = 0\,, \\ 1, & \text{otherwise}\,, \end{cases} \qquad x, y \in [0,1]\,.$$

Fig. 2.7(a) gives the plot of the QL-implication obtained from the triple $(T, S_{\mathbf{D}}, N_{\mathbf{C}})$, where T is any positive t-norm, which is, in fact, the Dubois-Prade implication $I_{\mathbf{DP}}$ (see Tables 2.4 and 2.8).

Example 2.6.17. Let N be a strong negation. Consider the following t-conorm

$$S^N_{\mathbf{nM}}(x,y) = \begin{cases} 1, & \text{if } x \geq N(y)\,, \\ \max(x,y), & \text{if } x < N(y)\,, \end{cases} \qquad x, y \in [0,1]\,, \tag{2.40}$$

which is only right-continuous. Let N^* be any negation such that $N^* \geq N$. Hence, we have $S^N_{\mathbf{nM}}(N^*(x), x) = 1$.

(i) The QL-operation from the triple $(T_{\mathbf{M}}, S^N_{\mathbf{nM}}, N^*)$ is

$$I_{T_{\mathbf{M}},S^N_{\mathbf{nM}},N^*}(x,y) = \begin{cases} 1, & \text{if } x \leq y\,, \\ \max(N^*(x), y), & \text{if } x > y\,, \end{cases} \tag{2.41}$$

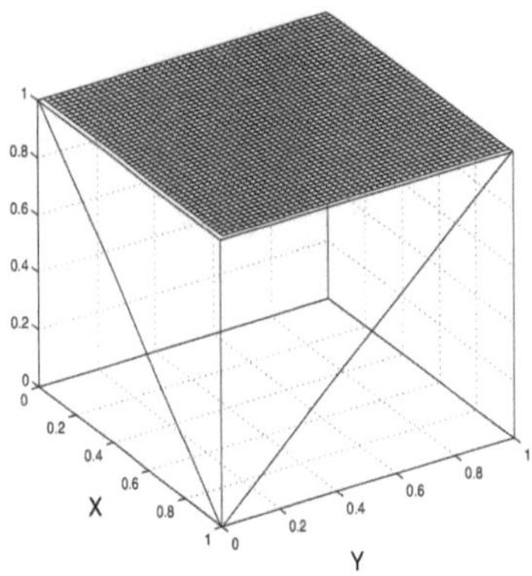

(a) QL-implication $I_{\mathbf{DP}}$ from Example 2.6.16 with $N = N_{\mathbf{C}}$

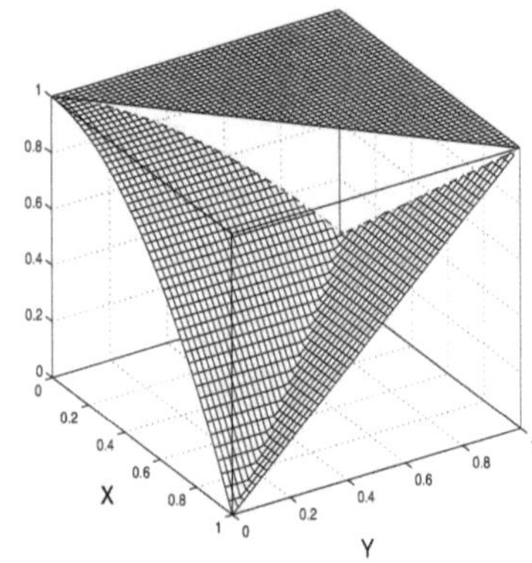

(b) QL-implication $I_{T_{\mathbf{M}},S_{\mathbf{nM}}^{N},N^*}$ from Example 2.6.17(i) with $N^* = N_{\mathbf{K}}$

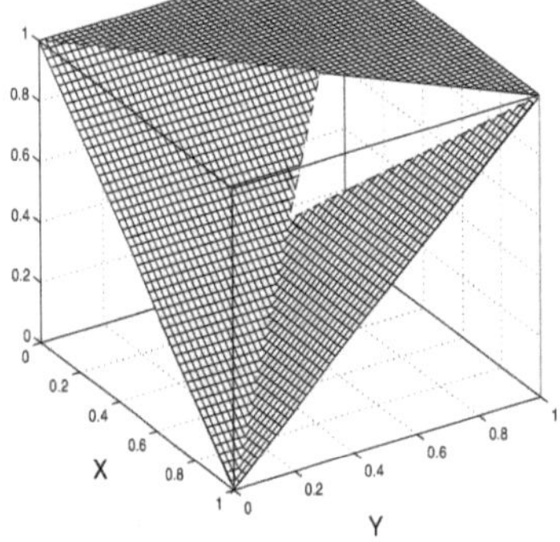

(c) QL-implication $I_{T_{\mathbf{nM}}^{N},S_{\mathbf{nM}}^{N},N^*}$ from Example 2.6.17(ii) with $N = N^* = N_{\mathbf{C}}$

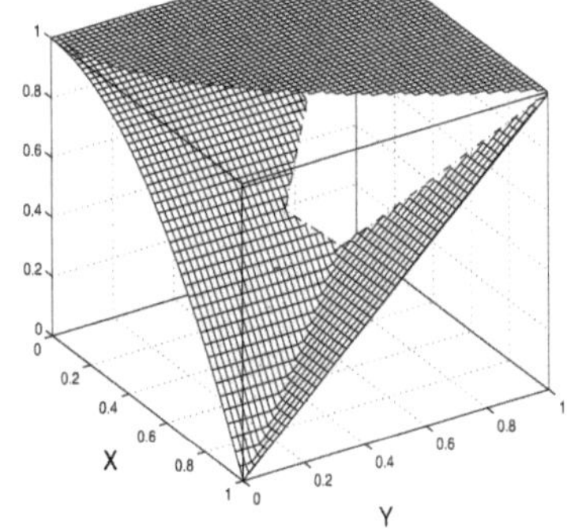

(d) QL-implication $I_{T_{\mathbf{nM}}^{N},S_{\mathbf{nM}}^{N},N^*}$ from Example 2.6.17(ii) with $N = N_{\mathbf{C}}$ and $N^* = N_{\mathbf{K}}$

Fig. 2.7. Plots of QL-implications from some non-continuous t-conorms

for $x, y \in [0,1]$. In the case $N = N^* = N_{\mathbf{C}}$ the QL-operation in (2.41) is a QL-implication, indeed, it is the Fodor implication $I_{\mathbf{FD}}$. Fig. 2.7(b) gives the plot of the QL-implication obtained from the triple $(T_{\mathbf{M}}, S_{\mathbf{nM}}^{N_{\mathbf{C}}}, N_{\mathbf{K}})$, where $N_{\mathbf{K}}(x) = 1 - x^2$.

(ii) Consider the following N-dual t-norm of $S_{\mathbf{nM}}^{N}$

$$T_{\mathbf{nM}}^{N}(x,y) = \begin{cases} 0, & \text{if } x \leq N(y)\,, \\ \min(x,y), & \text{if } x > N(y)\,, \end{cases} \qquad x, y \in [0,1]\,.$$

The QL-operation obtained from the triple $(T^N_{\mathbf{nM}}, S^N_{\mathbf{nM}}, N^*)$ is given by

$$I_{T^N_{\mathbf{nM}},S^N_{\mathbf{nM}},N^*}(x,y) = \begin{cases} N^*(x), & \text{if } x \le N(y) , \\ 1, & \text{if } N^*(x) \ge N(y) , \\ \max(N^*(x), y), & \text{if } N^*(x) < N(y) , \end{cases}$$

for $x, y \in [0,1]$. Figs. 2.7(c) and (d) give the plots of the QL-implication obtained from the triple $(T^N_{\mathbf{nM}}, S^N_{\mathbf{nM}}, N^*)$, when $N = N^* = N_{\mathbf{C}}$ and $N = N_{\mathbf{C}}$, $N^* = N_{\mathbf{K}}$, respectively.

In fact, the following result was proven by Mas et al. [174].

Proposition 2.6.18. *Let N be a strong negation with fixed point $e \in (0,1)$, T a continuous t-norm and $S^N_{\mathbf{nM}}$ the t-conorm obtained from N as given in* (2.40). *Let $I_{T,S^N_{\mathbf{nM}},N}$ be the QL-operation obtained from the triple $(T, S^N_{\mathbf{nM}}, N)$. Then the following statements are equivalent:*

(i) $I_{T,S^N_{\mathbf{nM}},N} \in \mathcal{FI}$.
(ii) $T(x,x) = x$ *for all* $x \in [e,1]$.

Moreover, the corresponding QL-implication is then given by

$$I_{T,S^N_{\mathbf{nM}},N}(x,y) = \begin{cases} 1, & \text{if } x, e \le y \text{ or } (x \le y < e \text{ and } T(x,y) = x) , \\ y, & \text{if } N(x) \le y < x , \\ N(x), & \text{otherwise} . \end{cases}$$

2.6.2 QL-Implications and the Exchange Principle

The main result concerning the exchange principle for QL-operations is the following, see also Mas et al. [174].

Theorem 2.6.19. *For a QL-implication $I_{T,S,N}$, with a continuous negation N, the following statements are equivalent:*

(i) $I_{T,S,N}$ *satisfies* (EP).
(ii) $I_{T,S,N}$ *is an (S,N)-implication generated from N.*

Proof. *(i)* $\Longrightarrow$ *(ii)* Let $I_{T,S,N}$ be a QL-implication with a continuous negation N. Firstly, observe that $I_{T,S,N}$ satisfies (I2) and $N_{I_{T,S,N}} = N$ is continuous. If $I_{T,S,N}$ satisfies (EP), then by virtue of Theorem 2.4.10 and Remark 2.4.13(ii), the function $I_{T,S,N}$ is an (S,N)-implication generated from N.

(ii) $\Longrightarrow$ *(i)* The reverse implication is obvious and follows from Proposition 2.4.3(i). □

Remark 2.6.20. (i) When N is a strict negation in Theorem 2.6.19, then $I_{T,S,N}$ is an (S,N)-implication generated from N and a t-conorm S^* given by $S^*(x,y) = S(x, T(N^{-1}(x), y))$.

(ii) Theorem 2.6.19 also gives a sufficient condition for a QL-operation obtained from the triple (T, S, N) with a continuous negation N to be a fuzzy implication.

(iii) However, the QL-implications $I_{\mathbf{WB}}$ and $I_{T,S_{\mathbf{D}},N}$ with an N that is discontinuous but non-vanishing (see Example 2.6.16), show that the continuity of N is not necessary for a QL-implication to satisfy (EP).
(iv) It is interesting to note that both $I_{\mathbf{WB}}$ and $I_{T,S_{\mathbf{D}},N}$, under the conditions of Example 2.6.16, are also (S,N)-implications. While $I_{\mathbf{WB}}$ is an (S,N)-implication obtained from any t-conorm S and $N = N_{\mathbf{D2}}$, i.e., $I_{\mathbf{WB}} = I_{S,N_{\mathbf{D2}}}$, $I_{T,S_{\mathbf{D}},N}$ is the (S,N)-implication $I_{S_{\mathbf{D}},N}$.

2.6.3 QL-Implications and the Identity Principle

The following result is immediate from Remark 2.6.6(iii), Examples 2.6.10 and 2.6.16.

Proposition 2.6.21. *A QL-implication $I_{T,S,N}$ satisfies* (IP) *if*

(i) $N = N_{\mathbf{D2}}$, S is any t-conorm and T any t-norm, or
(ii) $T = T_{\mathbf{M}}$, S is any t-conorm and N any negation such that they satisfy (LEM)*, or*
(iii) $S = S_{\mathbf{D}}$, N is any non-vanishing negation and T is a positive t-norm.

Remark 2.6.22. (i) If S is positive, then we know, from Proposition 2.6.7, that the QL-implication obtained is the Weber implication $I_{\mathbf{WB}}$, which satisfies (IP).
(ii) However, from Example 2.6.14 and Proposition 2.6.21(iii) we see that in the case when S is not positive, we can obtain QL-implications that satisfy the identity principle (IP) for many fuzzy negations N. In fact, we have the following easy to obtain result.

Proposition 2.6.23. *If a QL-implication $I_{T,S,N}$ satisfies* (IP)*, then $T(x,x) \geq N_S \circ N(x)$ for all $x \in [0,1]$.*

Proof. If $I_{T,S,N}$ satisfies (IP), then $I_{T,S,N}(x,x) = S(N(x),T(x,x)) = 1$, for any $x \in [0,1]$. From Remark 2.3.2(iii), we have that $T(x,x) \geq N_S \circ N(x)$, for all $x \in [0,1]$. □

In the case when the t-conorm considered in Proposition 2.6.23 is right-continuous, then the above condition is also sufficient.

Theorem 2.6.24. *For a QL-implication $I_{T,S,N}$ generated from a right-continuous t-conorm S the following statements are equivalent:*

(i) $I_{T,S,N}$ satisfies (IP).
(ii) $T(x,x) \geq N_S \circ N(x)$ for all $x \in [0,1]$.

Proof. *(i)* $\Longrightarrow$ *(ii)* It is obvious from Proposition 2.6.23.
(ii) $\Longrightarrow$ *(i)* By the right-continuity of S, from Proposition 2.3.3(iv) we have that, $S(N(x), N_S \circ N(x)) = 1$ for all $x \in [0,1]$. Now, by the monotonicity of the t-conorm S, we have that

$$I_{T,S,N}(x,x) = S(N(x),T(x,x)) \geq S(N(x), N_S \circ N(x)) = 1 ,$$

i.e., $I_{T,S,N}$ satisfies (IP). □

Example 2.6.25. Let us consider once again the Łukasiewicz t-conorm $S_{\mathbf{LK}}$ and the strict negation $N_{\mathbf{K}}$. The pair $(S_{\mathbf{LK}}, N_{\mathbf{K}})$ satisfies (LEM) and also $S_{\mathbf{LK}}$ is continuous, and hence is right-continuous. Let $T = T_{\mathbf{P}}$ be the product t-norm. Then the QL-operation obtained from the triple $(T_{\mathbf{P}}, S_{\mathbf{LK}}, N_{\mathbf{K}})$ is

$$I_{\mathbf{KP}}(x, y) = \min(1, 1 - x^2 + xy) , \qquad x, y \in [0, 1] .$$

Firstly, note that $I_{\mathbf{KP}}$ satisfies (I1) and hence is a fuzzy implication. Since $N_{\mathbf{S_{LK}}}(x) = 1 - x$, note also that, $N_S \circ N_{\mathbf{K}}(x) = 1 - N_{\mathbf{K}}(x) = 1 - (1 - x^2) = x^2$ and hence $T_{\mathbf{P}}(x, x) = N_S \circ N_{\mathbf{K}}(x)$ for all $x \in [0, 1]$. It is easy to observe that $I_{\mathbf{KP}}$ satisfies (IP).

Remark 2.6.26. Now, Proposition 2.6.21(i) follows directly from the above Theorem 2.6.24, since for $N = N_{\mathbf{D2}}$, we have $N_S \circ N(x) = 0$ if $x \in [0, 1)$ and $N_S \circ N(x) = 1$ if $x = 1$, for the natural negation N_S of any t-conorm S.

2.6.4 QL-Implications and the Ordering Property

From Proposition 2.6.7 and Remark 2.6.22(i) it is clear that if S is a positive t-conorm, then the QL-implication obtained from the triple (T, S, N) does not satisfy (OP).

Proposition 2.6.27. *If a QL-implication $I_{T,S,N}$ obtained from a non-positive t-conorm S satisfies* (OP), *then the negation N is strictly decreasing.*

Proof. To see this, if possible, let there exist $x, y \in [0, 1]$ such that $x < y$ but $N(x) = N(y)$. By (OP) we have

$$\begin{aligned} I_{T,S,N}(x, y) = 1 &\Longrightarrow S(N(x), T(x, y)) = 1 \\ &\Longrightarrow S(N(y), T(y, x)) = 1 \\ &\Longrightarrow I_{T,S,N}(y, x) = 1 \\ &\Longrightarrow y \leq x , \end{aligned}$$

a contradiction. □

Moreover, from Remark 2.6.6(iii) we require that $N \geq N_S$, which implies that the t-conorm S should be such that its natural negation N_S should be non-filling. From Definition 2.3.1, we see that this can happen only if every $x \in (0, 1)$ has a $y \in (0, 1)$ such that $S(x, y) = 1$. Noting that a fuzzy implication that satisfies (OP) also satisfies (IP), using also Theorem 2.6.24, we summarize the above discussion in the following result.

Theorem 2.6.28. *If a QL-implication $I_{T,S,N}$ satisfies* (OP), *then*

(i) $T(x, x) \geq N_S \circ N(x)$ for all $x \in [0, 1]$;
(ii) N is a strictly decreasing negation;
(iii) S is a non-positive t-conorm such that for every $x \in (0, 1)$ there exists a $y \in (0, 1)$ such that $S(x, y) = 1$.

Remark 2.6.29. (i) In fact, the QL-implication $I_{\mathbf{KP}}$ obtained from the triple $(T_{\mathbf{P}}, S_{\mathbf{LK}}, N_{\mathbf{K}})$ (see Example 2.6.25) not only satisfies (IP), but also - as it can be easily verified - (OP).

(ii) The fact that the above conditions are not sufficient can be seen from Example 2.6.16 where the drastic sum t-conorm $S_{\mathbf{D}}$ satisfies condition (iii) of Theorem 2.6.28. Since $N_{S_{\mathbf{D}}} = N_{\mathbf{D1}}$, if N is any strictly decreasing negation we have that N is both non-vanishing and $N > N_{\mathbf{D1}}$ for all $x \in (0,1)$. Notice also that any t-norm T satisfies condition (i) since $N_{\mathbf{D1}} \circ N(x) = 0$ for all $x \in (0,1)$. However, as can be seen from Example 2.6.16, the QL-implication $I_{T,S_{\mathbf{D}},N}$ obtained from such a triple $(T, S_{\mathbf{D}}, N)$ does not satisfy (OP).

(iii) Point (iii) of Theorem 2.6.28 is different from the pair (S,N) satisfying (LEM), in that, for an $x \in (0,1)$ the N may be such that $N(x) = 1$, but the y in Theorem 2.6.28(iii) has to be in $(0,1)$.

In the case the t-norm $T = T_{\mathbf{M}}$, we have the following stronger result.

Theorem 2.6.30. *Let S be a t-conorm and N a fuzzy negation such that the pair (S,N) satisfies conditions in Theorem 2.6.28. Further, for a t-norm T, let $I_{T,S,N}$ be a QL-implication which satisfies* (OP). *Then the following statements are equivalent:*

(i) T *is the minimum t-norm* $T_{\mathbf{M}}$.
(ii) $N_S \circ N = \mathrm{id}_{[0,1]}$.

Proof. Let (S,N) satisfy conditions in Theorem 2.6.28 and, for a t-norm T, let $I_{T,S,N} \in \mathcal{FI}$ satisfy (OP).

(i) $\Longrightarrow$ *(ii)* If $T = T_{\mathbf{M}}$, the minimum t-norm, then the QL-implication obtained from the triple $(T_{\mathbf{M}}, S, N)$ is the $I_{T_{\mathbf{M}},S,N}$ given in Example 2.6.10. From (2.38) we see that $x \leq y \Longrightarrow I_{T,S,N}(x,y) = 1$. The reverse implication is violated only if there exists a $y < x$ such that $S(N(x),y) = 1$. From Remark 2.3.2(iii), we know that for this to happen $y \geq N_S \circ N(x)$. However, from Lemma 2.3.9(ii), we see that $y \in [N_S \circ N(x), x)$. Now it is obvious that the reverse implication holds only if $x = N_S \circ N(x)$.

(ii) $\Longrightarrow$ *(i)* Now, let $N_S \circ N(x) = x$ for all $x \in [0,1]$. Since $I_{T,S,N}$ satisfies (OP), from Theorem 2.6.28(i), we get $x = N_S \circ N(x) \leq T(x,x) \leq x$, which implies that $T(x,x) = x$ for all $x \in [0,1]$, i.e., T is idempotent, or equivalently, $T = T_{\mathbf{M}}$. □

Remark 2.6.31. (i) From Example 2.6.16, we see that with the positive t-norm $T = T_{\mathbf{M}}$, if N is both non-vanishing, i.e., $N(x) = 0$ if and only if $x = 1$, and $N_S \circ N \neq \mathrm{id}_{[0,1]}$, then $I_{T,S,N}$ does not satisfy (OP).

(ii) Let S be a nilpotent t-conorm, i.e., Φ-conjugate with the Łukasiewicz t-conorm $S_{\mathbf{LK}}$. We know that the QL-implication obtained from the triple $(T, (S_{\mathbf{LK}})_\varphi, (N_{\mathbf{C}})_\varphi)$, where T is any t-norm, is $I_{\varphi,T}$ given by (2.39). Since $N_S = N = (N_{\mathbf{C}})_\varphi$ is a strong negation, from Theorem 2.6.30 and Theorem 2.6.24, we obtain the following result (cf. [174]).

Corollary 2.6.32. *For a QL-operation $I_{\varphi,T}$ given by* (2.39), *where T is any t-norm and $\varphi \in \Phi$, the following statements are equivalent:*

(i) $I_{\varphi,T}$ *satisfies* (IP).
(ii) $I_{\varphi,T}$ *satisfies* (OP).
(iii) $T = T_{\mathbf{M}}$.

Remark 2.6.33. The QL-implication $I_{\mathbf{KP}}$ in Example 2.6.25 shows that in the case $N \neq (N_{\mathbf{C}})_\varphi$ in Corollary 2.6.32, there do exist t-norms T other than $T_{\mathbf{M}}$ such that the QL-implication obtained from the triple $(T, (S_{\mathbf{LK}})_\varphi, N)$ satisfies (OP).

2.6.5 QL-Implications and the Law of Contraposition

Since every QL-operation satisfies (NP) (see Proposition 2.6.2), it is immediate from Lemma 1.5.4(v) that, if $I_{T,S,N}$ satisfies CP(N), then $N = N_I$ is strong. If S is a positive t-conorm, from Proposition 2.6.7 we see that a QL-operation $I_{T,S,N}$ is a fuzzy implication if and only if $N = N_{\mathbf{D2}}$, which is a non-strong negation. In fact, as should be familiar now, the QL-implication in this case is the Weber implication $I_{\mathbf{WB}}$, which does not satisfy (CP) with any negation N (see Table 1.9). Of course, if N is strong and $I_{T,S,N}$ satisfies (EP), then by Lemma 1.5.6, we have that $I_{T,S,N}$ satisfies CP(N).

Let S be a nilpotent t-conorm. Then it is non-positive, continuous and is Φ-conjugate with the Łukasiewicz t-conorm $S_{\mathbf{LK}}$. Consider, once again, the QL-implication $I_{\varphi,T}$ obtained from the triple $(T, (S_{\mathbf{LK}})_\varphi, (N_{\mathbf{C}})_\varphi)$, as given by (2.39). Since $(N_{\mathbf{C}})_\varphi$ is strong, we know that $I_{\varphi,T}$ satisfies CP only with $(N_{\mathbf{C}})_\varphi$. Now we have the following results firstly obtained by FODOR [103].

Theorem 2.6.34. *For a QL-implication $I_{\varphi,T}$ given by* (2.39), *where T is any t-norm and $\varphi \in \Phi$, the following statements are equivalent:*

(i) $I_{\varphi,T}$ *satisfies CP*($(N_{\mathbf{C}})_\varphi$).
(ii) T belongs to the family of Frank t-norms $T_{\mathbf{F}}^{\lambda}$.

Proof. Since $I_{\varphi,T} \in \mathcal{FI}$, from Remark 2.6.13 we see that T is continuous. Since φ is an increasing bijection, we have that for any arbitrary $x, y \in [0,1]$ there exist $a, b \in [0,1]$ such that $a = \varphi(x)$ and $b = \varphi(y)$. $I_{\varphi,T}$ satisfies CP($(N_{\mathbf{C}})_\varphi$) implies the following equivalences:

$$\begin{aligned}
I_{\varphi,T}(x,y) &= I_{\varphi,T}((N_{\mathbf{C}})_\varphi(y), (N_{\mathbf{C}})_\varphi(x)) \\
&\iff \varphi^{-1}(1 - \varphi(x) + \varphi(T(x,y))) \\
&\qquad\qquad = \varphi^{-1}(\varphi(y) + \varphi(T((N_{\mathbf{C}})_\varphi(y), (N_{\mathbf{C}})_\varphi(x)))) \\
&\iff 1 - \varphi(x) + \varphi(T(x,y)) = \varphi(y) + \varphi(T((N_{\mathbf{C}})_\varphi(y), (N_{\mathbf{C}})_\varphi(x))) \\
&\iff \varphi(T(x,y)) + 1 - \varphi(T((N_{\mathbf{C}})_\varphi(y), (N_{\mathbf{C}})_\varphi(x))) = \varphi(x) + \varphi(y) \\
&\iff T_{\varphi^{-1}}(\varphi(x), \varphi(y)) + 1 - T_{\varphi^{-1}}(1 - \varphi(x), 1 - \varphi(y)) = \varphi(x) + \varphi(y) \\
&\iff T_{\varphi^{-1}}(a,b) + 1 - T_{\varphi^{-1}}(1-a, 1-b) = a + b \,,
\end{aligned}$$

i.e., $T_{\varphi^{-1}}$ and its dual t-conorm satisfy the Frank equation. Now, by the continuity of T and from Theorem 2.2.11, we have that $I_{\varphi,T}$ satisfies $\mathrm{CP}((N_{\mathbf{C}})_\varphi)$ if and only if T belongs to the family of Frank t-norms. □

Theorem 2.6.35. *For the QL-implication $I_{T,S^N_{\mathbf{nM}},N}$ the following statements are equivalent:*

(i) $I_{T,S^N_{\mathbf{nM}},N}$ satisfies (CP) *with some fuzzy negation N^*.*
(ii) $N^* = N$ is strong and $T = T_{\mathbf{M}}$.

Proof. (i) $\Longrightarrow$ (ii) Let $I_{T,S^N_{\mathbf{nM}},N}$ satisfy (CP) with some fuzzy negation N^*. Since any QL-implication satisfies (NP), from Lemma 1.5.4(v) we see that $N^* = N$ and is a strong negation. Hence we have the following equality:

$$S^N_{\mathbf{nM}}(N(x), T(x,y)) = S^N_{\mathbf{nM}}(y, T(N(x), N(y))) \,, \qquad x, y \in [0,1] \,. \tag{2.42}$$

Next, we claim that $T = T_{\mathbf{M}}$, i.e., $T(x,x) = x$ for all $x \in (0,1)$. Since $T(x,x) \leq x$ for all $x \in (0,1)$ and for any t-norm T, let us assume on the contrary that for some $x \in (0,1)$, $T(x,x) < x$. This implies $N(T(x,x)) > N(x)$. Note that, if $x \in (0,1)$, then is so $N(x)$. Now, with $y = x$ in (2.42), from (2.40) for $S^N_{\mathbf{nM}}$, we have

$$S^N_{\mathbf{nM}}(N(x), T(x,x)) = \max(N(x), T(x,x)) < 1 \,,$$

which implies

$$S^N_{\mathbf{nM}}(x, T(N(x), N(x))) = \max(x, T(N(x), N(x))) < 1 \,,$$

i.e., $x < N(T(N(x), N(x)))$ and hence $N(x) > T(N(x), N(x))$. From the above, we see that $\max(N(x), T(x,x)) = \max(x, T(N(x), N(x)))$. From this equality, since $T(x,x) < x$, we have $\max(N(x), T(x,x)) = N(x)$ and also $\max(x, T(N(x), N(x))) = x$, for $N(x) > T(N(x), N(x))$. Thus we get $x = N(x)$.

Now, let us choose a $y \in (0,1)$ such that $x \neq y$ and once again consider (2.42) with the above $x = N(x)$, i.e., $S^N_{\mathbf{nM}}(x, T(x,y)) = S^N_{\mathbf{nM}}(y, T(x, N(y)))$. Since $T(x, N(y)) \leq N(y)$, we have

$$S^N_{\mathbf{nM}}(y, T(x, N(y))) = \max(y, T(x, N(y))) \leq \max(y, x) < 1 \,.$$

Hence

$$S^N_{\mathbf{nM}}(x, T(x,y)) = \max(x, T(x,y)) < 1 \,,$$

and $\max(x, T(x,y)) = x$, since $x \geq T(x,y)$. Thus $\max(y, T(x, N(y))) = x$. Noting the fact that $T(x, N(y)) \leq x$, we have $x = y$, a contradiction to our choice of y. Therefore, $T(x,x) = x$ for all $x \in (0,1)$, i.e., $T = T_{\mathbf{M}}$.

(ii) $\Longrightarrow$ (i) In the case $T = T_{\mathbf{M}}$ we have that the QL-implication obtained from the triple $(T_{\mathbf{M}}, S^N_{\mathbf{nM}}, N)$ is as given in (2.41) with $N^* = N$. By a straightforward verification, we see that $I_{T_{\mathbf{M}},S^N_{\mathbf{nM}},N}$ indeed satisfies $\mathrm{CP}(N)$. □

2.7 Bibliographical Remarks

We would like to underline here that some authors use the name S-implication, even if the negation N is not strong (see KLEMENT et al. [146], Definition 11.5). Since the name S-implication was firstly introduced in fuzzy logic framework by TRILLAS and VALVERDE (see [238, 239]) with restrictive assumptions, namely, S is a continuous t-conorm and N a strong negation, we use, in a general case, the name '(S,N)-implication' proposed by ALSINA and TRILLAS [5]. It is important to note that different assumptions on the function N are still considered (cf., e.g., KLEMENT et al. [146], Definition 11.5; GOTTWALD [113], Definition 5.4.1; ALCALDE et al. [4], p. 213).

The characterization of (S,N)-implications, where N is a continuous negation, is due to BACZYŃSKI and JAYARAM [15], which subsumes an earlier result of TRILLAS and VALVERDE [239]. From Remark 2.4.7, it should be noted that the above approach cannot be adopted for (S,N)-implications, where N is a non-continuous negation. Moreover, such a representation may not be unique. Hence we have:

Problem 2.7.1. What is the characterization of (S,N)-implications generated from non-continuous negations?

One of the earliest methods for obtaining implications was from conjunctions as their residuals, when no additional logical connectives are given. In fact, GÖDEL [117] extended the three-valued implication of Heyting [122], denoted $I_{\mathbf{GD}}$ in this treatise, while discussing the possible relationships between many-valued logic on the one hand, and intuitionistic logic on the other, where implication is obtained as a residuum of the conjunction. Residuals of conjunctions on a lattice $\mathcal{L}$, be it from t-norms (see DE BAETS and MESIAR [72]), uninorms (see DE BAETS and FODOR [69, 70, 71]), t-subnorms (see MESIAR and MESIAROVÁ [182]), copulas (see DURANTE et al. [95]), etc., have attracted the most attention from researchers, since they can transform the underlying lattice $\mathcal{L}$ into a residuated lattice.

R-implications also have a parallel origin other than its logical foundations. They were also obtained from the study of solutions of systems of fuzzy relational equations and have been known under different names, for example, as a Φ-operator in PEDRYCZ [201], as T-relative pseudocomplement and α_T-operator in [183] (see also the τ-operator in PEDRYCZ [203] and the α-operator in PEDRYCZ [202]). SANCHEZ [220] showed that the greatest solution of sup $-$ min composition of fuzzy relations is the relation obtained from the residual of min. In fact, MIYAKOSHI and SHIMBO [183] generalized this result to any (left-)continuous t-norm. They also showed that their α_T-operator is equivalent to the Φ-operator of PEDRYCZ [203]. Most importantly, they gave the first characterization of R-implications from (left-)continuous t-norms, requiring (NP) in addition to all the properties stated in Theorem 2.5.14.

One of the most important characterization results, not just in the context of R-implications, was that of SMETS and MAGREZ [226] (see Theorems 1 and 2, therein) wherein they showed any binary function I on the unit interval $[0, 1]$ that is continuous, satisfies (I1), (I2), (NP), (EP), (OP) and (CP) with some (strong) negation N is equivalent to the Łukasiewicz implication (up to a Φ-conjugation). FODOR and ROUBENS [105] showed that the conditions of (I1), (NP) and (CP) could be dropped from the above without affecting the validity of the result (see Theorem 1.15 therein). Further refinement was given by BACZYŃSKI [9], who showed the result was true even if (I2) was dispensed with. It should be remarked that this result can be seen as a characterization of the residuals from nilpotent t-norms and that a similar result for residuals from strict t-norms was given by BACZYŃSKI [9] (see Theorem 7.5.4 later). Still, as noted in Remark 2.5.18, the mutual independence of the properties in Theorem 2.5.14 are not known in full and we have the following:

Problem 2.7.2. Prove or disprove by giving a counter example:

> Let $I\colon [0,1]^2 \to [0,1]$ be any function that satisfies both (EP) and (OP). Then the following statements are equivalent:
> (i) I satisfies (I2).
> (ii) I is right-continuous in the second variable.

As given in Proposition 2.4.6 for (S,N)-implications, the equivalence between the different properties of a t-norm T and its residual I_T can be given, in the case the T considered is left-continuous, see, for instance, THIELE [230], DURANTE et al. [95]. Although such an approach in the case of R-implications would have ensured consistency in presentation, the current approach (compare Theorems 2.5.4, 2.5.7, 2.5.14 and Proposition 2.5.13) was taken so as not to obstruct the flow of contents.

However, it should be noted that given a t-norm T an equivalent condition for the R-implication I_T to satisfy the exchange principle (EP) has so far been given, once again, in terms of some conditions on I_T. In fact, such an equivalent condition is valid even for residuals obtained from more general conjunctions than t-norms (see DURANTE et al. [95]). In fact, the left-continuity of a t-norm T is only a sufficient condition for an I_T to satisfy (EP). For example, the Weber implication $I_{\mathbf{WB}}$ obtained from the non-left-continuous t-norm $T_{\mathbf{D}}$ satisfies (EP). See, also, the t-norm $T_{\mathbf{nM}^*}$ of Example 2.5.6 (iv). Thus we have the following question:

Problem 2.7.3. Give a necessary condition on a t-norm T for the corresponding I_T to satisfy (EP).

One of the first results on the contrapositive symmetry of R-implications was given in the framework of continuous t-norms, viz., an R-implication I_T from a continuous t-norm T satisfies (CP) with some (strong) negation N if and only if T is the Łukasiewicz t-norm and N is the classical strong negation (both up to a Φ-conjugation), and hence is Φ-conjugate with the Łukasiewicz implication (see, e.g., SMETS and MAGREZ [226], FODOR [103]). This also meant that the only R-implication I_T from a continuous t-norm T that was also an S-implication was the Łukasiewicz implication, once again up to a Φ-conjugation. It was FODOR [103] who provided a critical breakthrough by obtaining, what is perhaps the first left-continuous but non-continuous t-norm, the nilpotent minimum t-norm $T_{\mathbf{nM}}$, from his analysis of I_T and (CP). The residual of the nilpotent minimum t-norm, referred in this monograph as the Fodor implication $I_{\mathbf{FD}}$, was the first non-continuous fuzzy implication that was not only both an R-implication obtained from a left-continuous t-norm and an S-implication, but could also be shown to be a QL-implication. Of course, as we will show later in Chap. 4, the above is true, in general, for any left-continuous t-norm T whose natural negation is strong (see Theorems 4.1.2 and 4.3.2). Note that the Weber implication $I_{\mathbf{WB}}$ also has the same distinction as that of $I_{\mathbf{FD}}$, but is not obtained as a residual of a left-continuous t-norm.

Staying with intersections, one can further sub-divide the family of R-implications, for example, to consider the set of all R-implications obtained from border continuous t-norms. Such a division, though worthy of consideration, can only be justified if some newer insights can be gleaned therefrom. Hence, in this treatise we have only considered the major sub-families of t-norms that are relevant in the context of R-implications. As stated in the text, the following question still remains:

Problem 2.7.4. Does there exist a continuous R-implication generated from non-(left)-continuous t-norm?

QL-implications have not received as much attention as (S,N)- and R-implications within fuzzy logic. Perhaps, one of the reasons can be attributed to the fact that not all members of this family satisfy one of the main properties expected of a fuzzy implication, viz., left antitonicity. Moreover, in the earlier works, some conditions imposed on the fuzzy logic operations employed in the definition of QL-implications restricted both the class of operations from which QL-implications could be obtained and the properties these implications satisfied. However, the following question, as yet, remains unsolved:

Problem 2.7.5. Characterize triples (T, S, N) such that $I_{T,S,N}$ satisfies (I1).

In one of the earliest works on QL-implications Trillas and Valverde [238] (see also their recent work [235]), the authors required the negation N in Definition 2.6.1 to be strong. Moreover, the t-norm T and t-conorm S are continuous, and are expected to form a De Morgan triple with the negation N. In fact, in Theorem 3.2 of the same work, under these restrictions, condition (ii) of Theorem 2.6.24 has been obtained. From their proof, it is clear that the considered T and S are both continuous and Archimedean and hence either they are strict or nilpotent, in which case they show that the aforementioned condition is not satisfied and hence the claim that "QL-implications never satisfy (IP)". Whereas, from the QL-implications $I_{\mathbf{WB}}$ and $I_{\mathbf{KP}}$ (see Example 2.6.25) we see that $I_{T,S,N}$ satisfies (IP). Still, only some necessary conditions are known for a QL-implication to satisfy (OP) and hence the following question.

Problem 2.7.6. What extra sufficient condition(s) should we impose, other than the ones in Theorem 2.6.28, so that the QL-implication obtained from the triple (T, S, N) satisfies (OP)?

Although Theorem 2.6.19 gives an equivalence condition for a QL-implication $I_{T,S,N}$, with a continuous negation N, to satisfy (EP), it is obvious that its utility is quite limited. Nevertheless, all the examples, so far, seem to point that a QL-implication that satisfies (EP) also turns out to be an (S,N)-implication. The answers to the following posers will be of immense help in resolving the exact intersection between families of (S,N)- and QL-implications.

Problem 2.7.7. (i) Is Theorem 2.6.19 true even when N is not continuous, i.e., is any QL-implication $I_{T,S,N}$ that satisfies (EP) also an (S,N)-implication?
(ii) If not, give a counter-example and hence obtain an alternate necessary and sufficient condition for a QL-implication $I_{T,S,N}$ to satisfy (EP).

Interest on QL-implications has seen some rise in the recent past and some works have appeared on them. These can be broadly classified as follows:

(i) Studies that focus on QL-implications and their basic algebraic properties as in, for example, TRILLAS et al. [236], MAS et al. [174], JAYARAM and BACZYŃSKI [128], SHI et al. [224], and
(ii) Works that investigate QL-implications as part of determining which families of implications satisfy a property under consideration, for instance, TRILLAS et al. [235], JAYARAM [126], TRILLAS and ALSINA [233], FODOR [103], SHI et al. [223].

Once again, with the exception of JAYARAM and BACZYŃSKI [128], the former studies have all been done after restricting the underlying T, S, N to certain families. For example, in SHI et al. [223] though N is any fuzzy negation, T and S are assumed to be continuous, while in MAS et al. [174] N is strong, but they do consider the non-continuous t-conorm $S_{\mathbf{nM}}$, whereas in TRILLAS et al. [235] their investigations have been done only in the context of continuous T and S and strong N.

The reciprocal of a QL-implication, called *Dishkant implication*, has been investigated by MAS et al. [174, 175]. It is defined as

$$I_D(x,y) = S(y, T(N(x), N(y))) , \qquad x, y \in [0,1] ,$$

where T, S are a t-norm and a t-conorm, respectively, while N is (still) a strong negation. Once again these operations generally do not satisfy (I2). It can be shown that these become fuzzy implications if and only if the corresponding QL-operations obtained from (T, S, N) themselves are. Further, these have also been explored only under some restricted conditions on the underlying operations.

In this chapter, we have considered three ways of obtaining fuzzy implications from basic fuzzy logic operations, viz., t-norms, t-conorms and negations. A similar approach to obtaining newer families of fuzzy implications can be taken using the recently proposed uninorms or t-operators, which are generalizations of t-norms and t-conorms. This study will form the last chapter in the Part I of this monograph.

3 Fuzzy Implications from Generator Functions

It is not of the essence of mathematics to be occupied
with the ideas of number and quantity.
– George Boole (1815-1869)

Recently, YAGER [261] has introduced two new families of fuzzy implications, called the f- and g-generated implications, respectively, and discussed their desirable properties. In the next two sections we give the definitions of these newly proposed families of f- and g-generated implications and explore their algebraic properties.

3.1 f-Generated Implications

3.1.1 Definition and Examples

Definition 3.1.1. *Let $f\colon [0,1] \to [0,\infty]$ be a strictly decreasing and continuous function with $f(1) = 0$. The function $I\colon [0,1]^2 \to [0,1]$ defined by*

$$I(x,y) = f^{-1}(x \cdot f(y)) , \qquad x,y \in [0,1] , \tag{3.1}$$

with the understanding $0 \cdot \infty = 0$, is called an f-generated implication. *The function f itself is called an f-generator of the I generated as in (3.1). In such a case, to emphasize the apparent relation we will write I_f instead of I.*

Firstly, since for every $x, y \in [0,1]$ we have $x \cdot f(y) \leq f(y) \leq f(0)$, we see that the formula (3.1) is correctly defined.

Proposition 3.1.2. *If f is an f-generator, then $I_f \in \mathcal{FI}$.*

Proof. The fact that I_f defined by (3.1) is a fuzzy implication can be seen from the following:

- Let $x_1 \leq x_2$. Since f is strictly decreasing, so is f^{-1} and we have, for any $y \in [0,1]$, $x_1 \cdot f(y) \leq x_2 \cdot f(y)$ and hence

$$I_f(x_1,y) = f^{-1}(x_1 \cdot f(y)) \geq f^{-1}(x_2 \cdot f(y)) = I_f(x_2,y) , \tag{3.2}$$

  i.e., I_f satisfies (I1).

M. Baczyński and B. Jayaram: Fuzzy Implications, STUDFUZZ 231, pp. 109–125, 2008.
springerlink.com

- Once again, by the strictly decreasing nature of f, and hence f^{-1}, if $y_1 \leq y_2$, for any $x \in [0,1]$, we have

$$\begin{aligned} f(y_1) \geq f(y_2) &\Longrightarrow x \cdot f(y_1) \geq x \cdot f(y_2) \\ &\Longrightarrow f^{-1}(x \cdot f(y_1)) \leq f^{-1}(x \cdot f(y_2)) \\ &\Longrightarrow I_f(x, y_1) \leq I_f(x, y_2) \,, \end{aligned}$$

  i.e., I_f satisfies (I2).
- $I_f(0,0) = f^{-1}(0 \cdot f(0)) = f^{-1}(0) = 1$, i.e., I_f satisfies (I3).
- $I_f(1,1) = f^{-1}(1 \cdot f(1)) = f^{-1}(f(1)) = 1$, i.e., I_f satisfies (I4).
- $I_f(1,0) = f^{-1}(1 \cdot f(0)) = f^{-1}(f(0)) = 0$, i.e., I_f satisfies (I5). □

In the sequel, we will use the terms f-implication and f-generated implication interchangeably.

Example 3.1.3. From Theorem 2.1.5, the f-generators can be seen as continuous additive generators of continuous Archimedean t-norms. The following examples (see also YAGER [261]) illustrate this amply.

(i) If we take the f-generator $f(x) = -\ln x$, which is a continuous additive generator of the product t-norm $T_{\mathbf{P}}$, then we obtain the Yager implication $I_{\mathbf{YG}}$, which is neither an (S,N)-implication nor an R-implication (see Remark 2.4.4(i) and Remark 2.5.5).

(ii) If we take the f-generator $f(x) = 1 - x$, which is a continuous additive generator of the Łukasiewicz t-norm $T_{\mathbf{L}}$, then we obtain the Reichenbach implication $I_{\mathbf{RC}}$, which is an S-implication (see Table 2.4).

(iii) Let us consider the f-generator $f_{\mathbf{c}}(x) = \cos(\frac{\pi}{2}x)$, which is a continuous and strictly decreasing trigonometric function such that $f_{\mathbf{c}}(0) = \cos 0 = 1$ and $f_{\mathbf{c}}(1) = \cos\frac{\pi}{2} = 0$. Its inverse is given by $f_{\mathbf{c}}^{-1}(x) = \frac{2}{\pi} \cdot \cos^{-1} x$ and the corresponding f-generated implication is given by

$$I_{f_{\mathbf{c}}}(x,y) = \frac{2}{\pi} \cos^{-1}\left(x \cdot \cos\left(\frac{\pi}{2} y\right)\right) \,, \qquad x, y \in [0,1] \,.$$

Fig. 3.1(a) gives the plot of $I_{f_{\mathbf{c}}}$.

(iv) Let us consider the Frank's class of additive generators given by

$$f^s(x) = -\ln\left(\frac{s^x - 1}{s - 1}\right) \,,$$

where $s > 0, s \neq 1$, as the f-generators. Then $f^s(0) = \infty$, its inverse is given by $(f^s)^{-1}(x) = \log_s \left(1 + (s-1)e^{-x}\right)$ and the corresponding f-generated implication, for every s, is given by

$$I_{f^s}(x,y) = \log_s \left(1 + (s-1)^{1-x}(s^y - 1)^x\right) \,, \qquad x, y \in [0,1] \,.$$

Fig. 3.1(b) gives the plot of I_{f^s} for $s = 2$.

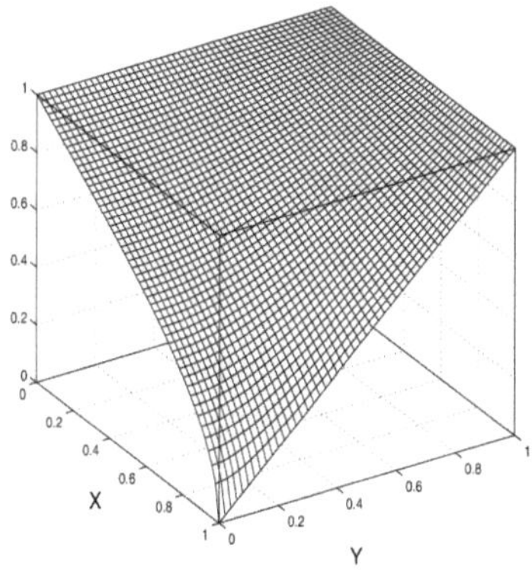

(a) $I_{f_{\mathbf{c}}}$ with $f_{\mathbf{c}}(x) = \cos(\frac{\pi}{2}x)$

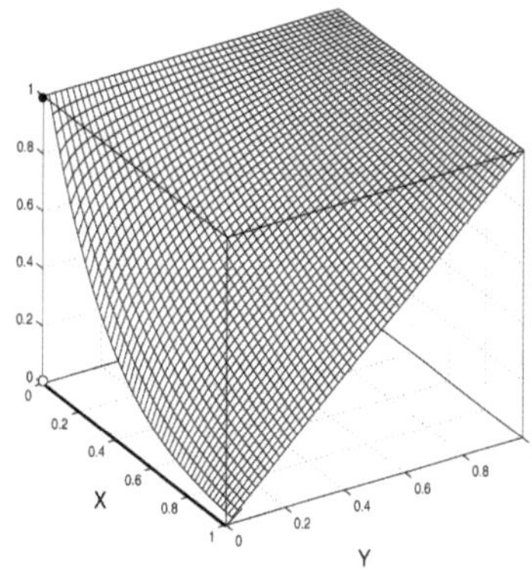

(b) I_{f^s} with $s = 2$

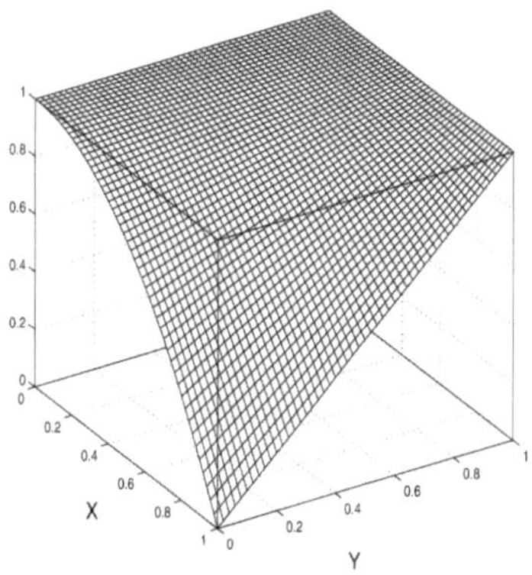

(c) I_{f^λ} with $\lambda = 2$

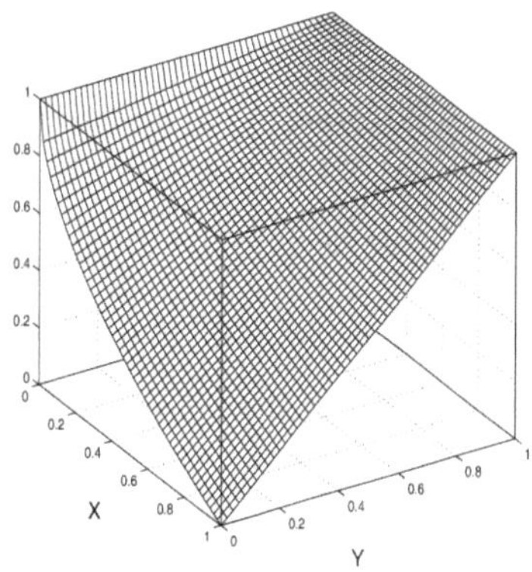

(d) I_{f^λ} with $\lambda = 0.5$

Fig. 3.1. Plots of some f-implications from Example 3.1.3

(v) If we take the Yager's class of additive generators, viz., $f^\lambda(x) = (1 - x)^\lambda$, where $\lambda \in (0, \infty)$, as the f-generators, then $f^\lambda(0) = 1$, its inverse is given by $(f^\lambda)^{-1}(x) = 1 - x^{\frac{1}{\lambda}}$ and the corresponding f-generated implication, for every $\lambda \in (0, \infty)$, is given by

$$I_{f^\lambda}(x, y) = 1 - x^{\frac{1}{\lambda}}(1 - y) , \qquad x, y \in [0, 1] .$$

Figs. 3.1(c) and (d) give the plots of I_{f^λ} for $\lambda = 2$ and $\lambda = 0.5$, respectively.

3.1.2 Properties of f-Implications

We know that the additive generators of t-norms are unique up to a positive multiplicative constant (see Theorem 2.1.5). Firstly, we show that this is also true for the f-generators. Following this, we investigate the natural negations of I_f and discuss its usual algebraic properties.

Theorem 3.1.4. *Let $f_1, f_2\colon [0,1] \to [0,\infty]$ be any two f-generators. Then the following statements are equivalent:*

(i) $I_{f_1} = I_{f_2}$.
(ii) There exists a constant $c \in (0,\infty)$ such that $f_2(x) = c \cdot f_1(x)$ for all $x \in [0,1]$.

Proof. (i) $\Longrightarrow$ (ii) Let f_1, f_2 be two f-generators of an f-generated implication, i.e., $I_{f_1}(x,y) = I_{f_2}(x,y)$ for all $x, y \in [0,1]$. Using (3.1) we get

$$f_1^{-1}(x \cdot f_1(y)) = f_2^{-1}(x \cdot f_2(y)) \,, \qquad x, y \in [0,1] \,.$$

If $f_1(0) = \infty$, then

$$I_{f_1}(x,0) = f_1^{-1}(x \cdot f_1(0)) = f_1^{-1}(x \cdot \infty) = f_1^{-1}(\infty) = 0 \,, \qquad x \in (0,1] \,.$$

Hence, for all $x \in (0,1]$, we have $0 = I_{f_1}(x,0) = I_{f_2}(x,0) = f_2^{-1}(x \cdot f_2(0))$, so $f_2(0) = x \cdot f_2(0)$. This implies that $f_2(0) = \infty$ or $f_2(0) = 0$. However, $f_2(0) = 0$ is impossible, since f_2 is a strictly decreasing function. By changing the role of f_1 and f_2 we obtain the following equivalence:

$$f_1(0) = \infty \Longleftrightarrow f_2(0) = \infty \,.$$

Now, we consider the following two cases:

(a) If $f_1(0) < \infty$, then $f_2(0) < \infty$ and we obtain, for every $x, y \in [0,1]$,

$$f_1^{-1}(x \cdot f_1(y)) = f_2^{-1}(x \cdot f_2(y)) \Longleftrightarrow f_2 \circ f_1^{-1}(x \cdot f_1(y)) = x \cdot f_2(y) \,.$$

In particular, for $y = 0$ and any $x \in [0,1]$, we get

$$\begin{aligned} & f_2 \circ f_1^{-1}(x \cdot f_1(0)) = x \cdot f_2(0) \\ & \qquad \Longleftrightarrow f_2 \circ f_1^{-1}(x \cdot f_1(0)) = x \cdot f_1(0) \cdot \frac{f_2(0)}{f_1(0)} \,. \end{aligned} \tag{3.3}$$

Let us fix arbitrarily $x \in [0,1]$ and consider $z = f_1(x)$. It immediately follows that $z \in [0, f_1(0)]$ and there exists $x_1 \in [0,1]$ such that $z = x_1 \cdot f_1(0)$. From (3.3), we obtain

$$f_2 \circ f_1^{-1}(z) = f_2 \circ f_1^{-1}(x_1 \cdot f_1(0)) = x_1 \cdot f_1(0) \cdot \frac{f_2(0)}{f_1(0)} = z \cdot \frac{f_2(0)}{f_1(0)} \,.$$

Since f_1 is a bijection on its range, substituting $c = \dfrac{f_2(0)}{f_1(0)} \in (0,\infty)$ we get

$$f_2(x) = f_1(x) \cdot \frac{f_2(0)}{f_1(0)} = c \cdot f_1(x) \,.$$

However, x was arbitrarily fixed, so we obtain the claim in this case.

(b) If $f_1(0) = \infty$, then $f_2(0) = \infty$. Firstly, see that $f_2(0) = c \cdot f_1(0)$ and $f_2(1) = c \cdot f_1(1)$ for every $c \in (0, \infty)$. Now, for every $x, y \in [0, 1]$ we have

$$f_1^{-1}(x \cdot f_1(y)) = f_2^{-1}(x \cdot f_2(y)) \Longleftrightarrow f_2 \circ f_1^{-1}(x \cdot f_1(y)) = x \cdot f_2 \circ f_1^{-1}(f_1(y)) .$$

By the substitution $h = f_2 \circ f_1^{-1}$ and $z = f_1(y)$ for any arbitrary $y \in [0, 1]$, we obtain the following equation

$$h(x \cdot z) = x \cdot h(z) , \qquad x \in [0, 1], z \in [0, \infty] , \tag{3.4}$$

where $h \colon [0, \infty] \to [0, \infty]$ is a continuous strictly increasing bijection. Let us substitute $z = 1$, we get

$$h(x) = x \cdot h(1) , \qquad x \in [0, 1] . \tag{3.5}$$

Now, fix arbitrarily $x \in (0, 1)$ and consider $z = f_1(x)$. Of course, $z \in (0, \infty)$. Hence there exists $x_1 \in (0, 1]$ such that $x_1 \cdot z \in (0, 1)$. From (3.4) and (3.5), we get

$$h(z) = \frac{h(x_1 \cdot z)}{x_1} = \frac{x_1 \cdot z \cdot h(1)}{x_1} = z \cdot h(1) .$$

Thus, by the definition of h, we have $f_2 \circ f_1^{-1}(z) = z \cdot f_2 \circ f_1^{-1}(1)$. Since f_1 is a bijection, substituting $c = f_2 \circ f_1^{-1}(1) \in (0, \infty)$ we get

$$f_2(x) = f_1(x) \cdot f_2 \circ f_1^{-1}(1) = c \cdot f_1(x) .$$

However, $x \in (0, 1)$ was arbitrarily fixed, so we have the proof in this direction.

(ii) $\Longrightarrow$ *(i)* Let f_1 be an f-generator and $c \in (0, \infty)$. Define $f_2(x) = c \cdot f_1(x)$ for all $x \in [0, 1]$. Firstly, note that f_2 is a well defined f-generator. Moreover, $f_2^{-1}(z) = f_1^{-1}\left(\frac{z}{c}\right)$ for every $z \in [0, f_2(0)]$. Now, for every $x, y \in [0, 1]$, we have

$$x \cdot c \cdot f_1(y) \leq c \cdot f_1(y) = f_2(y) \leq f_2(0) ,$$
$$\frac{x \cdot c \cdot f_1(y)}{c} = x \cdot f_1(y) \leq f_1(y) \leq f_1(0) ,$$

and thus

$$\begin{aligned} I_{f_2}(x, y) &= f_2^{-1}(x \cdot f_2(y)) = f_2^{-1}(x \cdot c \cdot f_1(y)) = f_1^{-1}\left(\frac{x \cdot c \cdot f_1(y)}{c}\right) \\ &= f_1^{-1}(x \cdot f_1(y)) = I_{f_1}(x, y) , \end{aligned}$$

for all $x, y \in [0, 1]$. □

Remark 3.1.5. From the above result it follows that if f is an f-generator such that $f(0) < \infty$, then the function $f_1 \colon [0, 1] \to [0, 1]$ defined by

$$f_1(x) = \frac{f(x)}{f(0)} , \qquad x \in [0, 1] , \tag{3.6}$$

is a well defined f-generator such that $I_f = I_{f_1}$ and $f_1(0) = 1$. In other words, it is enough to consider only decreasing generators for which $f(0) = \infty$ or $f(0) = 1$.

Proposition 3.1.6. *Let f be an f-generator.*

(i) If $f(0) = \infty$, then the natural negation N_{I_f} is the Gödel negation $N_{\mathbf{D1}}$, which is non-continuous.
(ii) The natural negation N_{I_f} is a strict negation if and only if $f(0) < \infty$.
(iii) The natural negation N_{I_f} is a strong negation if and only if $f(0) < \infty$ and f_1 defined by (3.6) is a strong negation.

Proof. Let f be an f-generator. We get

$$N_{I_f}(x) = I_f(x, 0) = f^{-1}(x \cdot f(0)), \qquad x \in [0,1].$$

(i) If $f(0) = \infty$, then for every $x \in [0,1]$ we have

$$N_{I_f}(x) = f^{-1}(x \cdot \infty) = \begin{cases} f^{-1}(0), & \text{if } x = 0 \\ f^{-1}(\infty), & \text{if } x > 0 \end{cases} = \begin{cases} 1, & \text{if } x = 0 \\ 0, & \text{if } x > 0 \end{cases}$$
$$= N_{\mathbf{D1}}(x).$$

(ii) If $f(0) < \infty$, then N_{I_f} is a composition of real continuous functions, so it is continuous. Moreover, if $x_1 < x_2$, then $x_1 \cdot f(0) < x_2 \cdot f(0)$ and by the strictness of f^{-1} we get that N_{I_f} is a strict negation. The converse implication is a consequence of point (i) of this proposition.

(iii) If $f(0) < \infty$, then because of Remark 3.1.5, the function f_1 defined by (3.6) is a well defined f-generator such that $I_f = I_{f_1}$ and $f_1(0) = 1$. In particular,

$$N_{I_f}(x) = N_{I_{f_1}}(x) = f_1^{-1}(x), \qquad x \in [0,1].$$

If N_{I_f} is a strong negation, then also f_1^{-1} is a strong negation, so $f_1 = f_1^{-1}$. Conversely, if f_1 is a strong negation, then $f_1^{-1} = f_1$, so N_{I_f} is also a strong negation. □

Theorem 3.1.7. *If f is an f-generator, then*

(i) I_f satisfies (NP) *and* (EP),
(ii) $I_f(x, x) = 1$ if and only if $x = 0$ or $x = 1$, i.e., I_f does not satisfy (IP),
(iii) $I_f(x, y) = 1$ if and only if $x = 0$ or $y = 1$, i.e., I_f does not satisfy (OP),
(iv) I_f satisfies (CP) *with some fuzzy negation N if and only if $f(0) < \infty$, f_1 defined by* (3.6) *is a strong negation and $N = N_{I_f}$,*
(v) I_f is continuous if and only if $f(0) < \infty$,
(vi) I_f is continuous except at the point $(0,0)$ if and only if $f(0) = \infty$.

Proof. (i) For any $x \in [0,1]$, we have

$$I_f(1, x) = f^{-1}(1 \cdot f(x)) = f^{-1}(f(x)) = x, \tag{3.7}$$

i.e., I_f satisfies (NP). If $x, y, z \in [0,1]$, then $y \cdot f(z) \le f(0)$ and we have

$$\begin{aligned} I_f(x, I_f(y, z)) &= f^{-1}(x \cdot f(I_f(y, z))) \\ &= f^{-1}(x \cdot f(f^{-1}(y \cdot f(z)))) = f^{-1}(x \cdot y \cdot f(z)) \\ &= I_f(y, I_f(x, z)), \end{aligned}$$

i.e., I_f satisfies (EP).

(ii) Let $I_f(x,x) = 1$ for some $x \in [0,1]$. This implies that $f^{-1}(x \cdot f(x)) = 1$, thus $x \cdot f(x) = f(1) = 0$, hence $x = 0$ or $f(x) = 0$, which by the strictness of f means $x = 1$. The reverse implication is obvious.

(iii) The proof is similar to that for (ii).

(iv) Since I_f satisfies (NP) and (EP), by Corollaries 1.5.5 and 1.5.9, it can satisfy (CP) with some fuzzy negation N if and only if $N = N_{I_f}$ is a strong negation. Now, from Proposition 3.1.6(iii), we obtain the thesis in the first direction. Conversely, if $f(0) < \infty$ and f_1 defined by (3.6) is a strong negation, then again from Proposition 3.1.6(iii), the natural negation N_{I_f} is strong, hence I_f satisfies CP(N_{I_f}).

(v) If $f(0) < \infty$, then I_f given by (3.1) is the composition of real continuous functions, so it is continuous. If $f(0) = \infty$, then because of previous proposition, the natural negation is not continuous and therefore I_f is also non-continuous.

(vi) The forward implication is a result of the previous point. Conversely, if $f(0) = \infty$, then I_f is continuous for every $x, y \in (0,1]$. Further, for every $y \in [0,1]$, we get $I_f(0,y) = 1$ and for every $x \in (0,1]$, we have $I_f(x,0) = 0$ and hence, I is not continuous at the point $(0,0)$. In addition, for every fixed $y \in (0,1]$, we have $f(y) < \infty$ and

$$\lim_{x \to 0^+} I_f(x,y) = \lim_{x \to 0^+} f^{-1}(x \cdot f(y)) = f^{-1}(0) = 1 = I_f(0,y) .$$

Finally, for every $x \in (0,1]$, we have

$$\lim_{y \to 0^+} I_f(x,y) = \lim_{y \to 0^+} f^{-1}(x \cdot f(y)) = f^{-1}(\infty) = 0 = I_f(x,0) . \qquad \square$$

Lemma 3.1.8. *If f is an f-generator, then I_f is one-to-one in the first variable, while the second variable lies in* $(0,1)$.

Proof. Let us assume that there exist $x_1, x_2 \in [0,1]$ and $y \in (0,1)$ such that $I_f(x_1,y) = I_f(x_2,y)$. Now, since $f(y) \in (0,\infty)$, we have

$$\begin{aligned} I_f(x_1,y) = I_f(x_2,y) &\Longrightarrow f^{-1}(x_1 \cdot f(y)) = f^{-1}(x_2 \cdot f(y)) \\ &\Longrightarrow x_1 \cdot f(y) = x_2 \cdot f(y) \\ &\Longrightarrow x_1 = x_2 . \end{aligned} \qquad \square$$

Remark 3.1.9. From Table 1.4, Theorem 3.1.7 and Lemma 3.1.8 we see that $I_{\mathbf{LK}}$, $I_{\mathbf{GD}}$, $I_{\mathbf{KD}}$, $I_{\mathbf{GG}}$, $I_{\mathbf{RS}}$, $I_{\mathbf{WB}}$ nad $I_{\mathbf{FD}}$ are not f-implications.

3.2 g-Generated Implications

YAGER [261] has also proposed another class of implications called the g-generated implications. Unlike f-generated implications, which are obtained from strictly decreasing functions, the g-generated implications are obtained from strictly increasing functions. Our discussion in this section will mirror the approach taken in the previous section.

3.2.1 Definition and Examples

Definition 3.2.1. *Let $g\colon [0,1] \to [0,\infty]$ be a strictly increasing and continuous function with $g(0) = 0$. The function $I\colon [0,1]^2 \to [0,1]$ defined by*

$$I(x,y) = g^{(-1)}\left(\frac{1}{x} \cdot g(y)\right), \qquad x, y \in [0,1], \tag{3.8}$$

with the understanding $\frac{1}{0} = \infty$ and $\infty \cdot 0 = \infty$, is called a g-generated implication, *where the function $g^{(-1)}$ in (3.8) is the pseudo-inverse of g given by*

$$g^{(-1)}(x) = \begin{cases} g^{-1}(x), & \text{if } x \in [0, g(1)], \\ 1, & \text{if } x \in [g(1), \infty]. \end{cases}$$

The function g itself is called a g-generator of the I generated as in (3.8). Once again, we will often write I_g instead of I.

Proposition 3.2.2. *If g is a g-generator, then I_g is a fuzzy implication.*

Proof. Once again, that I_g defined by (3.8) is a fuzzy implication can be seen from the following:

- Let $x_1, x_2 \in [0,1]$ and $x_1 \le x_2$. Then $\frac{1}{x_1} \ge \frac{1}{x_2}$. Since g is increasing, so is $g^{(-1)}$ and we have, for any $y \in [0,1]$,

$$I_g(x_1, z) = g^{(-1)}\left(\frac{1}{x_1} \cdot g(y)\right) \ge g^{(-1)}\left(\frac{1}{x_2} \cdot g(y)\right) = I_g(x_2, y),$$

  i.e., I_g satisfies (I1).
- Once again, by the increasing nature of g, and hence $g^{(-1)}$, if $y_1 \le y_2$, for any $x \in [0,1]$, it can be easily shown that $I_g(x, y_1) \le I_g(x, y_2)$, i.e., I_g satisfies (I2).
- $I_g(0,0) = g^{(-1)}\left(\frac{1}{0} \cdot g(0)\right) = g^{(-1)}(\infty \cdot 0) = g^{(-1)}(\infty) = 1$, i.e., I_g satisfies (I3).
- $I_g(1,1) = g^{(-1)}\left(\frac{1}{1} \cdot g(1)\right) = g^{(-1)}(1 \cdot g(1)) = 1$, i.e., I_g satisfies (I4).
- $I_g(1,0) = g^{(-1)}\left(\frac{1}{1} \cdot g(0)\right) = g^{(-1)}(1 \cdot g(0)) = 0$, i.e., I_g satisfies (I5). □

Remark 3.2.3. (i) Notice that the formula (3.8) can also be written in the following form without explicitly using the pseudo-inverse of g:

$$I(x,y) = g^{-1}\left(\min\left(\frac{1}{x} \cdot g(y), g(1)\right)\right), \qquad x, y \in [0,1]. \tag{3.9}$$

(ii) In the sequel, we will use the terms g-implication and g-generated implication interchangeably.

Example 3.2.4. Similar to f-generators, the g-generators can be seen as continuous additive generators of continuous Archimedean t-conorms. Once again, the following examples illustrate this notion.

(i) If we take the g-generator $g(x) = -\ln(1-x)$, which is a continuous additive generator of the probabilistic sum t-conorm $S_{\mathbf{P}}$, then we obtain the following fuzzy implication:

$$I(x,y) = \begin{cases} 1, & \text{if } x = 0 \text{ and } y = 0\,, \\ 1-(1-y)^{\frac{1}{x}}, & \text{otherwise}\,, \end{cases} \qquad x,y \in [0,1]\,.$$

Fig. 3.2(a) gives the plot of this fuzzy implication. Observe that since the I above does not satisfy (IP), from Theorem 2.5.4, we see that it is not an R-implication. Moreover, since the natural negation of the I above is the Gödel negation $N_{\mathbf{D1}}$, from Remark 2.4.4(i), we see that it is not an (S,N)-implication either. In fact, as we show later on, in Theorem 4.5.4, no g-generated implication I_g is an (S,N)-implication.

(ii) If we take the g-generator $g(x) = x$, which is a continuous additive generator of the Łukasiewicz t-conorm $S_{\mathbf{LK}}$, then we obtain the Goguen implication $I_{\mathbf{GG}}$, which is an R-implication but not an (S,N)-implication by Remark 2.4.4(i). In fact, we show in Theorem 4.6.1 that $I_{\mathbf{GG}}$ is the only g-generated implication which is also an R-implication obtained from some left-continuous t-norm.

(iii) One can easily calculate that for the g-generator $g(x) = -\dfrac{1}{\ln x}$ we obtain the Yager implication $I_{\mathbf{YG}}$, which is also an f-implication.

(iv) If we take the trigonometric function $g_{\mathbf{t}}(x) = \tan\left(\frac{\pi}{2}x\right)$, which is a continuous function with $g_{\mathbf{t}}(0) = 0; g_{\mathbf{t}}(1) = \infty$, as the g-generator, then its inverse is $g_{\mathbf{t}}^{-1}(x) = \frac{2}{\pi}\tan^{-1}(x)$ and we obtain the following g-generated implication:

$$I_{g_{\mathbf{t}}}(x,y) = \frac{2}{\pi}\tan^{-1}\left(\frac{1}{x}\cdot\tan\left(\frac{\pi}{2}y\right)\right)\,, \qquad x,y \in [0,1]\,.$$

Fig. 3.2(b) gives the plot of $I_{g_{\mathbf{t}}}$.

(v) If we take the Yager's class of additive generators, $g^{\lambda}(x) = x^{\lambda}$, where the parameter $\lambda \in (0,\infty)$, as the g-generators, then $g^{\lambda}(1) = 1$ for every λ, its pseudo-inverse is given by $(g^{\lambda})^{(-1)}(x) = \min(1, x^{\frac{1}{\lambda}})$ and the g-generated implication is given by

$$I_{g^{\lambda}}(x,y) = \min\left(1, \frac{y}{x^{\frac{1}{\lambda}}}\right) = \begin{cases} 1, & \text{if } x^{\frac{1}{\lambda}} \le y\,, \\ \dfrac{y}{x^{\frac{1}{\lambda}}}, & \text{otherwise}\,, \end{cases} \qquad x,y \in [0,1]\,.$$

Fig. 3.2(c) gives the plot of $I_{g^{\lambda}}$ for $\lambda = 2$.

(vi) If we take the Frank's class of additive generators,

$$g^{s}(x) = -\ln\left(\frac{s^{1-x}-1}{s-1}\right)\,, \qquad s > 0\,, s \neq 1\,,$$

as the g-generators, then for every s, we have $g^{s}(1) = \infty$,

$$(g^{s})^{-1}(x) = 1 - \log_s\left(1 + (s-1)e^{-s}\right)\,,$$

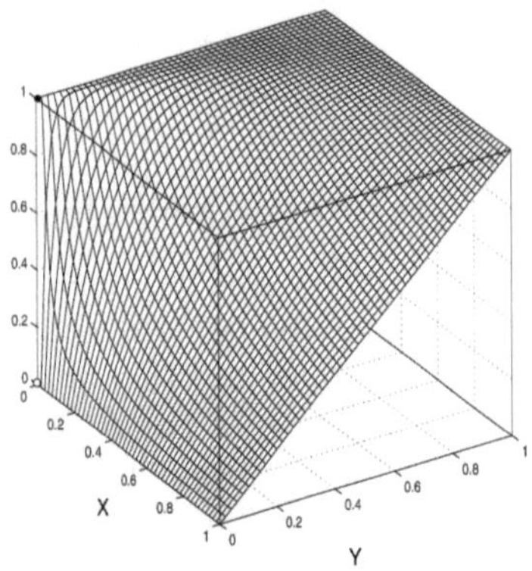

(a) I_g with $g(x) = -\ln(1-x)$

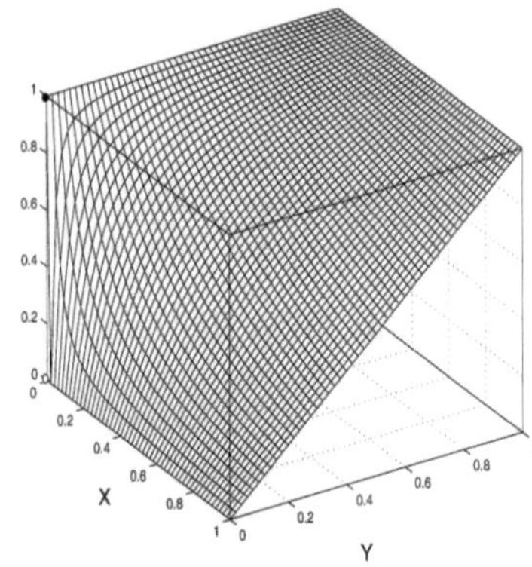

(b) $I_{g_{\mathbf{t}}}$ with $g_{\mathbf{t}}(x) = \tan\left(\frac{\pi}{2}x\right)$

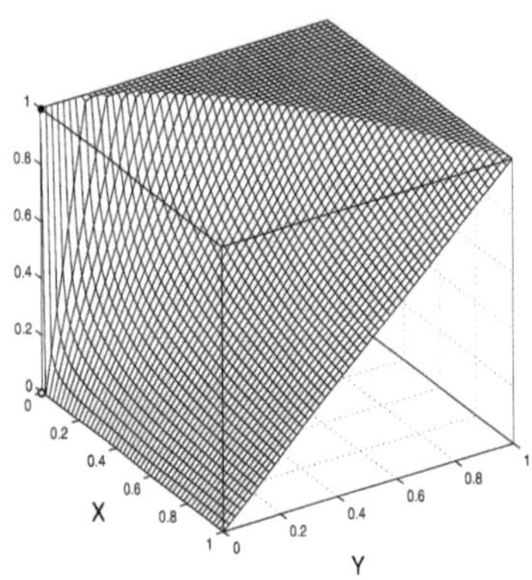

(c) $I_{g^{\lambda}}$ with $\lambda = 2$

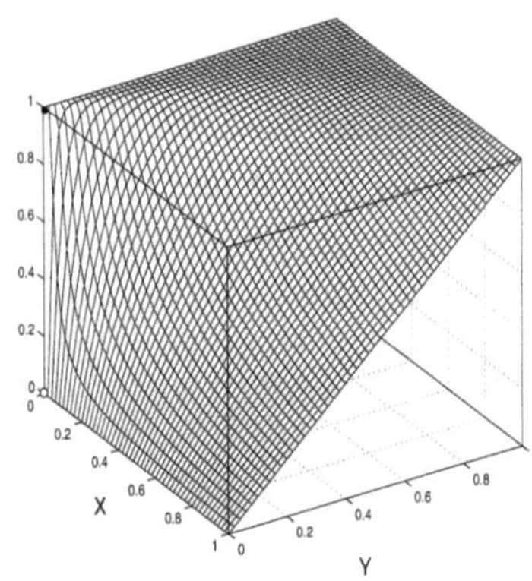

(d) I_{g^s} with $s = 2$

Fig. 3.2. Plots of some g-implications from Example 3.2.4

and the corresponding g-generated implication is given by

$$I_{g^s}(x,y) = 1 - \log_s\left(1 + (s-1)^{\frac{x-1}{x}}(s^{1-y}-1)^{\frac{1}{x}}\right), \qquad x, y \in [0,1] .$$

Fig. 3.2(d) gives the plot of I_{g^s} for $s = 2$.

3.2.2 Properties of g-Implications

In this section, similar to Sect. 3.1.2, we show that the g-generators of the g-generated implications are unique up to a positive multiplicative constant. Following this, we explore the properties of I_g starting with their natural negations.

Theorem 3.2.5. *Let $g_1, g_2\colon [0,1] \to [0,\infty]$ be any two g-generators. Then the following statements are equivalent:*

(i) $I_{g_1} = I_{g_2}$.
(ii) There exists a constant $c \in (0,\infty)$ *such that* $g_2(x) = c \cdot g_1(x)$ *for all* $x \in [0,1]$.

Proof. (i) $\Longrightarrow$ (ii) Let g_1, g_2 be two g-generators of a g-generated implication, i.e., assume that $I_{g_1}(x,y) = I_{g_2}(x,y)$ for all $x, y \in [0,1]$. Using (3.8), we get

$$g_1^{(-1)}\left(\frac{1}{x}\cdot g_1(y)\right) = g_2^{(-1)}\left(\frac{1}{x}\cdot g_2(y)\right), \qquad x, y \in [0,1]\,.$$

If $g_1(1) = \infty$, then $g_2(1) = \infty$. Indeed, let us assume that $g_2(1) < \infty$ and fix arbitrarily $y_0 \in (0,1)$. Since $\lim_{x\to 0^+} \frac{1}{x}\cdot g_2(y_0) = \infty$, there exists $x_0 \in (0,1)$ such that $\infty > \frac{1}{x_0}\cdot g_2(y_0) > g_2(1)$. Hence $g_2^{(-1)}\left(\frac{1}{x_0}\cdot g_2(y_0)\right) = 1$, but $g_1^{(-1)}\left(\frac{1}{x_0}\cdot g_1(y_0)\right) = g_1^{-1}\left(\frac{1}{x_0}\cdot g_1(y_0)\right) < 1$, a contradiction to the assumption that $I_{g_1} = I_{g_2}$. By changing the role of g_1 and g_2, we obtain the following equivalence:

$$g_1(1) = \infty \Longleftrightarrow g_2(1) = \infty\,.$$

Now, we consider the following two cases:

(a) Let $g_1(1) = \infty$. Hence $g_2(1) = \infty$. Firstly, note that $g_2(0) = c\cdot g_1(0)$ and $g_2(1) = c\cdot g_1(1)$, for every $c \in (0,\infty)$. Now, for every $x, y \in [0,1]$, we have

$$\begin{aligned} I_{g_1}(x,y) = I_{g_2}(x,y) &\Longleftrightarrow g_1^{-1}\left(\frac{1}{x}\cdot g_1(y)\right) = g_2^{-1}\left(\frac{1}{x}\cdot g_2(y)\right) \\ &\Longleftrightarrow g_2\circ g_1^{-1}\left(\frac{1}{x}\cdot g_1(y)\right) = \frac{1}{x}\cdot g_2(y) \\ &\Longleftrightarrow g_2\circ g_1^{-1}\left(\frac{1}{x}\cdot g_1(y)\right) = \frac{1}{x}\cdot g_2\circ g_1^{-1}(g_1(y))\,. \end{aligned}$$

By the substitution $h = g_2 \circ g_1^{-1}$ and $z = g_1(y)$ for any arbitrary $y \in [0,1]$, we obtain the following equation

$$h\left(\frac{1}{x}\cdot z\right) = \frac{1}{x}\cdot h(z)\,, \qquad x \in [0,1], z \in [0,\infty]\,, \tag{3.10}$$

where $h\colon [0,\infty] \to [0,\infty]$ is a continuous strictly increasing bijection such that $h(0) = 0$ and $h(\infty) = \infty$. Substituting $z = 1$, we get

$$h\left(\frac{1}{x}\right) = \frac{1}{x}\cdot h(1)\,, \qquad x \in [0,1]\,. \tag{3.11}$$

Fix arbitrarily $x \in (0,1)$ and consider $z = g_1(x)$. Obviously, $z \in (0,\infty)$ and there exists $x_1 \in (0,1)$ such that $x_1 \cdot \dfrac{1}{z} \in (0,1)$. From (3.10) and (3.11), we get

$$h(z) = x_1\cdot h\left(\frac{1}{x_1}\cdot z\right) = x_1\cdot h\left(\frac{1}{\frac{x_1}{z}}\right) = x_1\cdot\frac{1}{\frac{x_1}{z}}\cdot h(1) = z\cdot h(1)\,.$$

Now, by the definition of h, we have $g_2 \circ g_1^{-1}(z) = z \cdot g_2 \circ g_1^{-1}(1)$, and hence

$$g_2(x) = g_1(x) \cdot g_2 \circ g_1^{-1}(1) .$$

However, $x \in (0,1)$ was arbitrarily fixed. Thus, letting $c = g_2 \circ g_1^{-1}(1)$, we obtain the result in this case.

(b) In the case $g_1(1) < \infty$, we also have $g_2(1) < \infty$. Now, for every $x, y \in [0,1]$, we have

$$\begin{aligned} &I_{g_1}(x,y) = I_{g_2}(x,y) \\ &\iff g_1^{-1}\left(\min\left(\frac{1}{x} \cdot g_1(y), g_1(1)\right)\right) = g_2^{-1}\left(\min\left(\frac{1}{x} \cdot g_2(y), g_2(1)\right)\right) \\ &\iff g_2 \circ g_1^{-1}\left(\min\left(\frac{1}{x} \cdot g_1(y), g_1(1)\right)\right) = \min\left(\frac{1}{x} \cdot g_2(y), g_2(1)\right) . \end{aligned}$$

By the substitution $h = g_2 \circ g_1^{-1}$, $u = \frac{1}{x}$ and $v = g_1(y)$ for $x, y \in [0,1]$, we obtain the following equation

$$h\left(\min\left(u \cdot v, g_1(1)\right)\right) = \min\left(u \cdot h(v), g_2(1)\right) ,$$

for $u \in [1,\infty], v \in [0, g_1(1)]$, where the function $h\colon [0, g_1(1)] \to [0, g_2(1)]$ is a continuous and strictly increasing function such that $h(0) = 0$ and $h(g_1(1)) = g_2(1)$. Let us fix arbitrarily $x \in (0,1)$. Then $x \cdot v < g_1(1)$ for all $v \in (0, g_1(1))$. Since h is strictly increasing, $h(x \cdot v) < g_2(1)$ and $h(v) < g_2(1)$ for all $v \in (0, g_1(1))$. Therefore,

$$\begin{aligned} h(v) &= h\left(\frac{1}{x} \cdot x \cdot v\right) = h\left(\min\left(\frac{1}{x} \cdot x \cdot v, g_1(1)\right)\right) \\ &= \min\left(\frac{1}{x} \cdot h(x \cdot v), g_2(1)\right) = \frac{1}{x} \cdot h(x \cdot v) , \end{aligned}$$

for every $v \in (0, g_1(1))$. Hence, from the continuity of h, we have

$$\begin{aligned} g_2(1) = h(g_1(1)) &= \lim_{v \to g_1(1)^-} h(v) = \lim_{v \to g_1(1)^-} \frac{1}{x} \cdot h(x \cdot v) \\ &= \frac{1}{x} \cdot h\left(x \cdot \lim_{v \to g_1(1)^-} v\right) = \frac{1}{x} \cdot h(x \cdot g_1(1)) , \end{aligned}$$

Since $x \in (0,1)$ was arbitrarily fixed, we get

$$h(x \cdot g_1(1)) = x \cdot g_2(1) , \qquad x \in (0,1) . \tag{3.12}$$

Now, for any fixed $v \in (0, g_1(1))$, there exists $x_1 \in (0,1)$ such that $v = x_1 \cdot g_1(1)$ and the previous equality implies

$$h(v) = h(x_1 \cdot g_1(1)) = x_1 \cdot g_2(1) = x_1 \cdot g_1(1) \cdot \frac{g_2(1)}{g_1(1)} = v \cdot \frac{g_2(1)}{g_1(1)} ,$$

for all $v \in (0, g_1(1))$. Note that this formula is also correct for $v = 0$ and $v = g_1(1)$. Therefore, by the definition of h and from (3.12), we get

$$g_2 \circ g_1^{-1}(v) = v \cdot \frac{g_2(1)}{g_1(1)}, \qquad v \in [0, g_1(1)],$$

and hence,

$$g_2(y) = g_1(y) \cdot \frac{g_2(1)}{g_1(1)}, \qquad y \in [0,1].$$

Putting $c = \dfrac{g_2(1)}{g_1(1)}$ we obtain the result.

(*ii*) $\Longrightarrow$ (*i*) Let g_1 be a g-generator and $c \in (0, \infty)$. Define $g_2(x) = c \cdot g_1(x)$ for all $x \in [0,1]$. Evidently, g_2 is a well defined g-generator. Moreover, for any $z \in [0, \infty]$,

$$g_2^{(-1)}(z) = \begin{cases} g_1^{-1}\left(\dfrac{z}{c}\right), & \text{if } z \in [0, c \cdot g_1(1)], \\ 1, & \text{if } z \in [c \cdot g_1(1), \infty]. \end{cases}$$

This implies that, for every $x, y \in [0,1]$, we get

$$\begin{aligned} I_{g_2}(x,y) &= g_2^{-1}\left(\min\left(\frac{1}{x} \cdot g_2(y), g_2(1)\right)\right) \\ &= g_1^{-1}\left(\frac{1}{c}\min\left(\frac{1}{x} \cdot c \cdot g_1(y), c \cdot g_1(1)\right)\right) \\ &= g_1^{-1}\left(\min\left(\frac{1}{x} \cdot g_1(y), g_1(1)\right)\right) = I_{g_1}(x,y). \end{aligned}$$ □

Remark 3.2.6. From the above result it follows that, if g is a g-generator such that $g(1) < \infty$, then the function $g_1 \colon [0,1] \to [0,1]$ defined by

$$g_1(x) = \frac{g(x)}{g(1)}, \qquad x \in [0,1], \tag{3.13}$$

is a well defined g-generator such that $I_g = I_{g_1}$ and $g_1(1) = 1$. In other words, it is enough to consider only increasing generators for which $g(1) = \infty$ or $g(1) = 1$.

Proposition 3.2.7. *If g is a g-generator, then the natural negation of I_g is the Gödel negation $N_{\mathbf{D1}}$, which is not continuous.*

Proof. Let g be a g-generator. For every $x \in [0,1]$, we get

$$\begin{aligned} N_{I_g}(x) &= I_g(x,0) = g^{(-1)}\left(\frac{1}{x} \cdot g(0)\right) = g^{(-1)}\left(\frac{1}{x} \cdot 0\right) \\ &= \begin{cases} g^{(-1)}(\infty), & \text{if } x = 0 \\ g^{(-1)}(0), & \text{if } x > 0 \end{cases} = \begin{cases} 1, & \text{if } x = 0 \\ 0, & \text{if } x > 0 \end{cases} = N_{\mathbf{D1}}(x). \end{aligned}$$ □

Theorem 3.2.8. *Let g be a g-generator.*

(i) I_g *satisfies* (NP),
(ii) I_g *satisfies* (EP),
(iii) I_g *satisfies* (IP) *if and only if* $g(1) < \infty$ *and* $x \leq g_1(x)$ *for every* $x \in [0,1]$, *where* g_1 *is defined by* (3.13),
(iv) if $g(1) = \infty$, *then* $I_g(x,y) = 1$ *if and only if* $x = 0$ *or* $y = 1$, *i.e.,* I_g *does not satisfy* (OP) *when* $g(1) = \infty$,
(v) I_g *does not satisfy* (CP) *with any fuzzy negation,*
(vi) I_g *is continuous except at the point* $(0,0)$.

Proof. (i) The fact that I_g defined by (3.8) satisfies (NP) can be seen from the following:

$$I_g(1,y) = g^{(-1)}\left(\frac{1}{1} \cdot g(y)\right) = y\,, \qquad y \in [0,1]\,.$$

(ii) For any $x, y, z \in [0,1]$, using (3.9), we have

$$I_g(x, I_g(y,z)) = g^{(-1)}\left(\frac{1}{x} \cdot g \circ g^{-1}\left(\min\left(\frac{1}{y} \cdot g(z), g(1)\right)\right)\right)\,.$$

Firstly note that we have the following condition:

$$\min\left(\frac{1}{y}, \frac{1}{x}\right) \cdot g(z) > g(1) \Longrightarrow \left(\frac{1}{x} \cdot \frac{1}{y} \cdot g(z)\right) > g(1)\,. \tag{3.14}$$

Now we consider the following two cases.

- If $\left(\frac{1}{y} \cdot g(z)\right) \leq g(1)$, then by (3.14), we have

$$I_g(x, I_g(y,z)) = g^{(-1)}\left(\frac{1}{x} \cdot \frac{1}{y} \cdot g(z)\right) = I_g(y, I_g(x,z))\,.$$

- If $\left(\frac{1}{y} \cdot g(z)\right) > g(1)$, then by (3.14), we have

$$I_g(x, I_g(y,z)) = g^{(-1)}\left(\frac{1}{x} \cdot g(1)\right) = 1\,,$$

$$I_g(y, I_g(x,z)) = g^{(-1)}\left(\frac{1}{y} \cdot g \circ g^{-1}\left(\min\left(\frac{1}{x} \cdot g(z), g(1)\right)\right)\right) = 1\,.$$

Hence I_g satisfies (EP).

(iii) Let us assume that $g(1) = \infty$. This implies that $g^{(-1)} = g^{-1}$. Let us suppose that $I_g(x,x) = 1$ for some $x \in [0,1]$. This implies that $g^{-1}\left(\frac{1}{x} \cdot g(x)\right) = 1$, thus $\frac{1}{x} \cdot g(x) = g(1) = \infty$, hence $x = 0$ or $g(x) = \infty$, which by the strictness of g means $x = 1$. Therefore, I_g does not satisfy (IP) when $g(1) = \infty$. Let us assume now, that I_g satisfies the identity property (IP). Therefore, it should be $g(1) < \infty$. By Theorem 3.2.5, the function g_1 defined by (3.13) is a well defined g-generator such that $I_g = I_{g_1}$ and $g_1(1) = 1$. Now (IP) implies that, for every $x \in (0,1]$, we have

$$I_g(x,x) = 1 \iff I_{g_1}(x,x) = 1 \iff g_1^{(-1)}\left(\frac{1}{x} \cdot g_1(x)\right) = 1$$
$$\iff g_1^{-1}\left(\min\left(\frac{1}{x} \cdot g_1(x), g_1(1)\right)\right) = 1$$
$$\iff \frac{1}{x} \cdot g_1(x) \geq g_1(1) \iff \frac{1}{x} \cdot g_1(x) \geq 1$$
$$\iff x \leq g_1(x) .$$

The converse implication is a direct consequence of the above equivalences.

(iv) Let us assume that $g(1) = \infty$. This implies that $g^{(-1)} = g^{-1}$. Once again, let us suppose that $I_g(x,y) = 1$ for some $x, y \in [0,1]$. This implies that $g^{-1}(\frac{1}{x} \cdot g(y)) = 1$, thus $\frac{1}{x} \cdot g(y) = g(1) = \infty$ and hence, $x = 0$ or $g(y) = \infty$, which by the strictness of g means $y = 1$. The reverse implication is obvious.

(v) By points (i) and (ii) above, the g-generated implication I_g satisfies (NP) and (EP) and hence it can satisfy the contrapositive symmetry only with N_{I_g} which should be a strong negation. However, from Proposition 3.2.7, we see that the natural negation N_{I_g} is not strong.

(vi) By the formula (3.9), the implication I_g is continuous for every $x, y \in (0,1]$. Further, for every $y \in [0,1]$ we get $I_g(0,y) = 1$ and for every $x \in (0,1]$ we have $I_g(x,0) = 0$, so I_g is not continuous in the point $(0,0)$. In addition, for every fixed $y \in (0,1]$ we have $g(y) > 0$ and, consequently,

$$\lim_{x \to 0^+} I_g(x,y) = \lim_{x \to 0^+} g^{-1}\left(\min\left(\frac{1}{x} \cdot g(y), g(1)\right)\right)$$
$$= g^{-1}(g(1)) = 1 = I_g(0,y) .$$

Finally, for every $x \in (0,1]$ we have $\frac{1}{x} < \infty$, thus

$$\lim_{y \to 0^+} I_f(x,y) = \lim_{y \to 0^+} g^{\ 1}\left(\min\left(\frac{1}{x} \cdot g(y), g(1)\right)\right)$$
$$= g^{-1}(0) = 0 = I_g(x,0) .$$

□

In the last theorem in this section, we will show that I_g satisfies the ordering property (OP) only for a rather special class of g-generators.

Theorem 3.2.9. *If g is a g-generator, then the following statements are equivalent:*

(i) I_g satisfies (OP).

(ii) $g(1) < \infty$ and there exists a constant $c \in (0,\infty)$ such that $g(x) = c \cdot x$ for all $x \in [0,1]$.

(iii) I_g is the Goguen implication $I_{\mathbf{GG}}$.

Proof. (*i*) $\Longrightarrow$ (*ii*) Let us assume that I_g satisfies (OP). By Theorem 3.2.8(iii), we have $g(1) < \infty$. By Remark 3.2.6, the function g_1 defined by (3.13) is a well defined g-generator such that $I_g = I_{g_1}$ and $g_1(1) = 1$. Now (OP) implies that, for every $x, y \in (0,1]$, we have the following equivalences:

$$\begin{aligned} x \le y &\Longleftrightarrow I_g(x,y) = 1 \Longleftrightarrow I_{g_1}(x,y) = 1 \Longleftrightarrow g_1^{(-1)}\left(\frac{1}{x} \cdot g_1(y)\right) = 1 \\ &\Longleftrightarrow g_1^{-1}\left(\min\left(\frac{1}{x} \cdot g_1(y), g_1(1)\right)\right) = 1 \Longleftrightarrow \frac{1}{x} \cdot g_1(y) \ge g_1(1) \\ &\Longleftrightarrow \frac{1}{x} \cdot g_1(y) \ge 1 \Longleftrightarrow x \le g_1(y) \,. \end{aligned} \tag{3.15}$$

This equivalence can also be written in the following form

$$x > y \Longleftrightarrow x > g_1(y) \,, \qquad x, y \in (0,1] \,. \tag{3.16}$$

We show that $g_1(x) = x$ for all $x \in (0,1]$. Suppose that this does not hold, i.e., there exists $x_0 \in (0,1)$ such that $g_1(x_0) \neq x_0$. If $x_0 < g_1(x_0)$, then by the continuity and strict monotonicity of the generator g_1, there exists $y_0 \in (0,1)$ such that

$$x_0 < g_1(y_0) < g_1(x_0) \,. \tag{3.17}$$

However, by (3.15) we get $x_0 \le y_0$. Since g_1 is strictly increasing we have that $g_1(x_0) \le g_1(y_0)$, a contradiction to (3.17).

If $0 < g_1(x_0) < x_0$, then by the continuity and strict monotonicity of the generator g_1 there exists $y_0 \in (0,1)$ such that

$$g_1(x_0) < g_1(y_0) < x_0 \,. \tag{3.18}$$

Now, from (3.16), we get $y_0 < x_0$, which, by the strictly increasing nature of g_1, implies that $g_1(y_0) < g_1(x_0)$, a contradiction to (3.18). We have shown that $g_1(x) = x$ for all $x \in (0,1)$ and also that $g_1(0) = 0$ and $g_1(1) = 1$. By virtue of (3.13) we get that $g(x) = g(1) \cdot x$ for all $x \in [0,1]$.

(*ii*) $\Longrightarrow$ (*iii*) If $g(1) < \infty$ and $g(x) = c \cdot x$ for all $x \in [0,1]$, with some $c \in (0, \infty)$, then $g(1) = c$ and g-generator given by (3.13) is equal to $g_1(x) = x$. From Example 3.2.4(ii), we conclude that $I_{g_1} = I_g$ is the Goguen implication.

(*iii*) $\Longrightarrow$ (*i*) It is obvious that the Goguen implication satisfies (OP). □

Remark 3.2.10. From Table 1.4, Theorems 3.2.8 and 3.2.9 we see that $I_{\mathbf{LK}}$, $I_{\mathbf{GD}}$, $I_{\mathbf{RC}}$, $I_{\mathbf{KD}}$, $I_{\mathbf{RS}}$, $I_{\mathbf{WB}}$ and $I_{\mathbf{FD}}$ are not g-implications.

3.3 Bibliographical Remarks

Although it was YAGER [261] who formally proposed f- and g-generated implications, such an approach - of obtaining fuzzy implications from generators of t-norms - was firstly suggested by VILLAR and SANZ-BOBI in [250]. In fact, the first such example obtained in the aforementioned work was the Yager's implication $I_{\mathbf{YG}}$.

It should be noted that in the original definition of YAGER [261] the pseudo-inverse $f^{(-1)}$ of an f-generator (see Theorem 2.1.5) was employed. However, as was shown in the same work, it suffices to consider the actual inverse of f.

From Yager's work [261] it appears that the motivation to propose these two families stems from a desire to study and exploit the role of fuzzy implications in approximate reasoning (we will have more to say on this topic later on in Chap. 8).

So far very few works have dealt with these two families of fuzzy implications and hence many interesting questions need to be answered. Recently, BALASUBRAMANIAM [22] has investigated f-implications with respect to their distributivity over t-norms and t-conorms, the law of importation and contrapositive symmetry. In fact, the only major work is that of BACZYŃSKI and JAYARAM [16] wherein, after discussing the algebraic properties of these two families, they attempt to answer a significant question, viz., do these families of fuzzy implications intersect with the two well-known classes of fuzzy implications, viz., (S,N)-implications and R-implications? Their investigations have shown that, in general, they are different from the well established (S,N)- and R-implications. In the cases where they intersect with the above families of fuzzy implications, they have determined precisely the sub-families of such intersections. These investigations will form part of Chap. 4 wherein we investigate the intersections that exist among all the families of fuzzy implications dealt with so far in this treatise.

An important question is that of the characterization of these families. Of course, in the case when $f(0) < \infty$, as we show later in Theorem 4.5.1, the generated f-implication is an (S,N)-implication obtained from a continuous negation N and hence the characterization result in Theorem 2.4.10 is applicable. Moreover, based on some recent works of BACZYŃSKI and JAYARAM [18] on distributive equations involving fuzzy implications, rather not-so-elegant and partial characterizations can be given. However, the following still remains to be solved:

Problem 3.3.1. Characterize the families of f- and g-generated implications.

It is clear that the Yager's family of fuzzy implications were obtained from functions that were either strictly decreasing or increasing with the unit interval $[0,1]$ as their domain and $[0,\infty]$ as their codomain. Analogously, one can attempt to obtain fuzzy implications from such functions whose codomain is also $[0,1]$, in other words, from the multiplicative generators of t-norms and t-conorms (see Remarks 2.1.7(iv) and 2.2.7(iv)).

One such attempt was made by BALASUBRAMANIAM [21, 22] wherein a new class of fuzzy implications from the multiplicative generators of t-conorms called the h-generated implications has been proposed. However, as was shown in BACZYŃSKI and JAYARAM [16], this family of fuzzy implications is contained in the family of all (S,N)-implications obtained from continuous negations, and hence is not dealt with in this treatise.

4 Intersections between Families of Fuzzy Implications

Our similarities bring us to a common ground;
Our differences allow us to be fascinated by each other.
– Tom Robbins (1936-)

In the previous chapters, Chaps. 2 and 3, we have dealt with five main families of fuzzy implication operations, viz., (S,N)-, R-, QL-, f- and g-implications. This chapter presents results regarding the intersections that exist among the above families of fuzzy implications. Firstly, we discuss the pair-wise intersections of the families from Chap. 2 and the intersections that exist among the Yager's family of fuzzy implications in Chap. 3. Following this, we discuss the intersections that exist among the families of fuzzy implications across these two chapters, i.e., the intersections that exist among the Yager's family of fuzzy implications, viz., f- and g-implications, with (S,N)-, R- and QL–implications.

4.1 Intersections between (S,N)- and R-Implications

Firstly, note that since $I_{\mathbf{LK}}$ and $I_{\mathbf{WB}}$ are both (S,N)- and R-implications, we have that the intersection of (S,N)- and R-implications is non-empty:

$$\mathbb{I}_{\mathbb{S},\mathbb{N}} \cap \mathbb{I}_{\mathbb{T}} \neq \emptyset \, .$$

A complete characterization of the above intersection is as yet unknown, but some partial results do exist. From Theorem 2.5.4, it is obvious that if an $I_{S,N}$ belongs to the above intersection, then $I_{S,N}$ satisfies (IP). Let us denote by

- $\mathbb{I}_{\mathbb{S},\hat{\mathbb{N}}}$ – the family of all (S,N)-implications, where N is greater than or equal to the natural negation obtained from S, i.e., $N \geq N_S$.

From Proposition 2.3.3, Lemma 2.4.16 and properties of R-implications, we see that

$$\mathbb{I}_{\mathbb{S},\mathbb{N}} \cap \mathbb{I}_{\mathbb{T}} \subset \mathbb{I}_{\mathbb{S},\hat{\mathbb{N}}} \, .$$

Proposition 4.1.1. *(i) If T is a positive t-norm, then the R-implication I_T obtained from it is not an (S,N)-implication for any t-conorm S and any negation N.*
(ii) If I is an (S,N)-implication obtained from a non-involutive negation N, which is either strict, continuous or discontinuous but strictly decreasing, then it is not an R-implication obtained from any t-norm T.

M. Baczyński and B. Jayaram: Fuzzy Implications, STUDFUZZ 231, pp. 127–143, 2008.
springerlink.com

(iii) If T is a border continuous t-norm whose N_T is not strong, then the I_T obtained from it is not an (S,N)-implication for any t-conorm S and any negation N.

Proof. (i) Let T be a positive t-norm and, if possible, let $I_T = I_{S,N}$, for some t-conorm S and some negation N. By virtue of Proposition 2.4.3(ii) and Theorem 2.5.4, we get

$$N(x) = N_{I_{S,N}}(x) = I_{S,N}(x,0) = I_T(x,0) = N_T(x)\,, \qquad x \in [0,1]\,. \quad (4.1)$$

Since T is a positive t-norm, we have that $N_T = N_{\mathbf{D1}}$ (see Table 2.3). However, from Table 2.4, we note that the (S,N)-implication generated from $N_{\mathbf{D1}}$ is $I_{S,N_{\mathbf{D1}}} = I_{\mathbf{D1}}$, which does not satisfy (IP) and hence, by Theorem 2.5.4, is not an R-implication, a contradiction.

(ii) Let us suppose that with the given assumptions $I_{S,N} = I_T$. Then again (4.1) holds. The result now follows from Corollary 2.3.7.

(iii) Let T be a border continuous t-norm and $I_T = I_{S,N}$, for some t-conorm S and some negation N. Once again, (4.1) holds and N is discontinuous. We know, from Proposition 2.5.9, that I_T satisfies (OP). However, since $I_{S,N}$ satisfies (OP), from Theorems 2.4.19 and 2.5.4, we get that $N = N_T = N_S$ should be a strong negation, a contradiction to our hypothesis. □

However, note that, even if T is a border continuous t-norm whose N_T is strong, the I_T obtained from it need not be an (S,N)-implication for any t-conorm S and any negation N. Consider the Viceník t-norm $T_{\mathbf{VC}}$ from Remark 2.5.29 which is border continuous and whose natural negation $N_{T_{\mathbf{VC}}}$ is strong. Now, the R-implication $I_{\mathbf{VC}}$ obtained from it does not satisfy (EP) and hence is not an (S,N)-implication for any t-conorm S and any negation N.

The exact intersection of the family of (S,N)-implications $\mathbb{I}_{\mathbb{S},\mathbb{N}}$ with the family of R-implications obtained from left-continuous t-norms $\mathbb{I}_{\mathbb{T}_{\mathbb{LC}}}$ can be precisely determined. From Theorems 2.4.12, 2.5.7 and 2.5.33, we get

$$\mathbb{I}_{\mathbb{S}} \cap \mathbb{I}_{\mathbb{T}_{\mathbb{C}}} = \mathbb{I}_{\mathbb{LK}}\,.$$

From, by Theorems 2.4.20 and 2.5.17, we have

$$^{\mathbb{C}}\mathbb{I}_{\mathbb{S},\mathbb{N}} \cap \mathbb{I}_{\mathbb{T}_{\mathbb{LC}}} = \mathbb{I}_{\mathbb{LK}}\,.$$

The weaker version of the above two results is well known in the scientific literature and, in general, we say that the only continuous S-implication and R-implication (generated from a left-continuous t-norm) is the Łukasiewicz implication (up to a Φ-conjugation). However, there are many R-implications obtained from t-norms that are left-continuous but not continuous and are still S-implications, for example, the now familiar Fodor implication $I_{\mathbf{FD}}$.

Let us denote by

- $\mathbb{I}_{\mathbb{T}^*}$ – the family of all R-implications obtained from left-continuous t-norms having strong induced negations.

From Theorem 2.5.33 we see that every member of $\mathbb{I}_{\mathbb{LK}}$ is a continuous R-implication based on some continuous t-norm (and hence a left-continuous t-norm) T with a strong natural negation $N_{I_T} = N_T$. However, the Jenei family $I_{\mathbf{J}}^{\lambda}$ of fuzzy implications are R-implications based on the left-continuous t-norms $T_{\mathbf{J}}^{\lambda}$ with a strong natural negation and hence belong to $\mathbb{I}_{\mathbb{T}^*}$, but are non-continuous and hence by Proposition 1.2.4 do not belong to $\mathbb{I}_{\mathbb{LK}}$. It is clear now from our discussion, that

$$\mathbb{I}_{\mathbb{S}} \cap \mathbb{I}_{\mathbb{T}_{\mathbb{LC}}} \supseteq \mathbb{I}_{\mathbb{T}^*} \supsetneq \mathbb{I}_{\mathbb{LK}} \ .$$

A characterization of the fuzzy implications that fall under the above intersection was recently given in BACZYŃSKI and JAYARAM [17].

Theorem 4.1.2. *For a left-continuous t-norm T, a t-conorm S and a fuzzy negation N the following statements are equivalent:*

(i) The R-implication I_T is also an (S,N)-implication $I_{S,N}$, i.e., $I_T = I_{S,N}$.
(ii) $N = N_T$ is a strong negation and (T, S, N) form a De Morgan triple.

Proof. $(i) \Longrightarrow (ii)$ Assume that there exist a left-continuous t-norm T, a fuzzy negation N and a t-conorm S such that $I_T = I_{S,N}$. For notational simplicity, we denote this function by I. Since I is an R-implication from a left-continuous t-norm, by Theorem 2.5.17, it satisfies (OP). Since I is also an (S,N)-implication, by Theorem 2.4.19, we have that $N = N_S$ is a strong negation. Once again, by virtue of Proposition 2.4.3(ii), (4.1) is still valid and hence $N = N_T$. Further, since I satisfies (OP), (EP) and N_I is strong, from Theorem 2.5.30, we have

$$\begin{aligned} T(x, y) &= N_I(I(x, N_I(y))) = N_I(S(N_I(x), N_I(y))) \\ &= N_T(S(N_T(x), N_T(y))) = N(S(N(x), N(y))) \ , \end{aligned}$$

for all $x, y \in [0, 1]$, i.e., (T, S, N_T) form a De Morgan triple.

$(ii) \Longrightarrow (i)$ Let $N = N_T$ be a strong negation and the triple (T, S, N) form a De Morgan triple. Firstly, see that the R-implication I_T is an (S,N)-implication. Indeed, since I_T satisfies (I1), (EP) and its natural negation $N_{I_T} = N_T$ is a strong negation, by Theorem 2.4.12, we get that I_T is an S-implication, i.e., $I_T = I_{S',N'}$ for an appropriate t-conorm S' and a strong negation N'. Observe now that $I_T = I_{S,N_S}$. Indeed, if (T, N, S) form a De Morgan triple, then from our assumptions and Theorem 2.3.18, it follows that S is a right-continuous t-conorm such that $N = N_S$ is a strong negation. Therefore,

$$\begin{aligned} N_S(x) &= N(x) = N_T(x) = N_{I_T}(x) = I_T(x, 0) = I_{S',N'}(x, 0) \\ &= N_{I_{S',N'}}(x) = N'(x) \ , \end{aligned}$$

for all $x \in [0, 1]$. Hence $I_T = I_{S',N_S}$. Finally, from the proof of $(i) \Longrightarrow (ii)$ above, we know that T is the N_S dual of S' and by our assumption T is the N_S dual of S. Hence $S = S'$, i.e., $I_T = I_{S,N_S}$. □

In fact, using Theorems 2.3.17 and 2.3.18, the above result can be restated as follows.

Theorem 4.1.3. *For a left-continuous t-norm T and a t-conorm S the following statements are equivalent:*

(i) The R-implication I_T is also an (S,N)-implication $I_{S,N}$ with some fuzzy negation N.
(ii) The R-implication I_T is also an S-implication I_{S,N_T}, where N_T is a strong negation.
(iii) (T, N_T, S) form a De Morgan triple.

Remark 4.1.4. The left-continuity of T is very important in the above theorems.

(i) For example, consider any t-conorm S whose natural negation $N_S \neq N_{\mathbf{D2}}$. However, $I_{S,N_{\mathbf{D2}}} = I_{\mathbf{WB}}$, which is also an R-implication obtained from the non-left-continuous t-norm $T_{\mathbf{D}}$. It is obvious that the triple $(T_{\mathbf{D}}, N_{\mathbf{D2}}, S)$ do not form a De Morgan triple.
(ii) Even if T is a non-left-continuous t-norm, N_T can still be strong and the R-implication I_T obtained from it can be an S-implication, without the triple (T, N_T, S) forming a De Morgan triple. Once again, consider the t-norm $T_{\mathbf{nM}^*}$ from Example 2.5.6(iv). Now, $N_{T_{\mathbf{nM}^*}} = N_{\mathbf{C}}$ and $I_{T_{\mathbf{nM}^*}} = I_{\mathbf{FD}}$ which is an S-implication. However, it should be emphasized that the S-implication obtained from the N_C-dual t-conorm of $T_{\mathbf{nM}^*}$ is not equal to the Fodor implication $I_{\mathbf{FD}}$ (see Example 2.4.22).

Theorem 4.1.5. *For a right-continuous t-conorm S and a t-norm T the following statements are equivalent:*

(i) The (S,N)-implication I_{S,N_S} is also the R-implication I_T.
(ii) (S, N_S, T) form a De Morgan triple.

Let us denote by

- $\mathbb{I}_{\mathbb{N}_\mathbb{T}(\mathbb{T}),\mathbb{N}_\mathbb{T}}$ – the family of all (S,N)-implications obtained from the N_T-dual of the left-continuous t-norm T whose natural negation N_T is strong.

Recall that $\mathbb{I}_{\mathbb{S}^*,\mathbb{N}_\mathbb{S}^*}$ denotes the family of all (S,N)-implications obtained from right-continuous t-conorms and their natural negations which are strong (see Sect. 2.4.5). As a consequence of the presented facts, the following equalities hold:

$$\mathbb{I}_\mathbb{S} \cap \mathbb{I}_{\mathbb{T}_{\mathbb{LC}}} = \mathbb{I}_{\mathbb{S},\mathbb{N}} \cap \mathbb{I}_{\mathbb{T}_{\mathbb{LC}}} = \mathbb{I}_{\mathbb{N}_\mathbb{T}(\mathbb{T}),\mathbb{N}_\mathbb{T}} = \mathbb{I}_{\mathbb{T}^*} = \mathbb{I}_{\mathbb{S}^*,\mathbb{N}_\mathbb{S}^*} .$$

The results presented in this section are also diagrammatically represented in Fig. 4.1., for which the following example will be useful.

Example 4.1.6. Consider the following non-right-continuous t-conorm which is the $N_{\mathbf{C}}$-dual of the t-norm $T_{\mathbf{B}^*}$ given in Example 2.5.6(ii):

$$S_{\mathbf{B}^*}(x, y) = \begin{cases} 1, & \text{if } (x, y) \in (0.5, 1)^2 , \\ \max(x, y), & \text{otherwise} , \end{cases} \qquad x, y \in [0, 1] .$$

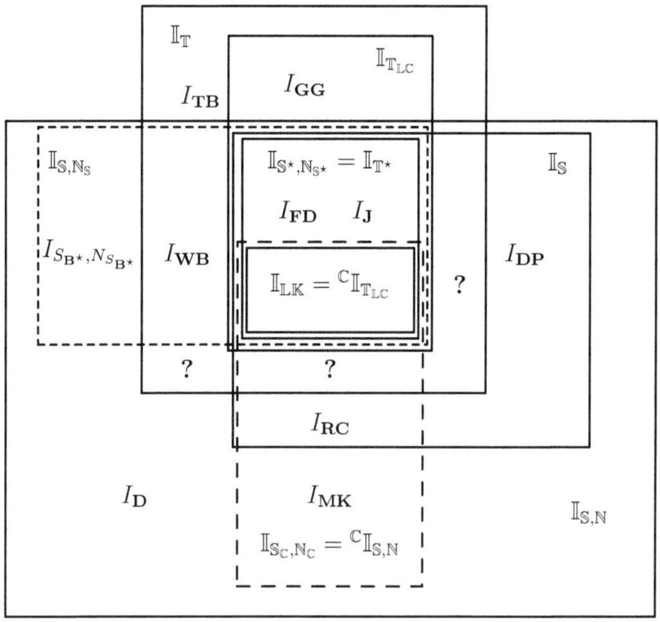

Fig. 4.1. Intersections between families of (S,N)- and R-implications

Then its natural negation is the discontinuous function given by

$$N_{S_{\mathbf{B}^*}}(x) = \begin{cases} 1, & \text{if } x \in [0, 0.5) , \\ 0.5, & \text{if } x \in (0.5, 1) , \\ 0, & \text{if } x = 1 . \end{cases}$$

The (S,N)-implication obtained from $S_{\mathbf{B}^*}$ and $N_{S_{\mathbf{B}^*}}$ is given by

$$I_{S_{\mathbf{B}^*},N_{S_{\mathbf{B}^*}}}(x, y) = \begin{cases} 1, & \text{if } x \in [0, 0.5] , \\ 0.5, & \text{if } x \in (0.5, 1) \text{ and } y \in [0, 0.5] , \\ y, & \text{otherwise} , \end{cases}$$

for all $x, y \in [0, 1]$. Clearly, $I_{S_{\mathbf{B}^*},N_{S_{\mathbf{B}^*}}}$ does not satisfy (IP). Therefore, from Theorem 2.5.4, we see that it is not an R-implication for any t-norm T.

4.2 Intersections between (S,N)- and QL-Implications

Let us denote by

- $\mathbb{I}_{\mathbb{QL}}$ – the family of all QL-implications.

We divide our investigation into two parts, based on whether the considered t-conorm S is positive or not. The next result is obvious from Proposition 2.6.7.

Theorem 4.2.1. *The QL-implication $I_{T,S,N}$, where S is a positive t-conorm, is an (S,N)-implication. In fact, $I_{T,S,N} = I_{\mathbf{WB}}$.*

Hence

$$\mathbb{I}_{\mathbb{S},\mathbb{N}} \cap \mathbb{I}_{\mathbb{QL}} \neq \emptyset \,.$$

From Theorem 2.6.19, we know that a QL-operation $I_{T,S,N}$ obtained from the triple (T, S, N), where N is a continuous negation, is an (S,N)-implication if it satisfies (EP). In the case, when T is the minimum t-norm $T_{\mathbf{M}}$, we have the following stronger results.

Proposition 4.2.2. *Let S be a t-conorm and N a fuzzy negation such that the pair (S, N) satisfies* (LEM). *Then the QL-operation $I_{T_{\mathbf{M}},S,N}$ is also an (S,N)-implication obtained from the same t-conorm S and negation N, i.e., $I_{T_{\mathbf{M}},S,N} = I_{S,N}$. In other words, $I_{S,N}$ can be represented as a QL-implication obtained from the triple $(T_{\mathbf{M}}, S, N)$.*

Proof. Consider the QL-operation generated from the minimum t-norm $T_{\mathbf{M}}$, a t-conorm S and a fuzzy negation N such that the pair (S, N) satisfies (LEM). Then $I_{T_{\mathbf{M}},S,N}$ is a fuzzy implication given by (2.38). Now, if $x \leq y$, then

$$I_{T,S,N}(x, y) = S(N(x), T_{\mathbf{M}}(x, y)) = S(N(x), x) = 1 \,,$$

and $I_{S,N}(x, y) = S(N(x), y) \geq S(N(x), x) = 1$. If $x > y$, then, of course, $I_{T,S,N}(x, y) = S(N(x), y) = I_{S,N}(x, y)$. □

Remark 4.2.3. In the above Proposition 4.2.2 the condition that the pair (S, N) satisfies (LEM) is essential. Otherwise, the QL-operation may not be a fuzzy implication, as can be seen for I_{-1} in Remark 2.6.3.

By virtue of Remark 2.1.4(viii) we see that $(T_{\mathbf{M}})_\varphi = T_{\mathbf{M}}$ for all $\varphi \in \Phi$, whence we get the following result.

Corollary 4.2.4. *The Φ-conjugate of the QL-implication $I_{T_{\mathbf{M}},S,N}$ is also the (S,N)-implication generated from the Φ-conjugate t-conorm of S and the Φ-conjugate fuzzy negation of N, i.e., if $\varphi \in \Phi$, then*

$$(I_{T_{\mathbf{M}},S,N})_\varphi(x, y) = I_{S_\varphi,N_\varphi}(x, y) \,, \qquad x, y \in [0, 1] \,.$$

As an interesting consequence of the above fact we obtain the following characterizations of some special classes of QL-implications.

Theorem 4.2.5. *For a function $I\colon [0, 1]^2 \to [0, 1]$ the following statements are equivalent:*

(i) I is a QL-implication obtained from the triple $(T_{\mathbf{M}}, S, N)$ with a continuous negation N.
(ii) I satisfies (I2), (IP), (EP) *and N_I is a continuous negation.*

Proof. (*i*) $\Longrightarrow$ (*ii*) From Proposition 4.2.2 we see that I is an (S,N)-implication obtained from the t-conorm S and the continuous negation N. Hence it satisfies (I2) and (EP) too. Since I is a QL-implication we know that the pair (S, N) satisfies (LEM) and hence, by Lemma 2.4.16 I satisfies (IP).

(*ii*) $\Longrightarrow$ (*i*) Since I satisfies (I2), (EP) and N_I is a continuous negation, I is an (S,N)-implication obtained from some t-conorm S and $N = N_I$. Consider the QL-operation J obtained from the triple $(T_{\mathbf{M}}, S, N)$ with $N = N_I$ and the above S. Since I satisfies (IP), once again by Lemma 2.4.16, we know that the pair (S, N) satisfies (LEM). Now with $T = T_{\mathbf{M}}$ we know from Proposition 4.2.2, that J is the (S,N)-implication obtained from the above t-conorm S and $N = N_I$, i.e., $J = I$. □

Theorem 4.2.6. *For a function* $I\colon [0,1]^2 \to [0,1]$ *the following statements are equivalent:*

(i) I is a QL-implication obtained from the triple $(T_{\mathbf{M}}, S, N_S)$ *where* N_S *is a strong negation and the pair* (S, N_S) *satisfies* (LEM).
(ii) I satisfies (I2), (OP), (EP) *and* N_I *is a strong negation.*

Proof. (*i*) $\Longrightarrow$ (*ii*) From Proposition 4.2.2 we see that $I = I_{S,N_S}$, the (S,N)-implication obtained from the t-conorm S and the strong negation N_S. Hence it satisfies (I2) and (EP). Moreover, $N_I = N_{I_{S,N_S}} = N_S$ is a strong negation. Finally, from Theorem 2.4.19 we see that I satisfies (OP).

(*ii*) $\Longrightarrow$ (*i*) Since I satisfies (I2), (EP) and N_I is a strong negation it is an S-implication obtained from the t-conorm $S(x, y) = I(N_I(x), y)$ and $N = N_I$. Once again, from Theorem 2.4.19 we see that $N = N_I = N_S$ and the pair (S, N_S) satisfies (LEM). Consider the QL-operation J obtained from the triple $(T_{\mathbf{M}}, S, N_S)$. From Proposition 4.2.2 we get that I' is the (S,N)-implication obtained from the above t-conorm S and $N = N_S$, i.e., $J = I$. □

Corollary 4.2.7. *For a function* $I\colon [0,1]^2 \to [0,1]$ *the following statements are equivalent:*

(i) I is a QL-implication obtained from the triple $(T_{\mathbf{M}}, S, N_S)$ *where S is a right-continuous t-conorm and* N_S *is a strong negation.*
(ii) I satisfies (I2), (OP), (EP), *I is right-continuous in the second variable and* N_I *is a strong negation.*

Proof. (*i*) $\Longrightarrow$ (*ii*) Since S is right-continuous, from Proposition 2.3.3(iv) we have that the pair (S, N_S) satisfies (LEM). Then, from Theorem 4.2.6, we see that I satisfies (I2), (OP), (EP) and N_I is a strong negation. Obviously, I is right-continuous in the second variable.

(*ii*) $\Longrightarrow$ (*i*) Since $S(x, y) = I(N_I(x), y)$ for any $x, y \in [0, 1]$ and I is right-continuous in the second variable we have that S is right-continuous. Once again, from Theorem 4.2.6 the rest of the proof is obvious. □

Theorem 4.2.8. *Let $I_{T,S,N}$ be a QL-implication, where S is a non-positive t-conorm with strong induced negation N_S. Consider the following statements:*

(i) $I_{T,S,N}$ is an (S,N)-implication obtained from the same t-conorm S and negation N, i.e., $I_{T,S,N} = I_{S,N}$.
(ii) $N = N_S$.
(iii) $T = T_{\mathbf{M}}$.

Then the following relationships exist among the above statements:
(i) and (ii) $\Longrightarrow$ (iii),
(ii) and (iii) $\Longrightarrow$ (i).

Proof. Firstly, note that if $I_{T,S,N}$ is a fuzzy implication, then by virtue of Lemma 2.6.5 the pair (S,N) satisfies (LEM).

(i) and $(ii) \Longrightarrow (iii)$ We know that for any t-norm $T(x,x) \leq x$ for all $x \in [0,1]$. Moreover, $N_S(x) \neq 1$ for any $x \in (0,1)$. Let us assume that $I_{T,S,N_S} = I_{S,N_S}$ for some some non-positive t-conorm S with strong natural negation N_S and some t-norm T. Then $S(N_S(x), T(x,x)) = S(N_S(x), x) = 1$, for all $x \in [0,1]$, since the pair (S, N_S) satisfies (LEM). From Remark 2.3.2(iii), we obtain

$$T(x,x) \geq (N_S \circ N_S)(x) = x \ ,$$

from whence we obtain $T(x,x) = x$, for all $x \in [0,1]$, i.e., $T = T_{\mathbf{M}}$.

(ii) and $(iii) \Longrightarrow (i)$ Since the pair (S,N) satisfies (LEM), this follows from Proposition 4.2.2. □

Remark 4.2.9. Let us consider a t-conorm S whose natural negation N_S is discontinuous. Note that, in this remark, by points (i), (ii) and (iii) we refer to the items described in Theorem 4.2.8.

From Proposition 4.2.2, we always have that $(iii) \Longrightarrow (i)$. Let us define a lenient version of (i) as follows:

(i') $I_{T,S,N}$ is an (S,N)-implication obtained from a (possibly different) t-conorm S' and a negation N', i.e., $I_{T,S,N} = I_{S',N'}$.

Then, from Table 4.1, the following observations can be made:

(a) From the first entry, we notice that $N = N_S$ is not strong and $I_{T,S,N} = I_{S,N}$, but $T \neq T_{\mathbf{M}}$, i.e., (i) and $(ii) \not\Longrightarrow (iii)$, when N is not strong. Note that the t-conorm $S_{\mathbf{P}}$ can be replaced by any positive t-conorm.
(b) From the second and third entries, it is clear that even if $I_{T,S,N} = I_{S,N}$ and $T = T_{\mathbf{M}}$ we can have $N_S \neq N$, i.e., (i) and $(iii) \not\Longrightarrow (ii)$, when N is not strong. For the formulae for $S_{\mathbf{B}^*}$ and $N_{S_{\mathbf{B}^*}}$, see Example 4.1.6.
(c) From the fourth and fifth entries, we see that $I_{T,S,N} = I_{S',N'}$ and $N = N' = N_{\mathbf{C}}$, a strong negation, but $T \neq T_{\mathbf{M}}$, i.e., (i') *and* $(ii) \not\Longrightarrow (iii)$.

Summarizing the above discussion, we get

$$\mathbb{I}_{\mathbb{S}^*,\mathbb{N}^*_{\mathbb{S}}} \subsetneq \mathbb{I}_{\mathbb{S},\mathbb{N}_\mathbb{S}} \subsetneq \mathbb{I}_{\mathbb{S},\hat{\mathbb{N}}} \subsetneq \mathbb{I}_{\mathbb{QL}} \ .$$

Table 4.1. Some QL-implications that are also (S,N)-implications

S	T	N	N_S	$I_{T,S,N}$
$S_{\mathbf{P}}$	$T_{\mathbf{P}}$	$N_{\mathbf{D2}}$	$N_{\mathbf{D2}}$	$I_{\mathbf{WB}}$
$S_{\mathbf{B^*}}$	$T_{\mathbf{M}}$	$N_{\mathbf{D2}}$	$N_{S_{\mathbf{B^*}}}$	$I_{\mathbf{WB}}$
$S_{\mathbf{D}}$	$T_{\mathbf{M}}$	$N_{\mathbf{C}}$	$N_{\mathbf{D1}}$	$I_{\mathbf{DP}}$
$S_{\mathbf{LK}}$	$T_{\mathbf{LK}}$	$N_{\mathbf{C}}$	$N_{\mathbf{C}}$	$I_{\mathbf{KD}}$
$S_{\mathbf{LK}}$	$T_{\mathbf{P}}$	$N_{\mathbf{C}}$	$N_{\mathbf{C}}$	$I_{\mathbf{RC}}$

The following examples illustrate the above chain of inclusions.

- $I_{\mathbf{LK}}, I_{\mathbf{FD}} \in \mathbb{I}_{\mathbb{S}^*,\mathbb{N}_{\mathbb{S}}^*} \subsetneq \mathbb{I}_{\mathbb{QL}}$.
- $I_{\mathbf{WB}} \in \mathbb{I}_{\mathbb{S},\mathbb{N}_{\mathbb{S}}} \setminus \mathbb{I}_{\mathbb{S}^*,\mathbb{N}_{\mathbb{S}}^*} \subsetneq \mathbb{I}_{\mathbb{QL}}$.
- $I_{\mathbf{DP}} \in \mathbb{I}_{\mathbb{S},\hat{\mathbb{N}}} \setminus \mathbb{I}_{\mathbb{S},\mathbb{N}_{\mathbb{S}}} \subsetneq \mathbb{I}_{\mathbb{QL}}$.
- $I_{\mathbf{KD}}, I_{\mathbf{RC}} \in \mathbb{I}_{\mathbb{QL}} \setminus \mathbb{I}_{\mathbb{S},\hat{\mathbb{N}}}$.
- The QL-implications $I_{\mathbf{PC}}, I_{\mathbf{PR}}$ from Example 2.6.15 and the QL-implication $I_{\mathbf{KP}}$ from Example 2.6.25 do not satisfy the exchange principle (EP) and hence are not (S,N)-implications, i.e.,

$$I_{\mathbf{PC}}, I_{\mathbf{PR}}, I_{\mathbf{KP}} \in \mathbb{I}_{\mathbb{QL}} \setminus \mathbb{I}_{\mathbb{S},\mathbb{N}} .$$

- Similarly, the fuzzy implication $I_{\mathbf{D}}$ (see Table 2.4) is an (S,N)-implication obtained from the least negation $N_{\mathbf{D1}}$ and hence, by Remark 2.6.6(i), it is not a QL-implication, i.e.,

$$I_{\mathbf{D}} \in \mathbb{I}_{\mathbb{S},\mathbb{N}} \setminus \mathbb{I}_{\mathbb{QL}} .$$

4.3 Intersections between R- and QL-Implications

Firstly, if S is a positive t-conorm or if $N = N_{\mathbf{D2}}$, then the QL-implication $I_{T,S,N}$ is the R-implication $I_{\mathbf{WB}}$ obtained from the non-left-continuous t-norm $T_{\mathbf{D}}$. Hence

$$\mathbb{I}_{\mathbb{QL}} \cap \mathbb{I}_{\mathbb{T}} \neq \emptyset .$$

A complete characterization of the above intersection is as yet unknown. However, as we show below, the exact intersection of the family of QL-implications $\mathbb{I}_{\mathbb{QL}}$ with the family of R-implications obtained from left-continuous t-norms $\mathbb{I}_{\mathbb{T}_{\mathbb{LC}}}$ can be precisely determined.

Proposition 4.3.1. *If a QL-implication $I_{T,S,N}$ is an R-implication obtained from a left-continuous t-norm T^*, then*

(i) $N = N_{T^*}$ is strong;
(ii) $I_{T,S,N}$ is also an S-implication obtained from a t-conorm S^*, such that S^* is the N-dual of T^*, and $N = N_{S^*}$, i.e., $I_{T,S,N} = I_{S^*,N_{S^*}}$.

Proof. Let a QL-implication obtained from the triple (T, S, N) also be an R-implication obtained from a left-continuous t-norm T^*, i.e., let $I_{T,S,N} = I_{T^*}$.

(i) From Theorem 2.5.17, we see that I_{T^*} satisfies both (EP) and (OP). Now, from Propositions 2.6.27 and 1.4.19, we get that

$$N = N_{I_{T,S,N}} = N_{I_{T^*}} = N_{T^*} \ ,$$

is either strong or discontinuous but strictly decreasing. However, we know from Corollary 2.3.7, that the natural negation N_{T^*} of a (left-continuous) t-norm T^*, if discontinuous, is not strictly decreasing. Hence $N = N_{T^*}$ is strong.

(ii) Since $I_{T,S,N}$ satisfies (EP), Theorem 2.6.19 implies that $I_{T,S,N}$ is also an S-implication $I_{S^*,N}$ for some t-conorm S^*, i.e., $I_{T,S,N} = I_{S^*,N} = I_{T^*}$. Now, from Theorem 4.1.2, we see that (T^*, S^*, N) forms a DeMorgan triple, i.e., S^* is the N-dual of T^* and that $N = N_{S^*}$. □

Theorem 4.3.2. *For a function $I\colon [0,1]^2 \to [0,1]$ the following statements are equivalent:*

(i) I is both a QL-implication obtained from the triple (T, S, N) and an R-implication obtained from some left-continuous t-norm T^*.
(ii) I can be represented as a QL-implication obtained from $(T_{\mathbf{M}}, S^*, N_{S^*})$, where S^* is a right-continuous t-conorm with a strong natural negation N_{S^*}.
(iii) I is both an (S,N)- and an R-implication obtained from a left-continuous t-norm.

Proof. *(i)* $\Longrightarrow$ *(ii)* Let $I = I_{T,S,N} = I_{T^*}$. From Proposition 4.3.1, we have that $N = N_{I_{T^*}} = N_{T^*}$ is strong and $I = I_{S^*,N}$, where S^* is the right-continuous t-conorm that is the N-dual of T^*. Moreover, $N = N_{T^*} = N_{S^*}$. Now, since S^* is right-continuous, by Proposition 2.3.3(iv), we see that the pair (S^*, N_{S^*}) indeed satisfies (LEM). Further, by Proposition 4.2.2, we see that $I_{S^*,N_{S^*}}$ can also be represented as a QL-implication obtained from the triple $(T_{\mathbf{M}}, S^*, N_{S^*})$, i.e., $I = I_{T^*} = I_{S^*,N_{S^*}} = I_{T_{\mathbf{M}},S^*,N_{S^*}}$.

(ii) $\Longrightarrow$ *(iii)* Firstly, from Proposition 4.2.2, we see that such a QL-implication is also an (S,N)-implication. In fact, we have $I_{T_{\mathbf{M}},S^*,N_{S^*}} = I_{S^*,N_{S^*}}$. Since S^* is a right-continuous t-conorm with a strong natural negation N_{S^*}, we see that (S^*, N_{S^*}, T^*) form a De Morgan triple, where T^* is the left-continuous t-norm which is N_{S^*}-dual of S^*. Now, from Theorem 4.1.5, we see that $I_{S^*,N_{S^*}}$ is also the R-implication obtained from T^*, i.e., $I_{S^*,N_{S^*}} = I_{T^*}$.

(iii) $\Longrightarrow$ *(i)* If I is both an (S,N)- and an R-implication obtained from a left-continuous t-norm, then we know, from Theorem 4.1.2, that $(T, S, N_T = N_S)$ form a De Morgan triple and $I = I_{S,N_S} = I_T$. Once again, invoking Proposition 4.2.2, we see that I_{S,N_S} can also be represented as a QL-implication obtained from the triple $(T_{\mathbf{M}}, S, N_S)$, i.e., $I = I_T = I_{S,N_S} = I_{T_{\mathbf{M}},S,N_S}$. □

From Theorem 4.3.2, we see that

$$\begin{aligned}\mathbb{I}_{\mathbb{QL}} \cap \mathbb{I}_{\mathbb{T}_{\mathbb{LC}}} &= \mathbb{I}_{\mathbb{T}_{\mathbb{LC}}} \cap \mathbb{I}_{\mathbb{S},\mathbb{N}} \\ &= \mathbb{I}_{\mathbb{QL}} \cap \mathbb{I}_{\mathbb{T}_{\mathbb{LC}}} \cap \mathbb{I}_{\mathbb{S},\mathbb{N}} \\ &= \mathbb{I}_{\mathbb{S}^*,\mathbb{N}_{\mathbb{S}}^*} = \mathbb{I}_{\mathbb{T}^*} \ .\end{aligned}$$

Example 4.3.3. (i) Consider the R-implication obtained from the t-norm $T_{\mathbf{B}^*}$ given in Example 2.5.6(ii). It is clear that $I_{\mathbf{TB}}$ satisfies (OP), but its natural negation

$$N_{\mathbf{TB}^*}(x) = \begin{cases} 1, & x = 0 \ , \\ 0.5, & x \in [0, 0.5) \ , \\ 0, & x \in [0.5, 1] \ , \end{cases}$$

is not strictly decreasing and hence, by Proposition 2.6.27, we have that $I_{\mathbf{TB}^*}$ is not a QL-implication. For the same reason the Goguen implication $I_{\mathbf{GG}}$ cannot be obtained as a QL-implication for any triple (T, S, N).

(ii) Consider the QL-implications $I_{\mathbf{PC}}, I_{\mathbf{PR}}$ from Example 2.6.15. Since they do not satisfy (IP), from Theorem 2.5.4 we see that they cannot be represented as R-implications of any t-norm. Let us now consider the QL-implication $I_{\mathbf{KP}}$ in Example 2.6.25. Since a QL-operation is generated by a unique negation (see Proposition 2.6.2(ii)), if $I_{\mathbf{KP}}$ is also an R-implication I_T obtained from some t-norm T, then $N_{I_{\mathbf{KP}}} = N_{\mathbf{K}} = N_{I_T} = N_T$. However, $N_{\mathbf{K}}$ is a strict negation that is not involutive, hence by Corollary 2.3.7, we see that $I_{\mathbf{KP}}$ cannot be an R-implication.

To summarize, we have the following facts:

- $I_{\mathbf{PC}}, I_{\mathbf{PR}}, I_{\mathbf{KP}} \in \mathbb{I}_{\mathbb{QL}} \setminus \mathbb{I}_{\mathbb{T}}$,
- $I_{\mathbf{WB}} \in \mathbb{I}_{\mathbb{QL}} \setminus \mathbb{I}_{\mathbb{T}_{\mathbb{LC}}}$,
- $I_{\mathbf{TB}^*} \in \mathbb{I}_{\mathbb{T}} \setminus \mathbb{I}_{\mathbb{QL}}$,
- $I_{\mathbf{GG}} \in \mathbb{I}_{\mathbb{T}_{\mathbb{LC}}} \setminus \mathbb{I}_{\mathbb{QL}}$.

The results presented in this section and Sect. 4.2 are also diagrammatically represented in Fig. 4.2.

4.4 Intersections between Yager's f- and g-Implications

Let us denote by

- $\mathbb{I}_{\mathbb{F},\infty}$ – the family of all f-generated implications such that $f(0) = \infty$;
- $\mathbb{I}_{\mathbb{F},\aleph}$ – the family of all f-generated implications such that $f(0) < \infty$;
- $\mathbb{I}_{\mathbb{F}} = \mathbb{I}_{\mathbb{F},\infty} \cup \mathbb{I}_{\mathbb{F},\aleph}$;
- $\mathbb{I}_{\mathbb{G},\infty}$ – the family of all g-generated implications such that $g(1) = \infty$;
- $\mathbb{I}_{\mathbb{G},\aleph}$ – the family of all g-generated implications such that $g(1) < \infty$;
- $\mathbb{I}_{\mathbb{G}} = \mathbb{I}_{\mathbb{G},\infty} \cup \mathbb{I}_{\mathbb{G},\aleph}$.

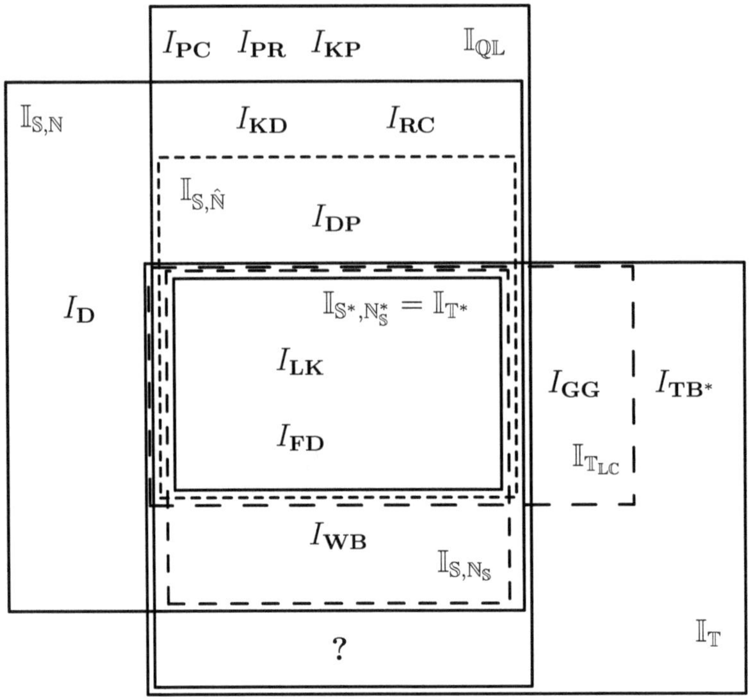

Fig. 4.2. Intersections between families of (S,N)-, R- and QL-implications

Proposition 4.4.1. *The following equalities are true:*

$$\mathbb{I}_{\mathbb{F},\aleph} \cap \mathbb{I}_{\mathbb{G}} = \emptyset \,, \tag{4.2}$$

$$\mathbb{I}_{\mathbb{F}} \cap \mathbb{I}_{\mathbb{G},\aleph} = \emptyset \,, \tag{4.3}$$

$$\mathbb{I}_{\mathbb{F},\infty} = \mathbb{I}_{\mathbb{G},\infty} \,. \tag{4.4}$$

Proof. (i) The equation (4.2) is the consequence of Theorem 3.1.6(ii) and Proposition 3.2.7.

(ii) Let $I \in \mathbb{I}_{\mathbb{F}}$. From Theorem 3.1.7(iii), we know that $I(x,y) = 1$ if and only if $x = 0$ or $y = 1$. However, if we assume that $I \in \mathbb{I}_{\mathbb{G},\aleph}$, then there exists a g-generator such that I has the form (3.8) and $g(1) < \infty$. Thus, for every $x \in (0,1)$ we get

$$I\left(\frac{g(x)}{g(1)}, x\right) = g^{(-1)}\left(\frac{g(1)}{g(x)} \cdot g(x)\right) = g^{(-1)}\left(g(1)\right) = 1 \,.$$

Therefore there exist $x, y \in (0,1)$ such that $I(x,y) = 1$, so we obtain (4.3).

(iii) Let us assume that $I \in \mathbb{I}_{\mathbb{F},\infty}$, i.e., there exists an f-generator f with $f(0) = \infty$ such that I has the form (3.1). Let us define the function $g\colon [0,1] \to [0,\infty]$ by

$$g(x) = \frac{1}{f(x)} \,, \qquad x \in [0,1] \,,$$

with the assumptions, that $\frac{1}{0} = \infty$ and $\frac{1}{\infty} = 0$. We see that g is a g-generator with $g(1) = \infty$. Moreover $g^{(-1)}(x) = g^{-1}(x) = f^{-1}(\frac{1}{x})$. Hence, for every $x,\ y \in [0,1]$, we have

$$\begin{aligned} I_g(x,y) &= g^{-1}\left(\frac{1}{x} \cdot g(y)\right) = g^{-1}\left(\frac{1}{x} \cdot \frac{1}{f(y)}\right) \\ &= f^{-1}\left(x \cdot f(y)\right) = I(x,y) \ . \end{aligned}$$

Conversely, if $I \in \mathbb{I}_{\mathbb{G},\infty}$, then there exists a g-generator g with $g(1) = \infty$ such that I has the form (3.8). Defining the function $f\colon [0,1] \to [0,\infty]$ by

$$f(x) = \frac{1}{g(x)} \ , \qquad x \in [0,1] \ ,$$

we get that f is an f-generator such that $I_f = I$. □

4.5 Intersections between Yager's and (S,N)-Implications

In this section we investigate whether any of the families $\mathbb{I}_{\mathbb{F}}, \mathbb{I}_{\mathbb{G}}$ intersect with $\mathbb{I}_{\mathbb{S},\mathbb{N}}$. We know from Proposition 2.4.3 that any (S,N)-implication satisfies (NP) and (EP). We know from Theorems 3.1.7 and 3.2.8 that the implications from the families $\mathbb{I}_{\mathbb{F}}, \mathbb{I}_{\mathbb{G}}$ satisfy (NP) and (EP). Hence, because of the characterization Theorems 2.4.10, 2.4.11 and 2.4.12 of some subclasses of (S,N)-implications we only need to check their natural negations. Once again, these have been done in Propositions 3.1.6 and 3.2.7 and we have the following results.

Theorem 4.5.1. *If f is an f-generator, then the following statements are equivalent:*

(i) I_f is an (S,N)-implication.
(ii) $f(0) < \infty$.

Proof. *(i)* $\Longrightarrow$ *(ii)* Let f be an f-generator such that $f(0) = \infty$ and assume that I_f is an (S,N)-implication generated from a t-conorm S and a fuzzy negation N. From Proposition 2.4.3 we get that $N_{I_f} = N$, but from Proposition 3.1.6(i) we see that N_{I_f} is the Gödel negation $N_{\mathbf{D1}}$. Hence, from Table 2.4.2, it follows that $I_f = I_{\mathbf{D}}$. Thus $f^{-1}(x \cdot f(y)) = y$ for all $x \in (0,1], y \in [0,1]$, which implies $x \cdot f(y) = f(y)$ for all $x \in (0,1], y \in [0,1]$, a contradiction.

(ii) $\Longrightarrow$ *(i)* Let f be an f-generator such that $f(0) < \infty$. Observe that from Proposition 3.1.6(ii) the natural negation N_{I_f} is a strict negation. Theorem 2.4.11 implies that I_f is an (S,N)-implication generated from some t-conorm and some strict negation. □

It is important to note that in this case we can fully describe t-conorms and strict negations from which the f-generated implications are obtained.

Corollary 4.5.2. *If $f(0) < \infty$, then the function $S\colon [0,1]^2 \to [0,1]$ defined by*

$$S(x,y) = I_f(N_{I_f}^{-1}(x), y)\,, \qquad x, y \in [0,1]$$

is a strict t-conorm, i.e., it is Φ-conjugate with the probabilistic sum t-conorm $S_{\mathbf{P}}$.

Proof. Let us assume that f is a decreasing generator such that $f(0) < \infty$. Then the function f_1 defined by the formula (3.6) is a strict negation. We know also that $I_f = I_{f_1}$, so for all x, $y \in [0,1]$ we get

$$\begin{aligned} S(x,y) &= I_f(N_{I_f}^{-1}(x), y) = I_{f_1}(N_{I_{f_1}}^{-1}(x), y) = I_{f_1}((I_{f_1}(x,0))^{-1}, y) \\ &= I_{f_1}((f_1^{-1})^{-1}(x), y) = I_{f_1}(f_1(x), y) \\ &= f_1^{-1}(f_1(x) \cdot f_1(y))\,. \end{aligned}$$

Let us define the function $\varphi\colon [0,1] \to [0,1]$ by $\varphi(x) = 1 - f_1(x)$ for all $x \in [0,1]$. Evidently, φ is an increasing bijection. Moreover $f_1^{-1}(x) = \varphi^{-1}(1-x)$ for all $x \in [0,1]$. This implies that

$$\begin{aligned} S(x,y) &= f_1^{-1}(f_1(x) \cdot f_1(y)) = f_1^{-1}((1-\varphi(x)) \cdot (1-\varphi(y))) \\ &= \varphi^{-1}(\varphi(x) + \varphi(y) - \varphi(x) \cdot \varphi(y)) \end{aligned}$$

for all x, $y \in [0,1]$, i.e., S is Φ-conjugate with the probabilistic sum t-conorm $S_{\mathbf{P}}$. Therefore, by virtue of Theorem 2.2.8, S is strict. □

This means that for $f(0) < \infty$ we have $I_f(x,y) = S(N_{I_f}(x), y)$ for x, $y \in [0,1]$, where S is Φ-conjugate with the probabilistic sum t-conorm $S_{\mathbf{P}}$. However,

$$N_{I_f}(x) = N_{I_{f_1}}(x) = f_1^{-1}(x) = \varphi^{-1}(1-x)\,, \qquad x \in [0,1]\,.$$

Therefore, if $f(0) < \infty$, then we do not obtain any new implication but only (S,N)-implication generated from Φ-conjugate probabilistic sum t-conorm for

$$\varphi(x) = 1 - \frac{f(x)}{f(0)}\,, \qquad x \in [0,1]\,,$$

and the strict negation $N(x) = \varphi^{-1}(1-x)$ for all $x \in [0,1]$.

Now, under the following restricted situation we can obtain S-implications.

Theorem 4.5.3. *Let f be an f-generator. Then the following statements are equivalent:*

(i) The function I_f is an S-implication.
(ii) $f(0) < \infty$ and the function f_1 defined by (3.6) is a strong negation.

Theorem 4.5.4. *If g is a g-generator, then I_g is not an (S,N)-implication.*

Proof. Assume that there exists a g-generator g such that I_g is an (S,N)-implication generated from a t-conorm S and a fuzzy negation N. We get $N_{I_g} = N$, but

Proposition 3.2.7 gives that N_{I_g} is the Gödel negation $N_{\mathbf{D1}}$. Hence, from Table 2.4, it follows that $I_g = I_{\mathbf{D}}$. Thus, for all $x \in (0,1], y \in [0,1]$,

$$g^{(-1)}\left(\frac{1}{x}\cdot g(y)\right) = y \Longrightarrow g^{-1}\left(\min\left(\frac{1}{x}\cdot g(y), g(1)\right)\right) = y\,.$$

Let us take any $x, y \in (0,1)$, we get $\frac{1}{x}\cdot g(y) = g(y)$, a contradiction. □

Denoting the sub-family of $\mathbb{I}_{\mathbb{F},\aleph}$ where the function f_1 defined by (3.6) is a strong negation by $\mathbb{I}_{\mathbb{F},\aleph^*}$, the above results can be summarized thus:

$$\mathbb{I}_{\mathbb{F}} \cap \mathbb{I}_{\mathbb{S},\mathbb{N}} = \mathbb{I}_{\mathbb{F},\aleph}\,,$$
$$\mathbb{I}_{\mathbb{F}} \cap \mathbb{I}_{\mathbb{S}} = \mathbb{I}_{\mathbb{F},\aleph^*}\,,$$
$$\mathbb{I}_{\mathbb{G}} \cap \mathbb{I}_{\mathbb{S},\mathbb{N}} = \emptyset\,.$$

4.6 Intersections between Yager's and R-Implications

In this section we investigate whether any of the families $\mathbb{I}_{\mathbb{F}}, \mathbb{I}_{\mathbb{G}}$ intersect with $\mathbb{I}_{\mathbb{T}}$. From Theorem 2.5.4, which states that an R-implication I_T obtained from any t-norm T, not necessarily left-continuous, satisfies (IP) and Theorem 3.1.7(ii), we get:

Theorem 4.6.1. *If f is an f-generator, then I_f is not an R-implication obtained from any t-norm.*

From Theorem 3.2.9 we have the following result:

Theorem 4.6.2. *If g is a g-generator of I_g, then the following statements are equivalent:*

(i) I_g is an R-implication obtained from a left-continuous t-norm.
(ii) $g(1) < \infty$ and there exists a constant $c \in (0,\infty)$ such that $g(x) = cx$ for all $x \in [0,1]$.
(iii) I_g is the Goguen implication $I_{\mathbf{GG}}$.

The above results can be summarized thus:

$$\mathbb{I}_{\mathbb{F}} \cap \mathbb{I}_{\mathbb{T}} = \emptyset\,,$$
$$\mathbb{I}_{\mathbb{G}} \cap \mathbb{I}_{\mathbb{T}_{\mathbb{LC}}} = \{I_{\mathbf{GG}}\}\,.$$

4.7 Intersections between Yager's and QL-Implications

In this section we investigate whether any of the families $\mathbb{I}_{\mathbb{F}}, \mathbb{I}_{\mathbb{G}}$ intersect with $\mathbb{I}_{\mathbb{QL}}$. Unfortunately, there does not exist any complete characterization of the family of QL-implications. Interestingly, some results can still be proven. Firstly, from Remark 2.6.6(i), we see that if the natural negation N_I of a fuzzy implication I is the Gödel negation $N_{\mathbf{D1}}$, then I is not a QL-implication. Now, from Propositions 3.1.6(i) and 3.2.7, we have the following result.

Theorem 4.7.1. *If I is either*

(i) a g-implication obtained from a g-generator, or
(ii) an f-implication obtained from an f-generator with $f(0) = \infty$,

then I is not a QL-implication.

Summarizing the above results we have:

$$\begin{aligned} \mathbb{I}_{\mathbb{F},\infty} \cap \mathbb{I}_{\mathbb{QL}} &= \emptyset , \\ \mathbb{I}_{\mathbb{G}} \cap \mathbb{I}_{\mathbb{QL}} &= \emptyset . \end{aligned}$$

The results presented in Sects. 4.4, 4.5 and 4.6 are also diagrammatically represented in Fig. 4.3.

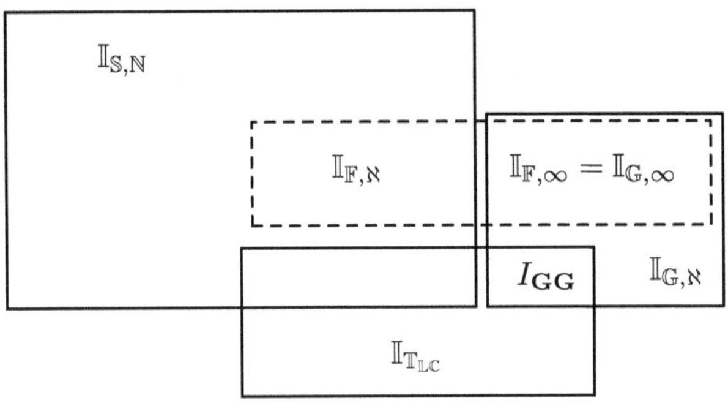

Fig. 4.3. Intersections between families of f-, g- and (S,N)-, R-implications

4.8 Bibliographical Remarks

Despite the many works that were done on (S,N)- and R-implications, an in-depth study of their overlaps was not done till recently. One of the first works on the intersection of S- and R-implications was done by Dubois and Prade [86, 87] wherein they have shown that S-implications and R-implications could be merged into a single family, provided that the class of triangular norms is enlarged to non-commutative conjunction operations. See also the follow-up works of Fodor [99, 100].

Perhaps, the only result known on this topic was that the only continuous (S,N)- and R-implication from a left-continuous t-norm is the Łukasiewicz implication, up to a Φ-conjugation (see, e.g., Smets and Magrez [226], Fodor [103]). But with the emergence of newer and interesting families of left-continuous t-norms with strong natural negations, it was obvious that there exist many other members in the above intersection. In fact, this was cited as one of the strengths of the nilpotent minimum $T_{\mathbf{nM}}$ by Fodor [103].

It was recently in Baczyński and Jayaram [17] that such a study was undertaken and the results presented here were obtained. Still, the following question remains (as already indicated in the text):

Problem 4.8.1. (i) Is there a fuzzy implication I, other than the Weber implication $I_{\mathbf{WB}}$, which is both an (S,N)-implication and an R-implication which is obtained from a non-left continuous t-norm and cannot be obtained as the residual of any other left-continuous t-norm, i.e., is the following equality true: $\mathbb{I}_{\mathbb{S},\mathbb{N}} \cap (\mathbb{I}_{\mathbb{T}} \setminus \mathbb{I}_{\mathbb{T}^*}) = \{I_{\mathbf{WB}}\}$?
(ii) If the answer to the above question is in the affirmative, characterize the non-empty intersection $\mathbb{I}_{\mathbb{S},\mathbb{N}} \cap \mathbb{I}_{\mathbb{T}}$.

We only remark that the phrasing of Problem 4.8.1(i) excludes the residual obtained from the non-left-continuous t-norm $T_{\mathbf{nM}^*}$ from Example 2.5.6(iv), since it is the Fodor implication $I_{\mathbf{FD}}$ and can be obtained as the residual of left-continuous t-norm $T_{\mathbf{nM}}$. See also Remark 2.5.29.

In the case of intersection between the families of $\mathbb{I}_{\mathbb{S},\mathbb{N}}$ and $\mathbb{I}_{\mathbb{QL}}$ only some partial results are known. In fact, only some examples of fuzzy implications that belonged to this intersection were known, viz., the QL-implications obtained from the Łukasiewicz t-conorm $S_{\mathbf{LK}}$, the classical strong negation $N_{\mathbf{C}}$ and the three basic t-norms $T_{\mathbf{M}}, T_{\mathbf{LK}}$ and $T_{\mathbf{P}}$, and to a lesser extent the Weber implication $I_{\mathbf{WB}}$ (see, e.g., TRILLAS and VALVERDE [239, 240, 241], FODOR [103]). In TRILLAS et al. [236] it was shown that every continuous QL-implication was an (S,N)-implication. MAS et al. [174] obtained a result similar to Proposition 4.2.2 but without explicitly stating that such QL-implications were also (S,N)-implications.

Once again, the only work known to the authors on the intersection between QL- and R-implications is that of JAYARAM and BACZYŃSKI [128]. Let us recall that the family $\mathbb{I}_{\mathbb{QL}}$ contains all of the R-implications obtained from left-continuous t-norms with strong natural negations and also many subfamilies of $\mathbb{I}_{\mathbb{S},\mathbb{N}}$. However, it is not yet clear if there is a QL-implication I which can be represented as an R-implication of some non-left-continuous t-norm, but which is not an (S,N)-implication. Note that Theorem 4.3.2 assumes the left-continuity of the t-norm T^* (see Fig. 4.2). The importance of the following questions need hardly be emphasized:

Problem 4.8.2. (i) Characterize the non-empty intersection $\mathbb{I}_{\mathbb{S},\mathbb{N}} \cap \mathbb{I}_{\mathbb{QL}}$.
(ii) Is the Weber implication $I_{\mathbf{WB}}$ the only QL-implication that is also an R-implication obtained from a non-left continuous t-norm? If not, give other examples from the above intersection and hence, characterize the non-empty intersection $\mathbb{I}_{\mathbb{QL}} \cap \mathbb{I}_{\mathbb{T}}$.
(iii) Prove or disprove by giving an example: $(\mathbb{I}_{\mathbb{QL}} \cap \mathbb{I}_{\mathbb{T}}) \setminus \mathbb{I}_{\mathbb{S},\mathbb{N}} = \emptyset$.

The results on the intersection of Yager's families of f- and g-implications among themselves and with (S,N)- and R-implications are from BACZYŃSKI and JAYARAM [16]. In this context, the only question that remains to be solved is

Problem 4.8.3. (i) Is the intersection $\mathbb{I}_{\mathbb{F},\aleph} \cap \mathbb{I}_{\mathbb{QL}}$ non-empty?
(ii) If yes, then characterize the intersection $\mathbb{I}_{\mathbb{F},\aleph} \cap \mathbb{I}_{\mathbb{QL}}$.

5 Fuzzy Implications from Uninorms

One should always generalize.
– Carl Gustav Jacobi (1804-1851)

In Chap. 2, we had seen the different ways of defining fuzzy implications based on the basic fuzzy logic connectives, viz., fuzzy negations, t-norms and t-conorms. Uninorms were introduced recently by YAGER and RYBALOV [263] (see also FODOR et al. [106]) as generalizations of t-norms and t-conorms and are thus another fertile source based on which one can define fuzzy implications. In this chapter, after giving the necessary introduction to uninorms, taking a similar approach as was done in the previous chapter we define fuzzy implications from uninorms and discuss their basic properties.

5.1 Uninorms

5.1.1 Definitions and Examples

In this section we recall basic definitions and facts related to uninorms. The relevant results are presented here without proofs. For more details, see, for example, FODOR et al. [106].

Definition 5.1.1. *An associative, commutative and increasing operation* $U\colon [0,1]^2 \to [0,1]$ *is called a* uninorm *if it has a neutral element* $e \in [0,1]$, *i.e.,* $U(e,x) = x$, *for all* $x \in [0,1]$.

Obviously, if $e = 0$, then U is a t-conorm and if $e = 1$, then U is a t-norm.

Proposition 5.1.2. *Let* U *be a uninorm with neutral element* $e \in [0,1]$.

(i) If $e \in (0,1]$, *then the function* $T_U\colon [0,1]^2 \to [0,1]$ *defined by*

$$T_U(x,y) = \frac{U(e \cdot x, e \cdot y)}{e}, \qquad x,y \in [0,1],$$

is a t-norm.

(ii) If $e \in [0,1)$, *then the function* S_U *defined by*

$$S_U(x,y) = \frac{U(e + (1-e)\cdot x, e + (1-e)\cdot y) - e}{1-e}, \qquad x,y \in [0,1],$$

is a t-conorm.

M. Baczyński and B. Jayaram: Fuzzy Implications, STUDFUZZ 231, pp. 145–177, 2008.
springerlink.com

Remark 5.1.3. (i) For a uninorm U with neutral element $e \in [0,1]$, if $x \le e \le y$ or $x \ge e \ge y$, then $\min(x,y) \le U(x,y) \le \max(x,y)$.

(ii) Any uninorm U with neutral element $e \in [0,1]$ is bounded above and below by $\underline{U}_e \le U \le \overline{U}_e$, where

$$\underline{U}_e(x,y) = \begin{cases} 0, & \text{if } 0 \le x,y \le e\,, \\ \max(x,y), & \text{if } e \le x,y \le 1\,, \\ \min(x,y), & \text{otherwise}\,, \end{cases}$$

$$\overline{U}_e(x,y) = \begin{cases} \min(x,y), & \text{if } 0 \le x,y \le e\,, \\ 1, & \text{if } e \le x,y \le 1\,, \\ \max(x,y), & \text{otherwise}\,. \end{cases}$$

(iii) One can easily observe that $U(0,1) = U(1,0) \in \{0,1\}$ for any uninorm U, and hence, no uninorm is continuous on the whole of $[0,1]^2$.

(iv) A uninorm U such that $U(0,1) = 0$ is called *conjunctive* and if $U(0,1) = 1$, then it is called *disjunctive.*

5.1.2 Pseudo-continuous Uninorms

Uninorms verifying that both functions $U(x,0)$ and $U(x,1)$ are continuous except at the point e, also referred to as *pseudo-continuous* uninorms, were characterized, again by Fodor et al. [106], as follows.

Theorem 5.1.4. *Let U be a uninorm with neutral element $e \in (0,1)$, such that the functions $U(x,1)$ and $U(x,0)$ are continuous except at the point $x = e$. If $U(0,1) = 0$, then there exist a t-norm T and a t-conorm S such that*

$$U(x,y) = \begin{cases} e \cdot T\left(\dfrac{x}{e}, \dfrac{y}{e}\right), & \text{if } x,y \in [0,e]\,, \\ e + (1-e) \cdot S\left(\dfrac{x-e}{1-e}, \dfrac{y-e}{1-e}\right), & \text{if } x,y \in [e,1]\,, \\ \min(x,y), & \text{otherwise}\,. \end{cases} \tag{5.1}$$

Similarly, if $U(0,1) = 1$, then there exist a t-norm T and a t-conorm S such that

$$U(x,y) = \begin{cases} e \cdot T\left(\dfrac{x}{e}, \dfrac{y}{e}\right), & \text{if } x,y \in [0,e]\,, \\ e + (1-e) \cdot S\left(\dfrac{x-e}{1-e}, \dfrac{y-e}{1-e}\right), & \text{if } x,y \in [e,1]\,, \\ \max(x,y), & \text{otherwise}\,. \end{cases} \tag{5.2}$$

The class of uninorms of the form (5.1) will be denoted by $\mathcal{U}_{\mathbf{Min}}$, while the class of uninorms of the form (5.2) will be denoted by $\mathcal{U}_{\mathbf{Max}}$. Note that, even if a t-norm T, a t-conorm S and $e \in (0,1)$ are fixed, a pseudo-continuous uninorm is not uniquely defined - it can be conjunctive or disjunctive. If U is a conjunctive (disjunctive), then we will write $U^{\mathbf{c}}_{T,S,e}$ ($U^{\mathbf{d}}_{T,S,e}$, respectively).

5.1.3 Idempotent Uninorms

Definition 5.1.5. *A uninorm U such that $U(x,x) = x$ for all $x \in [0,1]$ is said to be an* idempotent uninorm. *The class of all idempotent uninorms will be denoted by $\mathcal{U}_{\mathbf{Idem}}$.*

In fact, the first kind of uninorms, with neutral element $e \in (0,1)$, considered by YAGER and RYBALOV [263] were the following conjunctive $U^{\mathbf{c},e}_{\mathbf{YR}}$ and disjunctive $U^{\mathbf{d},e}_{\mathbf{YR}}$ idempotent uninorms:

$$U^{\mathbf{c},e}_{\mathbf{YR}}(x,y) = \begin{cases} \max(x,y), & \text{if } x,y \in [e,1]\,, \\ \min(x,y), & \text{otherwise}\,, \end{cases} \qquad x,y \in [0,1]\,,$$

$$U^{\mathbf{d},e}_{\mathbf{YR}}(x,y) = \begin{cases} \min(x,y), & \text{if } x,y \in [0,e]\,, \\ \max(x,y), & \text{otherwise}\,, \end{cases} \qquad x,y \in [0,1]\,.$$

MARTIN et al. [167] have characterized all idempotent uninorms, which subsumes the results of DE BAETS [67], who first characterized the class of left-continuous and right-continuous idempotent uninorms.

Theorem 5.1.6. *For a function $U\colon [0,1]^2 \to [0,1]$ the following statements are equivalent:*

(i) U is an idempotent uninorm with neutral element $e \in [0,1]$.
(ii) There exists a decreasing function $g\colon [0,1] \to [0,1]$ with a fixed point e, i.e., $g(e) = e$, satisfying

$$\begin{aligned} g(x) &= 0, && \textit{if } x \in (g(0),1]\,, \\ g(x) &= 1, && \textit{if } x \in [0,g(1))\,, \end{aligned}$$

and, for all $x \in [0,1]$

$$\inf\{y \mid g(y) = g(x)\} \le g(g(x)) \le \sup\{y \mid g(y) = g(x)\}\,, \tag{5.3}$$

such that U has the following form, for any $x,y \in [0,1]$:

$$U(x,y) = \begin{cases} \min(x,y), & \textit{if } y < g(x) \textit{ or } (y = g(x) \textit{ and } x < g(g(x)))\,, \\ \max(x,y), & \textit{if } y > g(x) \textit{ or } (y = g(x) \textit{ and } x > g(g(x)))\,, \\ \max(x,y) & \\ \quad \textit{or} & \textit{if } y = g(x) \textit{ and } x = g(g(x))\,. \\ \min(x,y), & \end{cases}$$

Moreover, U is commutative on the set $\{(x,y) \mid y = g(x) \text{ and } x = g(g(x))\}$.

Using the above result some interesting corollaries can be given, which use the notions of sub-involutive and super-involutive functions (see DE BAETS [67]).

Definition 5.1.7. *A function* $g\colon [0,1] \to [0,1]$ *is called*

(i) sub-involutive if $g(g(x)) \leq x$ *for all* $x \in [0,1]$,
(ii) super-involutive if $g(g(x)) \geq x$ *for all* $x \in [0,1]$.

Corollary 5.1.8. *For a function* $U\colon [0,1]^2 \to [0,1]$ *the following statements are equivalent:*

(i) U *is a conjunctive left-continuous idempotent uninorm with neutral element* $e \in (0,1]$.
(ii) There exists a super-involutive decreasing function g *with a fixed point* $e \in (0,1]$ *and* $g(0) = 1$ *such that* U *has the following form:*

$$U(x,y) = \begin{cases} \min(x,y), & \text{if } y \leq g(x)\,, \\ \max(x,y), & \text{otherwise}\,, \end{cases} \qquad x,y \in [0,1]\,.$$

Corollary 5.1.9. *For a function* $U\colon [0,1]^2 \to [0,1]$ *the following statements are equivalent:*

(i) U *is a disjunctive right-continuous idempotent uninorm with neutral element* $e \in [0,1)$.
(ii) There exists a sub-involutive decreasing function g *with* $e \in [0,1)$ *as its fixed point and* $g(1) = 0$ *such that* U *has the following form:*

$$U(x,y) = \begin{cases} \max(x,y), & \text{if } y \geq g(x)\,, \\ \min(x,y), & \text{otherwise}\,, \end{cases} \qquad x,y \in [0,1]\,. \tag{5.4}$$

Example 5.1.10. Let $e \in [0,1]$ be fixed and let us consider the functions g_c and g_d as defined below:

$$g_c(x) = \begin{cases} 1, & \text{if } x < e\,, \\ e, & \text{if } x \geq e\,, \end{cases} \qquad g_d(x) = \begin{cases} e, & \text{if } x \leq e\,, \\ 1, & \text{if } x > e\,. \end{cases}$$

It can be easily seen from the above corollaries that the corresponding idempotent uninorms generated by them are $U_{\mathbf{YR}}^{\mathbf{c},e}$ and $U_{\mathbf{YR}}^{\mathbf{d},e}$, respectively.

Remark 5.1.11. (i) The set $\{(x,y) | y = g(x) \text{ and } x = g(g(x))\}$ consists of points (x,y) and $(y, g(y))$ on the graph that are equidistant from the main diagonal $y = x$. Commutativity of U on the above set ensures that segments on the graph of g that are equidistant from the main diagonal either both belong to maximum, or, both belong to minimum.

(ii) In Theorem 5.1.6, if U is left-continuous, then it must be equal to minimum for all points on the graph of g, in which case $g(g(x)) \geq x$ for all $x \in [0,1]$. We will refer to the function g in Theorem 5.1.6 as the *associated function* of U (see [167]).

(iii) Moreover, note that the property (5.3) implies that $g(g(x)) = x$ whenever g is strictly decreasing and, in any interval (a,b) where g is constant, we have $a \leq g(g(x)) \leq b$.

5.1.4 Representable Uninorms

Analogous to the representation theorems for continuous Archimedean t-norms and t-conorms (see Theorems 2.1.5 and 2.2.6), FODOR et al. [106] have proven the following result.

Theorem 5.1.12. *For a function $U\colon [0,1]^2 \to [0,1]$ the following statements are equivalent:*

(i) U is a strictly increasing uninorm with neutral element $e \in (0,1)$ and continuous on $(0,1)^2$ such that U is self-dual with respect to a strong negation N with the fixed point e.
(ii) U has a continuous additive generator, i.e., there exists a continuous and strictly increasing function $h\colon [0,1] \to [-\infty,\infty]$, such that $h(0) = -\infty$, $h(e) = 0$ for an $e \in (0,1)$ and $h(1) = \infty$, which is uniquely determined up to a positive multiplicative constant, such that

$$U(x,y) = \begin{cases} 0, & \text{if } (x,y) \in \{(0,1),(1,0)\}\,, \\ h^{-1}(h(x)+h(y)), & \text{if } (x,y) \in [0,1]^2 \setminus \{(0,1),(1,0)\}\,, \end{cases}$$

or

$$U(x,y) = \begin{cases} 1, & \text{if } (x,y) \in \{(0,1),(1,0)\}\,, \\ h^{-1}(h(x)+h(y)), & \text{if } (x,y) \in [0,1]^2 \setminus \{(0,1),(1,0)\}\,. \end{cases}$$

Remark 5.1.13. (i) Uninorms that can be represented as in Theorem 5.1.12 are called *representable uninorms* and this class will be denoted by $\mathcal{U}_{\mathbf{Rep}}$.
(ii) Note that once the additive generator h is fixed, by its strictness e is also unique and hence h generates a unique (up to the constant value on the set $\{(0,1),(1,0)\}$) representable uninorm. A conjunctive representable uninorm generated by h will be denoted by $U_h^{\mathbf{c}}$. Similarly, a disjunctive representable uninorm generated by h will be denoted by $U_h^{\mathbf{d}}$.
(iii) It is interesting to note that every representable uninorm U_h (conjunctive or disjunctive) gives rise to a natural negation obtained as

$$N_{U_h}(x) = h^{-1}(-h(x))\,, \qquad x \in [0,1]\,, \tag{5.5}$$

which is a strong negation. Moreover, U_h is self-dual with respect to its natural negation.
(iv) For a representable uninorm U, we have $U(x,1) = 1$ for all $x > 0$ and $U(x,0) = 0$ for all $x < 1$. Hence, if U is conjunctive, then the function $U(\,\cdot\,,1)$ is not continuous at $x = 0$, while if U is disjunctive, then the function $U(\,\cdot\,,0)$ is not continuous at $x = 1$.

Example 5.1.14. The following are some well-known examples of representable uninorms (see [145] and [106]):

(i) For the additive generator $h_1(x) = \ln\left(\frac{x}{1-x}\right)$, we get the following conjunctive representable uninorm:

$$U^{\mathbf{c}}_{h_1}(x,y) = \begin{cases} 0, & \text{if } (x,y) \in \{(0,1),(1,0)\}\,, \\ \dfrac{xy}{(1-x)(1-y)+xy}, & \text{otherwise}\,. \end{cases}$$

In this case $e = \frac{1}{2}$.

(ii) Let $\beta > 0$. For the additive generator

$$h_\beta(x) = \ln\left(-\frac{1}{\beta}\cdot\ln(1-x)\right)\,, \qquad x \in (0,1)\,,$$

we get the following disjunctive representable uninorm:

$$U^{\mathbf{c}}_{h_\beta}(x,y) = \begin{cases} 1, & \text{if } (x,y) \in \{(0,1),(1,0)\}\,, \\ 1-\exp\left(-\dfrac{1}{\beta}\cdot\ln(1-x)\cdot\ln(1-y)\right), & \text{otherwise}\,. \end{cases}$$

In this case $e = 1-\exp(-\beta)$.

Remark 5.1.15. (i) From the above discussion it should be noted that no representable uninorm belongs to either $\mathcal{U}_{\mathbf{Min}}$ or $\mathcal{U}_{\mathbf{Max}}$.

(ii) From Example 5.1.10 and Theorem 5.1.4, it is obvious that an equivalent condition for an idempotent uninorm to belong to $\mathcal{U}_{\mathbf{Min}}$ or $\mathcal{U}_{\mathbf{Max}}$ is that its associated function g should have either of the following representations:

$$g_*(x) = \begin{cases} 1, & \text{if } x \in [0,e)\,, \\ e, & \text{if } x \in [e,1]\,, \end{cases} \qquad g^*(x) = \begin{cases} e, & \text{if } x \in [0,e]\,, \\ 0, & \text{if } x \in (e,1]\,. \end{cases}$$

Let us denote these sub-classes of idempotent uninorms by

$$\begin{aligned} \mathcal{U}_{I,G_*} &= \left\{U_{\langle e,g\rangle} \in \mathcal{U}_{\mathbf{Idem}} \mid g = g_*\right\}\,, \\ \mathcal{U}_{I,G^*} &= \left\{U_{\langle e,g\rangle} \in \mathcal{U}_{\mathbf{Idem}} \mid g = g^*\right\}\,. \end{aligned}$$

(iii) Clearly, the following relationships exist among the above families of uninorms:

$$\begin{aligned} &\mathcal{U}_{\mathbf{Min}} \cap \mathcal{U}_{\mathbf{Idem}} = \mathcal{U}_{I,G_*}\,, \\ &\mathcal{U}_{\mathbf{Max}} \cap \mathcal{U}_{\mathbf{Idem}} = \mathcal{U}_{I,G^*}\,, \\ &\mathcal{U}_{\mathbf{Min}} \cap \mathcal{U}_{\mathbf{Rep}} = \mathcal{U}_{\mathbf{Max}} \cap \mathcal{U}_{\mathbf{Rep}} = \mathcal{U}_{\mathbf{Idem}} \cap \mathcal{U}_{\mathbf{Rep}} = \emptyset\,. \end{aligned}$$

5.2 Natural Negations of Fuzzy Implications - Revisited

Definition 5.2.1. *Let $I \in \mathcal{FI}$. If $I(1,\alpha) = 0$ for some $\alpha \in [0,1)$, then the function $N_I^\alpha\colon [0,1] \to [0,1]$ given by*

$$N_I^\alpha(x) = I(x,\alpha)\,, \qquad x \in [0,1]\,,$$

is called the natural negation of I with respect to α.

It should be noted that for any $I \in \mathcal{FI}$ we have (I5), so for $\alpha = 0$ we have the natural negation N_I of I. Moreover, α should be less than 1, since $I(1,1) = 1$.

Lemma 5.2.2. *Let $I\colon [0,1]^2 \to [0,1]$ be any function and N_I^α be a fuzzy negation for an arbitrary but fixed $\alpha \in [0,1)$.*

(i) If I satisfies (I2)*, then I satisfies* (I5)*.*
(ii) If I satisfies (I2) *and* (EP)*, then I satisfies* (I3) *if and only if I satisfies* (I4)*.*
(iii) If I satisfies (EP)*, then I satisfies R-CP(N_I^α).*

Proof. (i) Since N_I^α is a fuzzy negation and I satisfies (I2) we get

$$I(1,0) \le I(1,\alpha) = N_I^\alpha(1) = 0\,.$$

(ii) Let I satisfy (I2) and (EP). If I satisfies (I4), then

$$\begin{aligned}1 &= I(1,1) = I(1, N_I^\alpha(0)) = I(1, I(0,\alpha)) = I(0, I(1,\alpha))\\ &= I(0, N_I^\alpha(1)) = I(0,0)\,,\end{aligned}$$

i.e., I satisfies (I3). The reverse implication can be shown similarly.

(iii) Since I satisfies (EP), we have

$$I(x, N_I^\alpha(y)) = I(x, I(y,\alpha)) = I(y, I(x,\alpha)) = I(y, N_I^\alpha(x))\,,$$

i.e., I satisfies R-CP(N_I^α). □

Lemma 5.2.3. *Let $I \in \mathcal{FI}$ and N_I^α be a fuzzy negation for an arbitrary but fixed $\alpha \in [0,1)$. If N is a fuzzy negation such that $N_I^\alpha \circ N = \mathrm{id}_{[0,1]}$ and I satisfies* (EP)*, then I satisfies L-CP(N).*

Proof. Firstly, observe that by Lemma 1.4.9 we see that N_I^α is continuous and N is strictly decreasing. Now, by our assumptions, we get

$$\begin{aligned}I(N(x),y) &= I(N(x), N_I^\alpha \circ N(y)) = I(N(x), I(N(y),\alpha))\\ &= I(N(y), I(N(x),\alpha)) = I(N(y), N_I^\alpha \circ N(x))\\ &= I(N(y),x)\,,\end{aligned}$$

for any $x, y \in [0,1]$. □

Remark 5.2.4. Under the assumptions of Lemma 5.2.3, we have:

(i) If N_I^α is a strict negation, then I satisfies L-CP($(N_I^\alpha)^{-1}$).
(ii) If N_I^α is a strong negation, then I satisfies L-CP (N_I^α), and hence, by Proposition 1.5.3, I also satisfies CP(N_I^α).

5.3 (U,N)-Implications

A natural generalization of (S,N)-implications in the uninorm framework is to consider a uninorm in the place of a t-conorm in Definition 2.4.1.

5.3.1 Definition and Basic Properties

Definition 5.3.1. *A function* $I\colon [0,1]^2 \to [0,1]$ *is called a* (U,N)-operation, *if there exist a uninorm* U *and a fuzzy negation* N *such that*

$$I(x,y) = U(N(x),y)\,, \qquad x,y \in [0,1]\,. \tag{5.6}$$

If I *is a (U,N)-operation generated from a uninorm* U *and a negation* N, *then we will often denote it by* $I_{U,N}$.

Proposition 5.3.2. *If* $I_{U,N}$ *is a (U,N)-operation, then*

(i) $I_{U,N}$ *satisfies* (I1), (I2), (I5), (NC) *and* (EP),
(ii) $N^e_{I_{U,N}} = N$ *and* $I_{U,N}$ *satisfies R-CP(N),*
(iii) if N *is strict, then* $I_{U,N}$ *satisfies L-CP*(N^{-1}),
(iv) if N *is strong, then* $I_{U,N}$ *satisfies CP(N).*

Proof. Let U be a uninorm with neutral element $e \in [0,1)$ and N any fuzzy negation.

(i) By the monotonicity of U and N we get that $I_{U,N}$ satisfies (I1) and (I2). Moreover, it can be easily verified that $I_{U,N}$ satisfies (I5) and (NC). Finally, from the associativity and the commutativity of U we have also (EP).
(ii) For any $x \in [0,1]$, we have

$$N^e_{I_{U,N}}(x) = I_{U,N}(x,e) = U(N(x),e) = N(x)\,.$$

Since $I_{U,N}$ satisfies (EP), from Lemma 5.2.2(iii) with $\alpha = e$, we have that $I_{U,N}$ satisfies R-CP(N).
(iii) If N is a strict negation, by Remark 5.2.4(i), we have that $I_{U,N}$ satisfies L-CP(N^{-1}).
(iv) If N is a strong negation, by Remark 5.2.4(ii), we have that $I_{U,N}$ satisfies CP(N). □

If $e = 0$, then U is a t-conorm and $I_{U,N}$, as an (S,N)-implication, is always a fuzzy implication. If $e = 1$, then U is a t-norm and $I_{U,N}$ is not a fuzzy implication, since (I3) is violated. If $e \in (0,1)$, then not for every uninorm the (U,N)-operation is a fuzzy implication. Next results characterize these (U,N)-operations, which satisfy (I3) and (I4).

Theorem 5.3.3. *Let* U *be a uninorm with neutral element* $e \in (0,1)$ *and* N *any fuzzy negation. Then the following statements are equivalent:*

(i) The (U,N)-operation $I_{U,N}$ *is a fuzzy implication.*
(ii) U *is a disjunctive uninorm.*

Proof. $(i) \implies (ii)$ If $I_{U,N}$ is a fuzzy implication, then from (I3) we obtain $U(0,1) = U(1,0) = I_{U,N}(0,0) = 1$.

$(ii) \implies (i)$ Assume that $U(0,1) = 1$. From Proposition 5.3.2, it is enough to show that $I_{U,N}$ satisfies (I3) and (I4):

$$I_{U,N}(0,0) = U(N(0),0) = U(1,0) = U(0,1) = 1\,,$$
$$I_{U,N}(1,1) = U(N(1),1) = U(0,1) = 1\,.$$
□

Lemma 5.3.4. *Let U be a uninorm. The function $I_{U,N}$ as defined in (5.6) is a fuzzy implication in the following cases:*

(i) $U \in \mathcal{U}_{\mathbf{Max}}$,

(ii) U *is a disjunctive representable uninorm,*

(iii) U *is a left-continuous idempotent uninorm with unary operator g such that $g(0) < 1$,*

(iv) U *is a right-continuous idempotent uninorm with unary operator g such that $g(1) = 0$.*

Proof. Points (i) and (ii) are obvious. Points (iii) and (iv) follow from Corollaries 5.1.8 and 5.1.9, since under these conditions U is disjunctive. □

Remark 5.3.5. As done before, only if the (U,N)-operation is a fuzzy implication we use the term *(U,N)-implication.*

Example 5.3.6. In the following, we give examples of (U,N)-implications obtained using the classical strong negation $N_{\mathbf{C}}$ and for different uninorms.

(i) Let us consider a disjunctive uninorm $U_{\mathbf{LK}}$ from the class $\mathcal{U}_{\mathbf{Max}}$ generated by the triple $(T_{\mathbf{LK}}, S_{\mathbf{LK}}, 0.5)$. Then

$$I_{U_{\mathbf{LK}},N_{\mathbf{C}}}(x,y) = \begin{cases} \max(y-x+0.5,0), & \text{if } \max(1-x,y) \leq 0.5 \,, \\ \min(y-x+0.5,1), & \text{if } \max(1-x,y) > 0.5 \,, \\ I_{\mathbf{KD}}(x,y), & \text{otherwise} \,. \end{cases}$$

Fig. 5.1(a) gives the plot of $I_{U_{\mathbf{LK}},N_{\mathbf{C}}}$.

(ii) Let us consider a disjunctive uninorm $U_{\mathbf{P}}$ from the class $\mathcal{U}_{\mathbf{Max}}$ generated by the triple $(T_{\mathbf{P}}, S_{\mathbf{P}}, 0.5)$. Then

$$I_{U_{\mathbf{P}},N_{\mathbf{C}}}(x,y) = \begin{cases} 2y-2xy, & \text{if } \max(1-x,y) \leq 0.5 \,, \\ 1-2x+2xy, & \text{if } \max(1-x,y) > 0.5 \,, \\ I_{\mathbf{KD}}(x,y), & \text{otherwise} \,. \end{cases}$$

Fig. 5.1(b) gives the plot of $I_{U_{\mathbf{P}},N_{\mathbf{C}}}$.

(iii) Let us consider a disjunctive uninorm $U_{\mathbf{M}}$ from the class $\mathcal{U}_{\mathbf{Max}}$ generated from the triple $(T_{\mathbf{M}}, S_{\mathbf{M}}, 0.5)$, which is also an idempotent uninorm. Then

$$I_{U_{\mathbf{M}},N_{\mathbf{C}}}(x,y) = \begin{cases} \min(1-x,y), & \text{if } \max(1-x,y) \leq 0.5 \,, \\ I_{\mathbf{KD}}(x,y), & \text{otherwise} \,. \end{cases}$$

In fact, $I_{U_{\mathbf{M}},N_{\mathbf{C}}}$ is the fuzzy implication $I_{\mathbf{MN}}$ given in Remark 1.3.5(iii), whose plot is given in Fig. 5.1(c).

(iv) Let us consider the disjunctive representable uninorm $U_{h_1}^{\mathbf{d}}$ from Example 5.1.14(i). Then

$$I_{U_{h_1}^{\mathbf{d}},N_{\mathbf{C}}}(x,y) = \begin{cases} 1, & \text{if } (x,y) \in \{(0,0),(1,1)\} \,, \\ \dfrac{(1-x)y}{x+y-2xy}, & \text{otherwise} \,. \end{cases}$$

Fig. 5.1(d) gives the plot of $I_{U_{h_1}^{\mathbf{d}},N_{\mathbf{C}}}$.

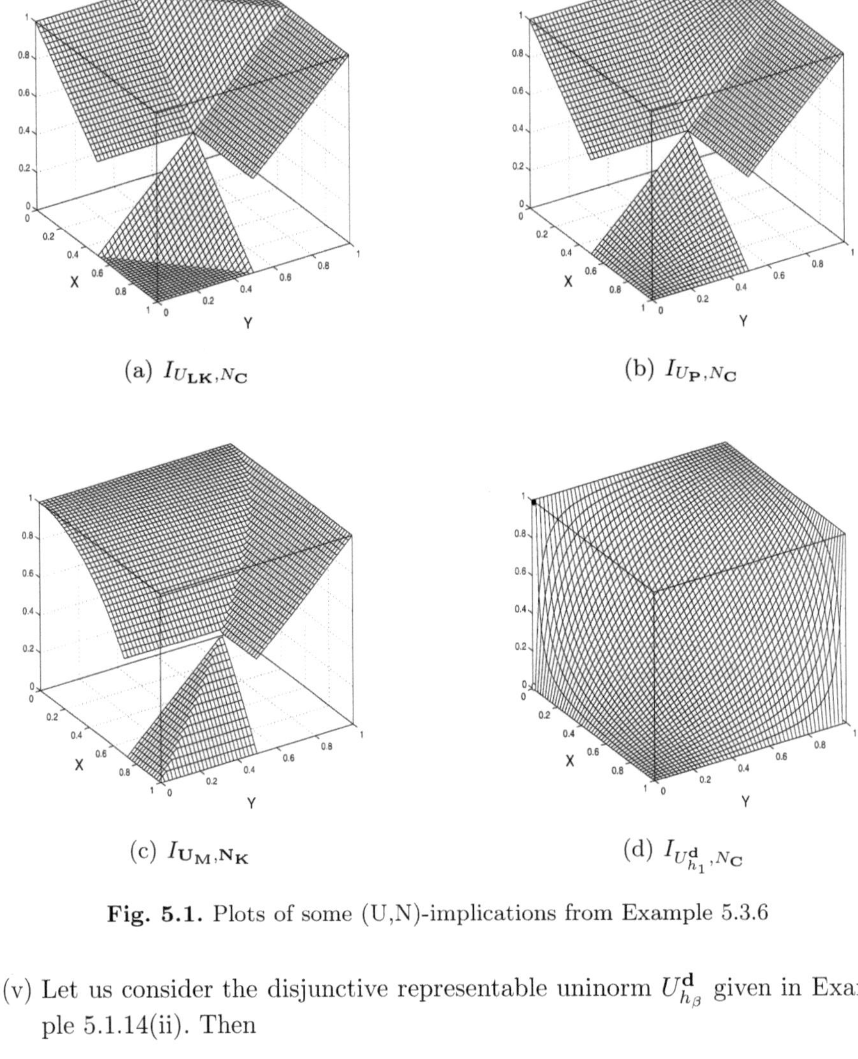

(a) $I_{U_{\mathbf{LK}},N_{\mathbf{C}}}$ (b) $I_{U_{\mathbf{P}},N_{\mathbf{C}}}$

(c) $I_{\mathbf{U_M},\mathbf{N_K}}$ (d) $I_{U^{\mathbf{d}}_{h_1},N_{\mathbf{C}}}$

Fig. 5.1. Plots of some (U,N)-implications from Example 5.3.6

(v) Let us consider the disjunctive representable uninorm $U^{\mathbf{d}}_{h_\beta}$ given in Example 5.1.14(ii). Then

$$I_{U^{\mathbf{d}}_{h_\beta},N_{\mathbf{C}}}(x,y) = \begin{cases} 1, & \text{if } (x,y) \in \{(0,0),(1,1)\}\,, \\ 1 - \exp\left(-\dfrac{1}{\beta} \cdot \ln(1+x-y)\right), & \text{otherwise}\,. \end{cases}$$

Lemma 5.3.7. *Let $I_{U,N}$ be a (U,N)-implication obtained from a uninorm U with $e \in (0,1)$ as its neutral element and a continuous negation N. Let $\alpha \in (0,1)$ be an arbitrary but fixed number. Then the following statements are equivalent:*

(i) $N^{\alpha}_{I_{U,N}} = N$.
(ii) $\alpha = e$.

Proof. Let $e \in (0,1)$ be the neutral element of U and $\alpha \in (0,1)$ be an arbitrary but fixed number.

$(i) \Longrightarrow (ii)$ If $N^{\alpha}_{I_{U,N}} = N$, then since N is continuous there exists an e' such that $e = N(e')$ and consequently we get

$$e = N(e') = N^{\alpha}_{I_{U,N}}(e') = I_{U,N}(e',\alpha) = U(N(e'),\alpha) = U(e,\alpha) = \alpha\,.$$

$(ii) \Longrightarrow (i)$ Conversely, if $\alpha = e$, then for all $x \in [0,1]$

$$N^{\alpha}_{I_{U,N}}(x) = I_{U,N}(x,\alpha) = I_{U,N}(x,e) = U(N(x),e) = N(x)\,. \qquad \square$$

Remark 5.3.8. It is interesting to note that, unlike (S,N)-implications, (U,N)-implications do not satisfy the neutrality property (NP). To see this, let U be a uninorm with neutral element $e \in (0,1)$ and N be a fuzzy negation. Then, $I_{U,N}(1,e) = U(N(1),e) = U(0,e) = 0 \neq e$ (see also Remark 5.4.6).

5.3.2 Characterizations of (U,N)-Implications

A first characterization of (U,N)-implications from continuous negations was given by JAYARAM and BACZYŃSKI [127].

Proposition 5.3.9. *Let $I \in \mathcal{FI}$ and N be a fuzzy negation. Let us define a binary operation U_I on $[0,1]$ as follows:*

$$U_{I,N}(x,y) = I(N(x),y)\,, \qquad x,y \in [0,1]\,. \tag{5.7}$$

Then, for all $x \in [0,1]$

(i) $U_{I,N}(x,1) = U_I(1,x) = 1$. In particular, $U_{I,N}(0,1) = 1$,
(ii) $U_{I,N}$ is increasing in both the variables,
(iii) $U_{I,N}$ is commutative if and only if I satisfies L-CP(N).

In addition, if I satisfies L-CP(N), then

(iv) $U_{I,N}$ is associative if and only if I satisfies the exchange property (EP),
(v) An $\alpha \in (0,1)$ is the neutral element of $U_{I,N}$ if and only if $N^{\alpha}_{I} \circ N = \mathrm{id}_{[0,1]}$.

Proof. Properties (i) to (iv) have been shown in Proposition 2.4.6. Therefore, we need to prove only point (v). Let us assume that some $\alpha \in (0,1)$ is the neutral element of $U_{I,N}$. Then, for any $x \in [0,1]$, we have

$$x = U_{I,N}(x,\alpha) = I(N(x),\alpha) = N^{\alpha}_{I}(N(x))\,.$$

Conversely, if $N^{\alpha}_{I} \circ N = \mathrm{id}_{[0,1]}$, then, for any $x \in [0,1]$ we get

$$U_{I,N}(\alpha,x) = U_{I,N}(x,\alpha) = I(N(x),\alpha) = N^{\alpha}_{I}(N(x)) = x\,,$$

and α is the neutral element of $U_{I,N}$. $\square$

If N_I^α is a continuous fuzzy negation for some $\alpha \in (0,1)$, then by Lemma 1.4.10 and previous results, we can consider the modified pseudo-inverse $\mathfrak{N}_I^\alpha$ given by

$$\mathfrak{N}_I^\alpha(x) = \begin{cases} \left(N_I^\alpha\right)^{(-1)}(x), & \text{if } x \in (0,1]\,, \\ 1, & \text{if } x = 0\,, \end{cases} \tag{5.8}$$

as the potential candidate for the fuzzy negation N in (5.7). Hence from Lemma 5.2.3 with $N = \mathfrak{N}_I^\alpha$ we obtain the following result.

Corollary 5.3.10. *If $I \in \mathcal{FI}$ satisfies* (EP) *and N_I^α, the natural negation of I with respect to an arbitrary but fixed $\alpha \in (0,1)$, is a continuous fuzzy negation, then I satisfies* (L-CP) *with $\mathfrak{N}_I^\alpha$ from* (5.8).

Hence, we obtain the following result:

Corollary 5.3.11. *If $I \in \mathcal{FI}$ satisfies* (EP) *and N_I^α is a continuous fuzzy negation with respect to an arbitrary but fixed $\alpha \in (0,1)$, then the function U_I defined by*

$$U_I(x,y) = I(\mathfrak{N}_I^\alpha(x), y)\,, \qquad x,y \in [0,1]\,, \tag{5.9}$$

is a disjunctive uninorm with neutral element α, where $\mathfrak{N}_I^\alpha$ is as defined in (5.8).

Theorem 5.3.12. *For a function $I\colon [0,1]^2 \to [0,1]$ the following statements are equivalent:*

(i) I is an (U,N)-operation generated from a disjunctive uninorm U with neutral element $e \in (0,1)$ and a continuous fuzzy negation N.
(ii) I is an (U,N)-implication generated from a uninorm U with neutral element $e \in (0,1)$ and a continuous fuzzy negation N.
(iii) I satisfies (I1), (I3), (EP) *and the function N_I^e is a continuous negation for some $e \in (0,1)$.*

Moreover, the representation (5.6) *of (U,N)-implication is unique in this case.*

Proof. *(i)* $\Longleftrightarrow$ *(ii)* This equivalence follows immediately from Theorem 5.3.3.

(ii) $\Longrightarrow$ *(iii)* Assume that I is an (U,N)-implication based on a uninorm U with neutral element $e \in (0,1)$ and a continuous negation N. Since every (U,N)-implication is a fuzzy implication, I satisfies (I1) and (I3). Moreover, by Proposition 5.3.2 it satisfies (EP) and $N_I^e = N$. In particular N_I^e is continuous.

(iii) $\Longrightarrow$ *(ii)* Firstly, from Lemma 5.2.2(iii), it follows that I satisfies (R-CP) with respect to the continuous negation N_I^e. Next, Lemma 1.5.20(i) implies that I satisfies (I2). Once again, from Lemma 5.2.2(i) and (ii), we have that I satisfies (I4) and (I5), and hence, $I \in \mathcal{FI}$. Further, by virtue of Corollary 5.3.10, the implication I satisfies L-CP($\mathfrak{N}_I^e$). By Corollary 5.3.11, the function U_I defined by (5.9) is a disjunctive uninorm with neutral element e. We will show that $I_{U_I,N_I^e} = I$. Fix arbitrarily $x,y \in [0,1]$. If $x \in \mathrm{Ran}(\mathfrak{N}_I^e)$, then by (1.11), we have

$$I_{U_I,N_I^e}(x,y) = U_I(N_I^e(x),y) = I(\mathfrak{N}_I^e \circ N_I^e(x),y) = I(x,y)\,.$$

If $x \notin \mathrm{Ran}(\mathfrak{N}_I^e)$, then, from the continuity of N_I^e, there exists $x_0 \in \mathrm{Ran}(\mathfrak{N}_I^e)$ such that $N_I^e(x) = N_I^e(x_0)$. Firstly, see that $I(x,y) = I(x_0,y)$ for all $y \in [0,1]$. Indeed, let us fix arbitrarily $y \in [0,1]$. From the continuity of N_I^e, there exists $y' \in [0,1]$ such that $N_I^e(y') = y$, and since I satisfies R-CP(N_I^e), we have

$$\begin{aligned} I(x,y) &= I(x, N_I^e(y')) = I(y', N_I^e(x)) = I(y', N_I^e(x_0)) \\ &= I(x_0, N_I^e(y')) = I(x_0, y) . \end{aligned}$$

From the above fact we get

$$I_{U_I,N_I^e}(x,y) = U_I(N_I^e(x), y) = U_I(N_I^e(x_0), y) = I(x_0, y) = I(x,y) .$$

Thus, I is an (U,N)-implication.

Finally, assume that there exist two continuous fuzzy negations N_1, N_2 and two uninorms U_1, U_2 with neutral elements $e, e' \in (0,1)$, respectively, such that $I(x,y) = U_1(N_1(x), y) = U_2(N_2(x), y)$ for all $x, y \in [0,1]$. Let $x_0, y_0 \in [0,1]$ be arbitrarily fixed. From Proposition 5.3.2, we get $N_1 = N_2 = N_I^e = N_I^{e'}$. By virtue of Lemma 5.3.7, we get that $e' = e$. Now, since N_I^e is a continuous negation, there exists $x_1 \in [0,1]$ such that $N_I^e(x_1) = x_0$. Thus, we have

$$U_1(x_0, y_0) = U_1(N_I^e(x_1), y_0) = U_2(N_I^e(x_1), y_0) = U_2(x_0, y_0) ,$$

i.e., $U_1 = U_2$. Hence N and U are uniquely determined. In fact, $U = U_I$ defined by (5.9). □

Now, the following result easily follows:

Theorem 5.3.13. *For a function $I\colon [0,1]^2 \to [0,1]$ the following statements are equivalent:*

(i) I is an (U,N)-implication generated from a disjunctive uninorm U with neutral element $e \in (0,1)$ and a strict (strong) fuzzy negation N.
(ii) I satisfies (I1), (I3), (EP) *and the function N_I^e is a strict (strong) negation for some $e \in (0,1)$.*

Moreover, the representation (5.6) *of an (U,N)-implication described above is unique.*

Remark 5.3.14. (i) In the above theorems the property (I1) can be substituted by (I2) and the property (I3) can be substituted by (I4).
(ii) In Table 5.1, we show the mutual independence of the properties from Theorem 5.3.12. The same examples can be considered for the mutual independence of axioms in Theorem 5.3.13. We recognize that the verification of the Table 5.1, vis-á-vis, the functions and the corresponding properties indicated, may not be obvious. Hence, in the following, we show that the presented examples in Table 5.1 are correct.

Firstly, observe that if a function F satisfies (NP), then $F(1,\alpha) = \alpha$ for all $\alpha \in (0,1)$, i.e., $N_F^\alpha(1) \neq 0$ and hence N_F^α is not a fuzzy negation for any $\alpha \in (0,1)$.

Table 5.1. Mutual independence of the properties in Theorem 5.3.12

Function F	(I1)	(I3)	(EP)	N_F^α is continuous for some $\alpha \in (0,1)$
$F_1(x,y) = \begin{cases} 1, & \text{if } x < y \\ 0, & \text{otherwise} \end{cases}$	✓	×	×	×
$F_2(x,y) = \begin{cases} 1, & \text{if } x = 0 \text{ and } y = 0 \\ y, & \text{if } x = 1 \\ 0, & \text{otherwise} \end{cases}$	×	✓	×	×
$F_3 = T_{\mathbf{M}}$	×	×	✓	×
$F_4(x,y) = \begin{cases} 0, & \text{if } (x,y) \in [0,0.5)^2 \cup (0.5,1]^2 \\ 1-x, & \text{if } y = 0.5 \\ 1, & \text{otherwise} \end{cases}$	×	×	×	✓
$F_5(x,y) = \begin{cases} 1, & \text{if } x, y \in [0,0.5] \\ 0, & \text{otherwise} \end{cases}$	✓	✓	×	×
$F_6 = 0$	✓	×	✓	×
$F_7(x,y) = \begin{cases} 1-x, & \text{if } y = 0.5 \\ \min(1-x,y), & \text{otherwise} \end{cases}$	✓	×	×	✓
$F_8(x,y) = \begin{cases} 1, & \text{if } (x,y) \in [0,0.5)^2 \cup (0.5,1]^2 \\ 0, & \text{otherwise} \end{cases}$	×	✓	✓	×
$F_9(x,y) = \begin{cases} 1, & \text{if } (x,y) \in [0,0.5)^2 \cup (0.5,1]^2 \\ 1-x, & \text{if } y = 0.5 \\ 0, & \text{otherwise} \end{cases}$	×	✓	×	✓
$F_{10} = \begin{cases} 0, & \text{if } (x,y) \in [0,0.5)^2 \cup (0.5,1]^2 \\ 1-x, & \text{if } y = 0.5 \\ y, & \text{if } x = 0.5 \\ 1, & \text{otherwise} \end{cases}$	×	×	✓	✓
$F_{11}(x,y) = \begin{cases} 1, & \text{if } (x,y) \in [0,0.5)^2 \cup (0.5,1]^2 \\ 1-x, & \text{if } y = 0.5 \\ y, & \text{if } x = 0.5 \\ 0, & \text{otherwise} \end{cases}$	×	✓	✓	✓
$F_{12}(x,y) = \begin{cases} \min(1-x,y), & \text{if } x \leq y \\ \max(1-x,y), & \text{if } x > y \end{cases}$	✓	×	✓	✓
$F_{13}(x,y) = 1-x$	✓	✓	×	✓
$F_{14} = I_{\mathbf{YG}}$	✓	✓	✓	×

(a) It is clear that F_1 satisfies (I1), but neither does it satisfy (I3) nor is $N_{F_1}^\alpha$ a continuous fuzzy negation for any $\alpha \in (0,1)$. Moreover, F_1 does not satisfy (EP), since $F_1(0.4, F_1(0.6, 0.5)) = 0 \neq 1 = F_1(0.6, F_1(0.4, 0.5))$.
(b) F_2 does not satisfy (I1), since $F_2(0,y) = 0 < y = F_2(1,y)$ for any $y \in (0,1)$. From the same equality we see that $N_{F_2}^\alpha$ is not a fuzzy negation for any

$\alpha \in (0,1)$. Since $F_2(0, F_2(0.5,1)) = 1 \neq 0 = F_2(0.5, F_2(0,1))$, F_2 does not satisfy (EP).

(c) Function F_3 is self-explanatory.

(d) F_4 does not satisfy (I1), since $F_4(0,0.3) = 0 < 1 = F_4(1,0.3)$. It is easy to see that $N_{F_4}^{0.5} = N_{\mathbf{C}}$ and also that F_4 does not satisfy (I3). F_4 does not satisfy (EP), since $F_4(0.5, F_4(0.6,0.5)) = 1 \neq 0.4 = F_4(0.6, F_4(0.5,0.5))$.

(e) It is clear that F_5 satisfies both (I1) and (I3). F_5 does not satisfy (EP), since $F_5(0.5, F_5(0.8,0.5)) = 1 \neq 0 = F_5(0.8, F_5(0.5,0.5))$. Moreover, $N_{F_5}^{\alpha}$ is not a continuous fuzzy negation for any $\alpha \in (0,1)$.

(f) Function F_6 is self-explanatory.

(g) Clearly, the function F_7 satisfies (I1), $N_{F_7}^{0.5} = N_{\mathbf{C}}$, but it does not satisfy (I3). Moreover, since $F_7(0.5, F_7(0.4,0.6)) = 0.5 \neq 0.6 = F_7(0.4, F_7(0.5,0.6))$, F_7 does not satisfy (EP).

(h) F_8 does not satisfy (I1), since $F_8(0,0.6) = 0 < 1 = F_8(1,0.6)$. $N_{F_8}^{\alpha}$ is not a continuous negation for any $\alpha \in (0,1)$. However, it can be easily verified that F_8 does satisfy both (I3) and (EP).

(i) F_9 does not satisfy (I1), since $F_9(0,0.6) = 0 < 1 = F_9(1,0.6)$. It is easy to see that $N_{F_9}^{0.5} = N_{\mathbf{C}}$ and also that F_9 does satisfy (I3). F_9 does not satisfy (EP), since $F_9(0.5, F_9(0.6,0.5)) = 0 \neq 0.4 = F_9(0.6, F_9(0.5,0.5))$.

(j) Although $N_{F_{10}}^{0.5} = N_{\mathbf{C}}$ and F_{10} satisfies (EP), it does not satisfy either (I1), since $F_{10}(0,0.3) = 0 < 1 = F_{10}(1,0.3)$, or (I3).

(k) F_{11} does not satisfy (I1), since $F_{11}(0,0.6) = 0 < 1 = F_{11}(1,0.6)$. Interestingly, $N_{F_{11}}^{0.5} = N_{\mathbf{C}}$ and F_{11} satisfies both (EP) and (I3).

(l) Clearly, the function F_{12} does not satisfy (I3), but it satisfies both (EP) and (I1). Moreover, $N_{F_{12}}^{0.5} = N_{\mathbf{C}}$.

(m) F_{13} satisfies both (I1) and (I3), and once again, $N_{F_{13}}^{0.5} = N_{\mathbf{C}}$. It does not satisfy (EP), since $F_{13}(0, F_{13}(0.5,0)) = 1 \neq 0 = F_{13}(0.5, F_{13}(0,0))$.

(n) Finally, F_{14} is the Yager implication which is a fuzzy implication that satisfies both (EP) and (NP).

The following characterization of (U,N)-operations can now be obtained along similar lines as above.

Theorem 5.3.15. *For a function $I\colon [0,1]^2 \to [0,1]$ the following statements are equivalent:*

(i) I is an (U,N)-operation generated from some uninorm U with neutral element $e \in (0,1)$ and some continuous fuzzy negation N.

(ii) I satisfies (I1), (EP) *and the function N_I^e is a continuous negation for some $e \in (0,1)$.*

Remark 5.3.16. (i) Let $U \in \mathcal{U}_{\mathbf{Max}}$ with the neutral element $e \in (0,1)$. If $\alpha > e$, then the natural negation of $I_{U,N}$ with respect to α is not a negation. To see this, observe that $N_{I_{U,N}}^{\alpha}(1) = I_{U,N}(1,\alpha) = U(0,\alpha) = \max(0,\alpha) = \alpha \neq 0$.

(ii) If $U = U_h$ is a disjunctive representable uninorm and N_U is its natural negation, then

$$I_{U,N_U}(x,y) = h^{-1}(h(y) - h(x)) .$$

In fact, it can be easily shown that $N = N_U$ if and only if $I_{U,N}(x,x) = e$ for any $x \in (0,1)$.

(iii) If N is any continuous fuzzy negation, then consider the disjunctive idempotent uninorm U generated from N. Moreover, let $N^* = \mathfrak{N}$ as obtained using (1.8). Then N^* is a strictly decreasing, possibly discontinuous, fuzzy negation. Now, the (U,N)-implication obtained from U and N^* is

$$I_{U,N^*}(x,y) = \begin{cases} \max(N^*(x), y), & \text{if } x \leq y \,, \\ \min(N^*(x), y), & \text{otherwise} \,. \end{cases}$$

It is interesting to note that, if N is any fuzzy negation with fixed point e, then the role of N in U is to demarcate the regions on the unit square $[0,1]^2$ on which U becomes min or max. However, in the (U,N)-implication obtained from the disjunctive idempotent uninorm U, it is the main diagonal $y = x$ that decides this demarcation, while N is relegated to acting on the individual values.

(iv) While in the idempotent uninorm U generated from N, different parts of the graph of N can belong to either min or max (See Remark 5.1.11), the I_{U,N^*} obtained from them is still the same for the given N (see also Remark 5.4.15).

5.4 RU-Implications

Analogous to the definition of R-implications from t-norms, one can also define residual implications from uninorms, the study of which forms the main focus of this section. We call the family of residual implications from uninorms as RU-implications. The discussion and analysis in this section will be parallel to that done in the section on the class of residual implications from t-norms, viz., Sect. 2.5.

5.4.1 Definition and Basic Properties

Definition 5.4.1. *A function $I\colon [0,1]^2 \to [0,1]$ is called an* RU-operation *if there exists a uninorm U such that*

$$I(x,y) = \sup\{t \in [0,1] \mid U(x,t) \leq y\} \,, \qquad x, y \in [0,1] \,. \tag{5.10}$$

If I is an RU-operation generated from a uninorm U, then we will often denote this by I_U.

Firstly, we examine the basic properties of RU-operations. In the specific case when the neutral element e of U is 1, U reduces to a t-norm, which has been dealt with in Sect. 2.5, while the case when $e = 0$, U reduces to a t-conorm and (5.10) leads to what are termed as co-implications, a discussion of which is beyond the scope of this monograph and the interested readers are referred to, for example, De Baets and Fodor [71].

Proposition 5.4.2. *If U is a uninorm with neutral element $e \in (0,1)$, then I_U satisfies* (I1), (I2), (I4), (I5), (NC) *and* (RB). *Moreover, $I_U(e,y) = y$ for all $y \in [0,1]$.*

Proof. Let U be a uninorm with neutral element $e \in (0,1)$. By the monotonicity of U we have that I_U defined as in (5.10) is increasing (decreasing) in the first (second) variable, i.e., I_U satisfies (I1) and (I2). Further, since $U(x,1) \leq U(1,1) = 1$ for any uninorm U, for any $x \in [0,1]$

$$I_U(x,1) = \sup\{t \in [0,1] \mid U(x,t) \leq 1\} = 1 ,$$

i.e., I_U satisfies (RB) and hence, both (I4) and (NC).

Now let us consider $I_U(1,0) = \sup\{t \in [0,1] \mid U(1,t) \leq 0\}$. For any uninorm U, we have that if $e \leq t \leq 1$, then $1 = U(e,1) \leq U(t,1)$ and when $e \geq t \geq 0$, $U(t,1) \geq \underline{U}_e(t,1) = \min(t,1) = t$. Hence the set $\{t \in [0,1] \mid U(1,t) \leq 0\}$ is either the singleton $\{0\}$, in the case of conjunctive uninorms, or is empty in the case of disjunctive uninorms. In either case, we have $\sup\{t \in [0,1] \mid U(1,t) \leq 0\} = 0$ and thus $I_U(1,0) = 0$, i.e., it satisfies (I5). Finally,

$$I_U(e,y) = \sup\{t \in [0,1] \mid U(e,t) \leq y\} = \sup\{t \in [0,1] \mid t \leq y\} = y . \qquad \square$$

Proposition 5.4.3. *Let U be a uninorm with neutral element $e \in (0,1)$. Then the following statements are equivalent:*

(i) I_U is a fuzzy implication.
(ii) For all $z \in [0,1)$, $U(0,z) = 0$.

Proof. Let U be a uninorm with neutral element $e \in (0,1)$.

(i) $\Longrightarrow$ *(ii)* By the monotonicity of U we have that $U(0,z) \leq U(0,e) = 0$ for all $z \in [0,e]$. Hence, we only need to prove that $U(0,z) = 0$, for all $z \in [e,1)$. If I_U is a fuzzy implication, then I_U satisfies (I3) and hence,

$$I_U(0,0) = \sup\{t \in [0,1] \mid U(0,t) \leq 0\} = 1 ,$$

which implies that for all $z \in [e,1)$, $U(0,z) = 0$.

(ii) $\Longrightarrow$ *(i)* From Proposition 5.4.2, it suffices to prove that I_U satisfies (I3). Indeed, for any $y \in [0,1]$, we have

$$I_U(0,y) = \sup\{t \in [0,1] \mid U(0,t) \leq y\} \geq \sup[0,1) = 1 ,$$

by our assumption. $\square$

Once again, only if the RU-operation is a fuzzy implication we use the term *RU-implication.*

Corollary 5.4.4. *If U is a uninorm with neutral element $e \in (0,1)$, then I_U is an RU-implication in the following cases:*

(i) U is a conjunctive uninorm,
(ii) U is a disjunctive representable uninorm,

(iii) U is an idempotent uninorm with associated function g such that $g(0) = 1$,
(iv) U is a disjunctive right-continuous idempotent uninorm whose associated function g is such that $g(z) = 0 \iff z = 1$, for all $z \in [0, 1]$.

Proof. (i) When U is a conjunctive uninorm we have $0 = U(1, 0) \geq U(z, 0)$, for all $z \in [0, 1)$ and hence I_U is a fuzzy implication.
(ii) For any representable uninorm U with generator h and for all $z \in [0, 1)$, we have $U(0, z) = h^{-1}(h(0) + h(z)) = 0$, .
(iii) Let U be an idempotent uninorm whose associated function g is such that $g(0) = 1$. From Theorem 5.1.6, we see that for all $z \in [0, 1)$, $U(0, z) = 0$ if and only if $g(0) = 1$.
(iv) If U is a disjunctive right-continuous idempotent uninorm with an associated function g, then by Corollary 5.1.9, we have that g is sub-involutive, in which case we have $g(g(0)) \leq 0$. By our assumption now, we have that $g(0) = 1$ and hence (iv) follows from (iii) above. □

The proof of the next result can be given along similar lines as that of Theorem 2.5.7.

Proposition 5.4.5. *If U is a conjunctive left-continuous uninorm with neutral element $e \in (0, 1)$, then I_U satisfies* (EP).

Remark 5.4.6. Similar to (U,N)-implications, RU-implications too do not satisfy (NP) (cf. Remark 5.3.8). To see this, let $1 > y > e$. By the monotonicity of U, for any $t \in [e, 1]$ we get $1 = U(1, e) \leq U(1, t)$, i.e., $U(1, t) > y$. Hence $I_U(1, y) \leq e \neq y$.

The block structure of a uninorm U precludes any further investigations on the basic properties of RU-implications unless the class to which U belongs is known. For example, there are $U \in \mathcal{U}_{\mathbf{Idem}}$ that are right-continuous and still their RU-implications satisfy (EP) (see Theorem 5.4.17). Moreover, a characterization of the family of all RU-implications is as yet unknown.

In the following sections we investigate RU-implications I_U, when U belongs to one of the three main classes. Our approach in the following three subsections is as follows. Firstly, we describe the structure of the RU-implication I_U and illustrate it with suitable examples and plots. Subsequently, we investigate the algebraic properties that these subfamilies possess.

5.4.2 RU-Implications from Pseudo-continuous Uninorms

Theorem 5.4.7. *If $U^{\mathbf{c}}_{T,S,e} \in \mathcal{U}_{\mathbf{Min}}$, then the RU-implication obtained from U is given by:*

$$I_{U^{\mathbf{c}}_{T,S,e}}(x, y) = \begin{cases} e \cdot I_T\left(\frac{x}{e}, \frac{y}{e}\right), & \text{if } x, y \in [0, e) \text{ and } x > y\ , \\ e + (1 - e) \cdot I_S\left(\frac{x-e}{1-e}, \frac{y-e}{1-e}\right), & \text{if } x, y \in [e, 1] \text{ and } x \leq y\ , \\ e, & \text{if } x, y \in [e, 1] \text{ and } x > y\ , \\ I_{\mathbf{GD}}(x, y), & \text{otherwise}\ , \end{cases}$$

for all $x, y \in [0,1]$, where I_S is the residual of the t-conorm S, obtained from (2.16) by employing S instead of a t-norm.

Proof. Let us fix arbitrarily $x, y \in [0,1]$. For notational simplicity, we will write U instead of $U^{\mathbf{c}}_{T,S,e}$. We divide our proof into several cases.

If $x \in [0, e)$, then

$$
\begin{aligned}
I_U(x,y) &= \max\left(\sup\left\{ t \in [0,e] \mid e \cdot T\left(\frac{x}{e}, \frac{t}{e}\right) \leq y \right\}, \right. \\
&\qquad \left. \sup\left\{ t \in (e,1] \mid \min(x,t) \leq y \right\} \right) \\
&= \max\left(\sup\left\{ t \in [0,e] \mid T\left(\frac{x}{e}, \frac{t}{e}\right) \leq \frac{y}{e} \right\}, \right. \\
&\qquad \left. \sup\left\{ t \in (e,1] \mid x \leq y \right\} \right)
\end{aligned}
$$

Now, if $x \leq y$, then $I_U(x,y) = 1 = I_{\mathbf{GD}}(x,y)$. If $x > y$, then putting $t' = \dfrac{t}{e}$, we have that if $t \in [0,e]$, then $t' \in [0,1]$. Hence

$$
\begin{aligned}
I_U(x,y) &= \sup\left\{ t \in [0,e] \mid T\left(\frac{x}{e}, \frac{t}{e}\right) \leq \frac{y}{e} \right\} \\
&= \sup\left\{ e \cdot t' \mid t' \in [0,1] \text{ and } T\left(\frac{x}{e}, t'\right) \leq \frac{y}{e} \right\} \\
&= e \cdot \sup\left\{ t' \in [0,1] \mid T\left(\frac{x}{e}, t'\right) \leq \frac{y}{e} \right\} \\
&= e \cdot I_T\left(\frac{x}{e}, \frac{y}{e}\right).
\end{aligned}
$$

If $x \in [e, 1]$, then

$$
\begin{aligned}
I_U(x,y) &= \max\left(\sup\left\{ t \in [e,1] \mid e + (1-e) \cdot S\left(\frac{x-e}{1-e}, \frac{t-e}{1-e}\right) \leq y \right\}, \right. \\
&\qquad \left. \sup\left\{ t \in [0,e) \mid \min(x,t) \leq y \right\} \right) \\
&= \max\left(\sup\left\{ t \in [e,1] \mid S\left(\frac{x-e}{1 \quad c}, \frac{t-e}{1 \quad c}\right) \leq \frac{y-e}{1 \quad c} \right\}, \right. \\
&\qquad \left. \sup\left\{ t \in [0,e) \mid t \leq y \right\} \right). \qquad (5.11)
\end{aligned}
$$

Obviously, $e + (1-e) \cdot S(p,q) \geq e$ for any $p, q \in [0,1]$. So, if $y \in [0,e)$, then $I_U(x,y) = \max(0,y) = y = I_{\mathbf{GD}}(x,y)$. In the case when $y \in [e,1]$, we have $\sup\{t \in [0,e) \mid t \leq y\} = e$. Now, observe that if $x > y$, then $\frac{x-e}{1-e} > \frac{y-e}{1-e}$, and so

$$
S\left(\frac{x-e}{1-e}, \frac{t-e}{1-e}\right) \geq \frac{x-e}{1-e} > \frac{y-e}{1-e},
$$

and hence $\left\{t \in [e,1] \mid S\left(\frac{x-e}{1-e}, \frac{t-e}{1-e}\right) \leq \frac{y-e}{1-e}\right\} = \emptyset$ and $I_U(x,y) = \max(0,e) = e$, in this case. Finally, if $x \leq y$, then putting $t' = \frac{t-e}{1-e}$, we get

$$\begin{aligned} I_U(x,y) &= \max\left(\sup\left\{t \in [e,1] \mid S\left(\frac{x-e}{1-e}, \frac{t-e}{1-e}\right) \leq \frac{y-e}{1-e}\right\}, e\right) \\ &= \sup\left\{e + (1-e)\cdot t' \mid t' \in [0,1] \text{ and } S\left(\frac{x-e}{1-e}, t'\right) \leq \frac{y-e}{1-e}\right\} \\ &= e + (1-e)\cdot \sup\left\{t' \in [0,1] \mid S\left(\frac{x-e}{1-e}, t'\right) \leq \frac{y-e}{1-e}\right\} \\ &= e + (1-e)\cdot I_S\left(\frac{x-e}{1-e}, \frac{y-e}{1-e}\right). \end{aligned}$$

□

Example 5.4.8. (i) Let us consider the uninorm $U_{\mathbf{LK}} = (T_{\mathbf{LK}}, S_{\mathbf{LK}}, 0.5) \in \mathcal{U}_{\mathbf{Min}}$. Then the RU-implication obtained from $U_{\mathbf{LK}}$ is given by:

$$I_{U_{\mathbf{LK}}}(x,y) = \begin{cases} 0.5 + y - x, & \text{if } (x,y \in [0,0.5) \text{ and } y < x) \\ & \quad \text{or } (x,y \in [0.5,1] \text{ and } y > x), \\ 0.5, & \text{if } x,y \in [0.5,1] \text{ and } y \leq x, \\ I_{\mathbf{GD}}(x,y), & \text{otherwise}. \end{cases}$$

(ii) Let us consider the uninorm $U_{\mathbf{P}} = (T_{\mathbf{P}}, S_{\mathbf{P}}, 0.5) \in \mathcal{U}_{\mathbf{Min}}$. Then the RU-implication obtained from $U_{\mathbf{P}}$ is given by:

$$I_{U_{\mathbf{P}}}(x,y) = \begin{cases} \frac{y}{2x}, & \text{if } x,y \in [0,0.5) \text{ and } y < x, \\ 0.5 + \frac{y-x}{2\cdot(1-x)}, & \text{if } x,y \in [0.5,1] \text{ and } x \leq y, \\ 0.5, & \text{if } x,y \in [0.5,1] \text{ and } y < x, \\ I_{\mathbf{GD}}(x,y), & \text{otherwise}. \end{cases}$$

(iii) Let us consider the uninorm $U_{\mathbf{M}} = (T_{\mathbf{M}}, S_{\mathbf{M}}, 0.5) \in \mathcal{U}_{\mathbf{Min}}$. Then the RU-implication obtained from $U_{\mathbf{M}}$ is given by:

$$I_{U_{\mathbf{M}}}(x,y) = \begin{cases} y, & \text{if } x,y \in [0.5,1] \text{ and } y > x, \\ 0.5, & \text{if } x,y \in [0.5,1] \text{ and } y \leq x, \\ I_{\mathbf{GD}}(x,y), & \text{otherwise}. \end{cases}$$

Fig. 5.2(a) gives the structure of an RU-implication I_U when $U \in \mathcal{U}_{\mathbf{Min}}$, while Figs. 5.2(b)–(d) give the plots of all the RU-implications in Example 5.4.8.

Remark 5.4.9. (i) From Proposition 5.4.5 it is obvious that the RU-implication obtained from any left-continuous uninorm $U \in \mathcal{U}_{\mathbf{Min}}$ satisfies (EP). Now, consider the uninorm $U_{\mathbf{M}} = (T_{\mathbf{M}}, S_{\mathbf{M}}, 0.5) \in \mathcal{U}_{\mathbf{Min}}$, whose associated function is g_c, see Example 5.1.10, with $e = 0.5$. It is easy to see that $U_{\mathbf{M}}$ is a right-continuous conjunctive uninorm (see [67], Example 3) and still its RU-implication I_U, given in Example 5.4.8(iii), satisfies (EP). A general result

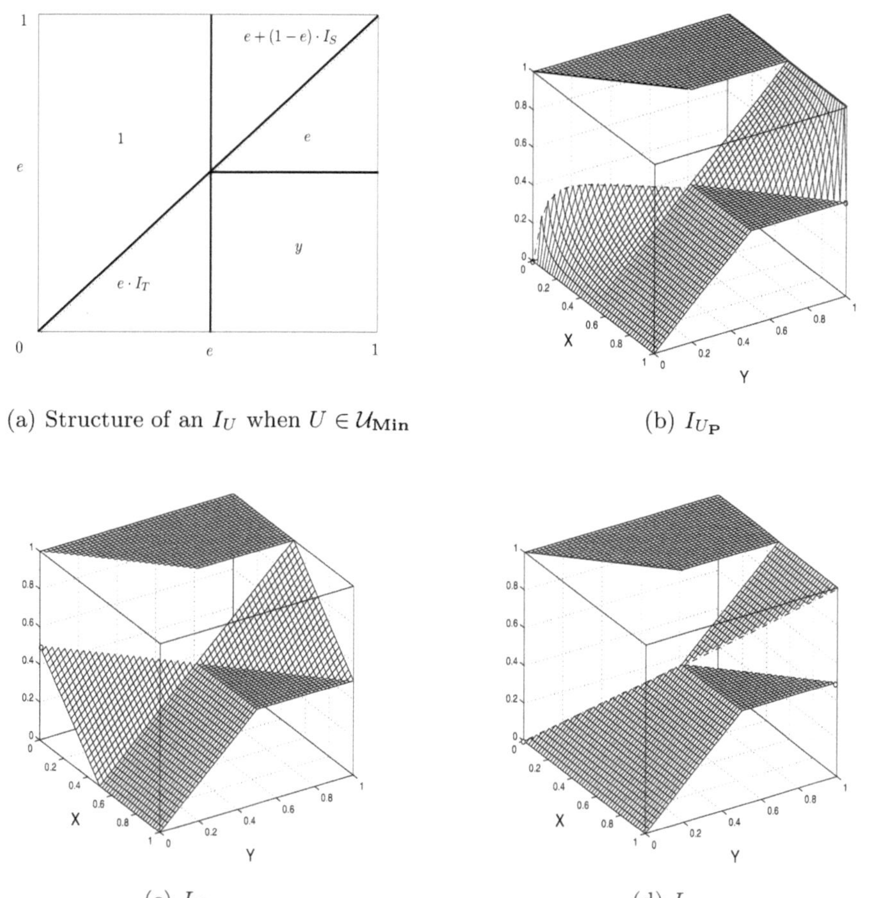

(a) Structure of an I_U when $U \in \mathcal{U}_{\mathbf{Min}}$ (b) $I_{U_{\mathbf{P}}}$

(c) $I_{U_{\mathbf{LK}}}$ (d) $I_{U_{\mathbf{M}}}$

Fig. 5.2. Plots of RU-implications when $U \in \mathcal{U}_{\mathbf{Min}}$ (see Example 5.4.8)

on the conditions under which an RU-implication obtained from $U \in \mathcal{U}_{\mathbf{Min}}$ satisfying (EP) is yet to be determined.

(ii) If $U \in \mathcal{U}_{\mathbf{Min}}$ with neutral element $e \in (0,1)$, then the natural negation of I_U with respect to e is the function

$$N_{I_U}^{e}(x) = \begin{cases} 1, & \text{if } x \in [0,e)\,, \\ e, & \text{if } x \in [e,1]\,, \end{cases}$$

which is not a fuzzy negation. In fact, from the formula of I_U given in Theorem 5.4.7, it can be seen that, for any $\alpha \in (0,1)$, the natural negation of I_U with respect to α is not a fuzzy negation, since $N_{I_U}^{\alpha}(1) = \alpha$ if $\alpha \in (0,e)$ and $N_{I_U}^{\alpha}(1) = e$ if $\alpha \in [e,1)$ (see also the 2D-plot given in Fig. 5.2(a)).

5.4.3 RU-Implications from Representable Uninorms

Let the uninorm U in Definition 5.4.1 be a representable uninorm. Then we have the following representation of the RU-implications from DE BAETS and FODOR [71].

Theorem 5.4.10. *If U_h is a representable uninorm with additive generator h, then the RU-implication obtained from U_h is given by*

$$I_{U_h}(x,y) = \begin{cases} 1, & \text{if } (x,y) \in \{(0,0),(1,1)\}, \\ h^{-1}(h(y)-h(x)), & \text{otherwise}, \end{cases} \tag{5.12}$$

for all $x, y \in [0,1]$.

Proof. Let us fix arbitrarily $x, y \in [0,1]$. We consider several cases now.

If $x \in (0,1)$ and $y \in [0,1]$, then by Theorem 5.1.12 and the strict monotonicity of h, we get

$$\begin{aligned} I_{U_h}(x,y) &= \sup\{t \in [0,1] \mid U(x,t) \leq y\} \\ &= \sup\left\{t \in [0,1] \mid h^{-1}(h(x)+h(t)) \leq y\right\} \\ &= \sup\{t \in [0,1] \mid h(x)+h(t) \leq h(y)\} \\ &= h^{-1}(h(y)-h(x)) . \end{aligned}$$

Further, if $x = 0$ and $y \in [0,1]$, then

$$\begin{aligned} I_{U_h}(0,y) &= \sup\{t \in [0,1] \mid U(0,t) \leq y\} \\ &\geq \sup\left\{t \in [0,1) \mid h^{-1}(h(0)+h(t)) \leq y\right\} \\ &= \sup\{t \in [0,1) \mid 0 \leq y\} = 1 . \end{aligned}$$

Observe that for $y \in [0,1]$, it also holds that $h^{-1}(h(y)-h(0)) = 1$.

If $x = 1$ and $y \in [0,1)$, then

$$\begin{aligned} I_{U_h}(1,y) &= \sup\{t \in [0,1] \mid U(1,t) \leq y\} \\ &= \max\left(\sup\left\{t \in (0,1] \mid h^{-1}(h(1)+h(t)) \leq y\right\}, 0\right) \\ &= \max\left(\sup\{t \in (0,1] \mid 1 \leq y\}, 0\right) = \max(0,0) = 0 . \end{aligned}$$

Once again, for $y \in [0,1)$, it also holds that $h^{-1}(h(y)-h(1)) = 0$. Finally,

$$\begin{aligned} I_{U_h}(1,1) &= \sup\{t \in [0,1] \mid U(1,t) \leq 1\} \\ &= \max\left(\sup\left\{t \in (0,1] \mid h^{-1}(h(1)+h(t)) \leq 1\right\}, 0\right) \\ &= \max\left(\sup\{t \in (0,1] \mid 1 \leq 1\}, 0\right) = \max(1,0) = 1 . \end{aligned}$$

□

Example 5.4.11. (i) Let us consider the conjunctive representable uninorm $U^{\mathbf{c}}_{h_1}$ given in Example 5.1.14(i). Its RU-implication given in Theorem 5.4.10 is also the (U,N)-implication $I_{U^{\mathbf{d}}_{h_1}, N_{\mathbf{C}}}$ given as in Example 5.3.6(iv).

(ii) Similarly the RU-implication of the conjunctive representable uninorm $U^{\mathbf{c}}_{h_\beta}$, given in Example 5.1.14(ii), is also the (U,N)-implication $I_{U^{\mathbf{d}}_{h_\beta},N_{\mathbf{C}}}$ as given in Example 5.3.6(v).

Lemma 5.4.12. *If U_h is a representable uninorm with additive generator h, then I_{U_h} satisfies* (EP).

Proof. Let us fix arbitrarily $x, y, z \in [0,1]$. We will consider several cases.

Since I_{U_h} in (5.12) is a fuzzy implication, it reduces to the classical implication when $x, y, z \in \{0,1\}$ and hence, it satisfies (EP).

If $x, y, z \in (0,1)$, then from (5.12) we have

$$\begin{aligned}
I_{U_h}(x, I_{U_h}(y,z)) &= h^{-1}\left(h(I_{U_h}(y,z)) - h(x)\right) \\
&= h^{-1}\left(h \circ h^{-1}(h(z) - h(y)) - h(x)\right) \\
&= h^{-1}\left(h(z) - h(y) - h(x)\right) \\
&= h^{-1}\left(h(z) - h(x) - h(y)\right) \\
&= h^{-1}\left(h \circ h^{-1}(h(z) - h(y)) - h(x)\right) \\
&= h^{-1}\left(h(I_{U_h}(x,z)) - h(x)\right) = I_U(y, I_U(x,z)) \,.
\end{aligned}$$

If $x, y \in (0,1)$ and $z = 0$, then $I_{U_h}(y,0) = h^{-1}(h(0) - h(y)) = 0$ and hence $I_{U_h}(x, I_{U_h}(y,z)) = 0$, in which case we get also $I_{U_h}(y, I_{U_h}(x,z)) = 0$.

If $x, y \in (0,1)$ and $z = 1$, then $I_{U_h}(x, I_{U_h}(y,z)) = I_{U_h}(y, I_{U_h}(x,z)) = 1$.

If $x, z \in (0,1)$ and $y = 0$, then $I_{U_h}(0,z) = h^{-1}(h(z) - h(0)) = 1$ and hence $I_{U_h}(x, I_{U_h}(y,z)) = I_{U_h}(y, I_{U_h}(x,z)) = 1$.

If $x, z \in (0,1)$ and $y = 1$, then $I_{U_h}(1,z) = h^{-1}(h(z) - h(1)) = 0$ and hence $I_{U_h}(x, I_{U_h}(y,z)) = I_{U_h}(y, I_{U_h}(x,z)) = 0$.

The cases when $x \in \{0,1\}$ and $y, z \in (0,1)$ are similar as above. Hence I_{U_h} satisfies (EP). □

Proposition 5.4.13. *If U_h is a representable uninorm with additive generator h and neutral element $e \in (0,1)$, then*

(i) $I_{U_h}(x,x) = e$ for all $x \in (0,1)$,
(ii) $N^e_{I_{U_h}} = N_{U_h}$ is a strong negation,
(iii) I_{U_h} satisfies (CP)*(N_{U_h}),*
(iv) $I_{U_h}(x,y) \geq y$ only if $x < e$.

Proof. Let U_h be a representable uninorm with additive generator h and neutral element $e \in (0,1)$.

(i) From (5.12), we see that $I_{U_h}(x,x) = h^{-1}(h(x) - h(x)) = h^{-1}(0) = e$, for all $x \in (0,1)$. However, note that $I_{U_h}(0,0) = I_{U_h}(1,1) = 1$.

(ii) The natural negation of I_{U_h} with respect to e is

$$\begin{aligned}
N^e_{I_{U_h}}(x) &= I_{U_h}(x,e) = h^{-1}(h(e) - h(x)) \\
&= h^{-1}(-h(x)) = N_{U_h}(x), \qquad x \in [0,1] \,,
\end{aligned}$$

by Remark 5.1.13(iii), and is a strong negation.

(iii) If $(x, y) \in \{(0,0),(1,1)\}$, then the contrapositivity is obvious. Now, with $N_{U_h}(x) = h^{-1}(-h(x))$, if $(x, y) \in [0,1]^2 \setminus \{(0,0),(1,1)\}$, then

$$\begin{aligned} I_{U_h}(N_{U_h}(y), N_{U_h}(x)) &= h^{-1}\left(h\left(h^{-1}(-h(x))\right) - h\left(h^{-1}(-h(y))\right)\right) \\ &= h^{-1}(-h(x) + h(y)) = I_{U_h}(x, y) \,. \end{aligned}$$

(iv) By the monotonicity of h, it is obvious that $I_{U_h}(x, y) \geq y$ only if $x < e$. □

5.4.4 RU-Implications from Idempotent Uninorms

From Corollary 5.4.4(iii), we see that an RU-implication can be constructed from an idempotent uninorm whenever the unary operator g is such that $g(0) = 1$. The following result of Ruiz and Torrens [215] gives the general structure of an I_U obtained from such idempotent uninorms, which subsumes an earlier result by De Baets and Fodor [71].

Theorem 5.4.14. *Let U be an idempotent uninorm whose associated function g is such that $g(0) = 1$. Then the RU-implication obtained from U is given by*

$$I_U(x, y) = \begin{cases} \max(g(x), y), & \text{if } x \leq y \,, \\ \min(g(x), y), & \text{if } x > y \,, \end{cases} \tag{5.13}$$

for all $x, y \in [0,1]$.

Proof. Let us fix arbitrarily $x, y \in [0,1]$. Using the representation of idempotent uninorms given in Theorem 5.1.6, we consider the following four cases now.

If $y < x$ and $y < g(x)$, then $U(x, y) = \min(x, y) = y$. Thus, for any $t > y$, we have $U(x, t) \in \{x, t\} > y$ and $I_U(x, y) = \sup\{t \in [0,1] \mid U(x, t) \leq y\} = y$.

If $y < x$ and $y > g(x)$, then $U(x, t) = \min(x, t) = t < y$, for any $t < g(x) \leq y$. If $t > g(x)$, then we get $U(x, t) = \max(x, t) \geq x > y$ and therefore, $I_U(x, y) = \sup\{t \in [0,1] \mid U(x, t) \leq y\} = g(x)$.

If $x \leq y$ and $y > g(x)$, then $U(x, y) = \max(x, y) = y$. Thus, for any $t > y$, we have $U(x, t) = t > y$. If $y > t > g(x)$, then we get $U(x, t) = \max(x, t) < y$ and hence, $I_U(x, y) = \sup\{t \in [0,1] \mid U(x, t) \leq y\} = y$.

Finally, if $x \leq y$ and $y \leq g(x)$, then $U(x, y) = \min(x, y) = x$. Therefore, for any $t > g(x)$, we have $U(x, t) = \max(x, t) = t > y$. If $t < g(x)$, then we get $U(x, t) = \min(x, t) = x < y$, hence $I_U(x, y) = \sup\{t \in [0,1] | U(x, t) \leq y\} = g(x)$. □

Remark 5.4.15. From Theorem 5.1.6, we know that for a uninorm $U \in \mathcal{U}_{\mathbf{Idem}}$, different parts of the graph of its associated function g can belong to either minimum or maximum. Whereas, from Theorem 5.4.14, we see that the RU-implication obtained from U is still the same for a given g. In other words, the RU-implications obtained from idempotent uninorms with the same neutral element e and associated function g are identical (see Remark 5.3.16(v)).

Example 5.4.16. (i) Let us consider the idempotent uninorm $U_{\mathbf{YR}}^{\mathbf{c},e} \in \mathcal{U}_{\mathbf{Min}}$, with the associated function $g_{\mathbf{c}}$ given in Example 5.1.10. Then its RU-implication is given by

$$I_{U_{\mathbf{YR}}^{\mathbf{c},e}}(x,y) = \begin{cases} y, & \text{if } (y < x \text{ and } y \leq e) \text{ or } (y \geq x \text{ and } x \geq e)\ , \\ e, & \text{if } y < x \text{ and } y > e\ , \\ 1, & \text{otherwise}\ . \end{cases}$$

(ii) Let us consider the right-continuous idempotent uninorm $U_{N_{\mathbf{C}}}$, whose associated function is the classical negation $N_{\mathbf{C}}$, which is obviously not in $\mathcal{U}_{\mathbf{Min}}$, given by

$$U_{N_{\mathbf{C}}}(x,y) = \begin{cases} \min(x,y), & \text{if } y < 1-x\ , \\ \max(x,y), & \text{if } y \geq 1-x\ . \end{cases}$$

Then its RU-implication is given by

$$I_{U_{N_{\mathbf{C}}}}(x,y) = \begin{cases} \min(1-x,y), & \text{if } y < x\ , \\ \max(1-x,y), & \text{if } y \geq x\ . \end{cases}$$

(iii) Let us consider the right-continuous idempotent uninorm $U_{N_{\mathbf{K}}}$, whose associated function is the strict negation $N_{\mathbf{K}}$, which is again not in $\mathcal{U}_{\mathbf{Min}}$, given by

$$U_{N_{\mathbf{K}}}(x,y) = \begin{cases} \min(x,y), & \text{if } y < N_{\mathbf{K}}(x)\ , \\ \max(x,y), & \text{if } y \geq N_{\mathbf{K}}(x)\ . \end{cases}$$

Then its RU-implication is given by

$$I_{U_{N_{\mathbf{K}}}}(x,y) = \begin{cases} \min(N_{\mathbf{K}}(x),y), & \text{if } y < x\ , \\ \max(N_{\mathbf{K}}(x),y), & \text{if } y \geq x\ . \end{cases}$$

Fig. 5.3(a) gives the structure of an RU-implication I_U when $U \in \mathcal{U}_{\mathbf{Idem}}$, while Figs. 5.3(b)–(d) give the plots of all the RU-implications in Example 5.4.16.

Characterization of RU-implications generated from idempotent uninorms that satisfy the exchange principle (EP) has been obtained by Ruiz and Torrens [215]. The proof of this next result is rather involved and can be found in the above reference.

Theorem 5.4.17. *Let U be an idempotent uninorm whose associated function g is such that $g(0) = 1$. Then the following statements are equivalent:*

(i) I_U satisfies (EP).
(ii) The following property is satisfied:

$$\textit{if } g(g(x)) < x \textit{ for some } x \in [0,1],\ \textit{then } x > e \textit{ and } g(x) = e. \tag{5.14}$$

Corollary 5.4.18. *Let U be an idempotent uninorm whose associated function g is such that $g(0) = 1$. If g is super-involutive, then I_U satisfies* (EP).

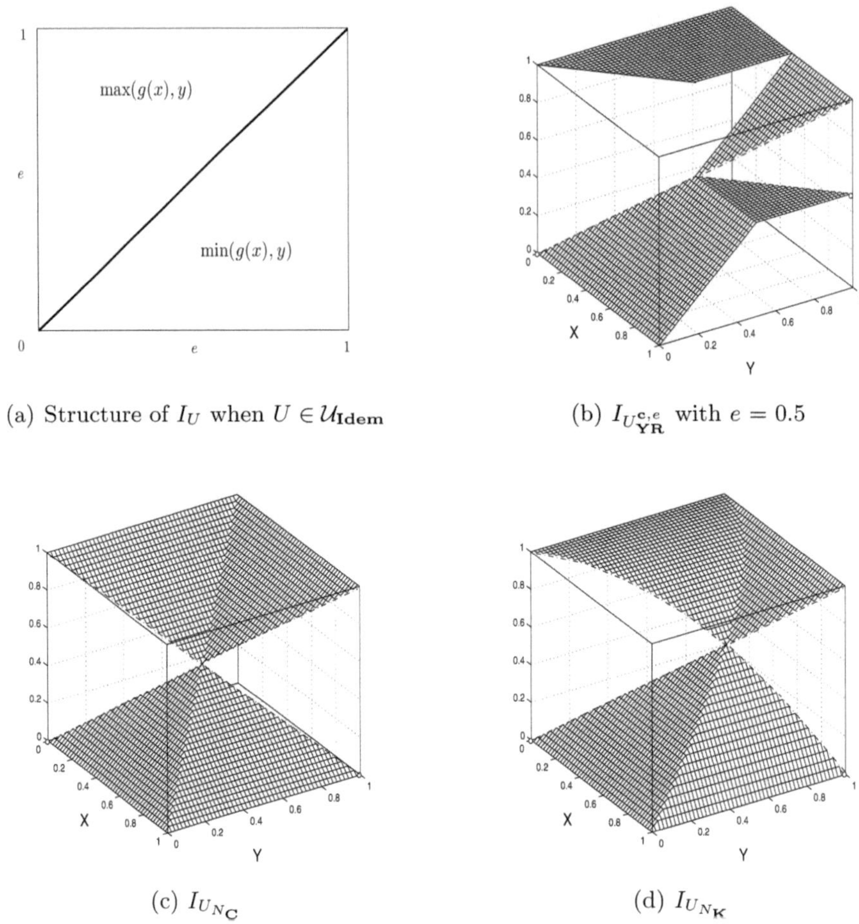

(a) Structure of I_U when $U \in \mathcal{U}_{\mathbf{Idem}}$

(b) $I_{U^{\mathbf{c},e}_{\mathbf{YR}}}$ with $e = 0.5$

(c) $I_{U_{N_{\mathbf{C}}}}$

(d) $I_{U_{N_{\mathbf{K}}}}$

Fig. 5.3. Plots of RU-implications when $U \in \mathcal{U}_{\mathbf{Idem}}$ (see Example 5.4.16)

Example 5.4.19. (i) Consider the RU-implication in Example 5.4.16(i). From the definition of g_c and Corollary 5.1.9 (see also [67], Example 3), it is obvious that the uninorm U_{g_c} is a right-continuous conjunctive uninorm and g_c satisfies (5.14). Hence the RU-implication $I_{U^{\mathbf{c},e}_{\mathbf{YR}}}$ in Example 5.4.16(i) does satisfy (EP).

(ii) Let us consider the unary operator g_n as defined below, with $g_n(0.75) = 0.75$:

$$g_n(x) = \begin{cases} 1, & \text{if } x \in [0, 0.5) , \\ 1.5 - x, & \text{if } x \in [0.5, 1] . \end{cases}$$

Now, if $x \in [0, 0.5)$, then $g_n(x) = 1$ and $g_n(g_n(x)) = g_n(1) = 0.5 \geq x$. If $x \in [0.5, 1]$, then so is $g_n(x)$, and $g_n(g_n(x)) = 1.5 - (1.5 - x) = x$. Hence, $g_n(g_n(x)) \geq x$ for all $x \in [0, 1]$, i.e., g_n is super-involutive and since

$g_n(0) = 1$, from Corollary 5.1.8, the uninorm U_{g_n} is a left-continuous conjunctive idempotent uninorm. Moreover, note that $U_{g_n} \notin \mathcal{U}_{\mathbf{Min}}$. Now, from Corollary 5.4.18, we have that the corresponding $I_{U_{g_n}}$ does have the exchange principle (EP). This can also be seen from Proposition 5.4.5.

Lemma 5.4.20. *Let $N\colon [0,1] \to [0,1]$ be a continuous negation. If, in addition, N is either sub-involutive or super-involutive, then N is involutive.*

Proof. We give the proof for the case when N is sub-involutive. Since a negation N is continuous on $[0,1]$ it is onto. Now, for any $x \in [0,1]$ there exists a $y \in [0,1]$ such that $x = N(y)$. Consequently, we get

$$x = N(y) \Longrightarrow N(x) = N(N(y)) \leq y \Longrightarrow N(N(x)) \geq N(y) = x \,,$$

i.e., $N(N(x)) \geq x$. Since N is sub-involutive, we have that $N(N(x)) \leq x$, whence N is involutive. □

Proposition 5.4.21. *Let N be a continuous negation and U be an idempotent uninorm whose associated function is N. Then the following statements are equivalent:*

(i) I_U *satisfies* (EP).
(ii) N *is super-involutive, and hence strong.*

Proof. (*i*) $\Longrightarrow$ (*ii*) Let N be a continuous negation and U be an idempotent uninorm obtained from N. If the RU-implication obtained from U satisfies (EP), then N satisfies the condition (5.14). We show that $N(N(x)) \geq x$ for all $x \in [0,1]$ by discussing the following two cases.

If $x \leq e$, then by (5.14) we see that $N(N(x)) \geq x$.

Let us suppose that there exists an $x > e$ such that $N(N(x)) < x$. By Theorem 5.4.17 and (5.14), we have $N(x) = e$. Since N is continuous, there exists a $y < e$ such that $N(y) = x > e = N(e) = N(x)$. Once again, by the continuity of N, if z' is such that $N(y) = x > z' > e = N(e)$, then there exists a z such that $y < z < e$ and $N(z) = z'$. Now, from these two inequalities we get

$$U(x,z) = \min(x,z) = z \,,$$

since $z < N(x) = e$. Finally, by the commutativity of U, $U(z,x) = z$, which implies $x \leq N(z) = z'$, a contradiction.

Hence there does not exist any $x \in [0,1]$ such that $N(N(x)) < x$, i.e., N is super-involutive. From Proposition 5.4.20, we have that N is strong.

(*ii*) $\Longrightarrow$ (*i*) This follows from Corollary 5.4.18. □

Proposition 5.4.22. *Let U be an idempotent uninorm with the generator g such that $g(0) = 1$. Then*

(i) $I_U(x,x) = \max(x, g(x))$,
(ii) $I_U(x,x) = e$ *for an* $x \in (0,1)$ *if and only if* $x \in \{t \in [0,e] \mid g(t) = e\}$,

(iii) $I_U(x, g(x)) = g(x)$,
(iv) $I_U(x, e) = g(x)$.

Proof. Observe that (i), (iii) and (iv) are obvious from (5.13). Therefore, we show (ii) now. Let $I_U(x, x) = e$, for some $x \in (0, 1)$. Then $I_U(x, x) = \max(g(x), x) = e$, which implies $x = e$ or $g(x) = e$ and $x \leq e$. In either case $x \in \{t \in [0, e] | g(t) = e\}$. Conversely, if $x \in \{t \in [0, e] \mid g(t) = e\}$, then the result is obvious. □

From (5.13) and Proposition 5.4.22 the following result can be obtained.

Lemma 5.4.23. *Let N be a fuzzy negation and U be an idempotent uninorm whose associated function g is such that $g(0) = 1$, $g(1) = 0$. Then the following statements are equivalent:*

(i) $N_{I_U}^e = N$.
(ii) $g = N$.

5.5 Intersections between (U,N)- and RU-Implications

In the previous sections we have discussed two families of fuzzy implications derived from uninorms, viz., (U,N)-implications and RU-implications. In the case of RU-implications from uninorms, we have specifically considered uninorms U from the three main families of $\mathcal{U}_{\mathbf{Min}}$, $\mathcal{U}_{\mathbf{Rep}}$ and $\mathcal{U}_{\mathbf{Idem}}$. In this section we discuss the intersections that exist among these families of fuzzy implications.

Let us denote by

- $\mathbb{I}_{\mathbb{U},\mathbb{N}}$ – the family of all (U,N)-implications.
- $\mathbb{I}_{\mathbb{U},\mathbb{N}_\mathbb{C}}$ – the family of all (U,N)-implications obtained from continuous negations.
- $\mathbb{I}_{\mathbb{U}_\mathbf{M}}$ – the family of all RU-implications generated from uninorms in $\mathcal{U}_{\mathbf{Min}}$.
- $\mathbb{I}_{\mathbb{U}_\mathbf{R}}$ – the family of all RU-implications generated from uninorms in $\mathcal{U}_{\mathbf{Rep}}$.
- $\mathbb{I}_{\mathbb{U}_\mathbf{I}}$ – the family of all RU-implications generated from uninorms in $\mathcal{U}_{\mathbf{Idem}}$.

Needless to state, in the case $e = 0$ we have $\mathbb{I}_{\mathbb{U},\mathbb{N}} = \mathbb{I}_{\mathbb{S},\mathbb{N}}$, while if $e = 1$ we have $\mathbb{I}_\mathbb{U} = \mathbb{I}_\mathbb{T}$. Hence, in the sequel, we consider only uninorms with neutral elements in $(0, 1)$.

5.5.1 Intersection between $\mathbb{I}_{\mathbb{U},\mathbb{N}}$ and $\mathbb{I}_{\mathbb{U}_\mathbf{M}}$

From Proposition 5.3.2(ii) and Remark 5.4.9(ii), we get

$$\mathbb{I}_{\mathbb{U},\mathbb{N}} \cap \mathbb{I}_{\mathbb{U}_\mathbf{M}} = \emptyset \,. \tag{5.15}$$

5.5.2 Intersection between $\mathbb{I}_{\mathbb{U},\mathbb{N}}$ and $\mathbb{I}_{\mathbb{U}_\mathbf{R}}$

Theorem 5.5.1. *Let N be a fuzzy negation and U_h be a representable uninorm with additive generator h. Then the following statements are equivalent:*

(i) The (U,N)-implication $I_{U_h,N}$ is also the RU-implication I_{U_h}.
(ii) $N = N_{U_h}$, the natural negation obtained from the generator h, i.e., N is given by the formula (5.5).

Proof. (i) $\Longrightarrow$ (ii) Let us assume that $I_{U_h,N} = I_{U_h}$. From Proposition 5.3.2(ii) and Theorem 5.4.10 we have, for all $x, y \in [0,1]$,

$$\begin{aligned} N(x) &= U_h(N(x), e) = I_{U_h,N}(x, e) \\ &= I_{U_h}(x, e) = h^{-1}(-h(x) + h(e)) \\ &= h^{-1}(-h(x)) = N_{U_h} . \end{aligned}$$

(ii) $\Longrightarrow$ (i) Obvious by retracing the above steps. □

From Proposition 5.4.13(iii) and Theorem 5.5.1 we have the following:

Corollary 5.5.2. *Let U_h be a representable uninorm with additive generator h. Then the RU-implication I_{U_h} is also the (U,N)-implication obtained from U_h and its natural negation N_{U_h}, i.e., $I_{U_h} = I_{U_h,N_{U_h}}$.*

From Corollary 5.5.2 we have

$$\mathbb{I}_{\mathbb{U}_{\mathbf{R}}} \subsetneq \mathbb{I}_{\mathbb{U},\mathbb{N}} . \tag{5.16}$$

Remark 5.5.3. (i) We know that an R-implication I_T obtained from a continuous t-norm is also an (S,N)-implication - in fact, an S-implication - if and only if the t-norm T is nilpotent. The above results seem to suggest that representable uninorms are generalizations of nilpotent t-norms and t-conorms, whereas their definition indicates that they are, in fact, obtained from generators of strict t-norms and t-conorms.
(ii) Observe now that Lemma 5.4.12 becomes a simple corollary of the above result.

5.5.3 Intersection between $\mathbb{I}_{\mathbb{U},\mathbb{N}}$ and $\mathbb{I}_{\mathbb{U}_\mathbf{I}}$

Ruiz and Torrens [215] have investigated the conditions under which the RU-implication from an idempotent uninorm is also an (U,N)-implication obtained from a strong N. In fact, it can be shown, using Proposition 5.4.21, that the strongness of N need not be assumed and is consequential. The next lemma will be handy in the main result.

Lemma 5.5.4. *Let N be a strong negation and let U be the disjunctive idempotent uninorm whose associated function is N. Then the corresponding (U,N)- and RU-implications are identical, i.e., $I_{U,N} = I_U$.*

Proof. Since U is disjunctive, it is right-continuous. From (5.4) and (5.13), we have, for any $x, y \in [0,1]$,

$$I_{U,N}(x,y) = U(N(x), y) = \begin{cases} \max(N(x), y), & \text{if } x \le y \\ \min(N(x), y), & \text{otherwise} \end{cases} = I_U(x,y) .$$ □

Theorem 5.5.5. *Let N be a continuous negation and U be an idempotent uninorm obtained with associated function g. Then the following statements are equivalent:*

(i) The (U,N)-implication $I_{U,N}$ is also the RU-implication I_U.
(ii) $g = N$, N is strong and U is right-continuous.

Proof. (i) $\Longrightarrow$ (ii) Let us assume that $I_{U,N} = I_U$. From Propositions 5.3.2(ii) and 5.4.22 we have

$$N(x) = U(N(x), e) = I_{U,N}(x, e) = I_U(x, e) = g(x) ,$$

for all $x \in [0,1]$. Since I_U is an (U,N)-implication it satisfies (EP) and from Proposition 5.4.21, we get that N is strong. Finally, from Proposition 5.4.22(i), with $g = N$, we see that

$$\max(N(x), x) = I_U(x, x) = I_{U,N}(x, x) = U(N(x), x) ,$$

for all $x \in [0,1]$. By virtue of Corollary 5.1.9, U is right-continuous.

(ii) $\Longrightarrow$ (i) Since N is a strong negation and U is right-continuous, U is disjunctive and from Lemma 5.5.4 we have that the corresponding (U,N)- and RU-implications are identical, i.e., $I_{U,N} = I_U$. □

Remark 5.5.6. It immediately follows that the RU-implication given in Example 5.4.16(ii) is a (U,N)-implication while the RU-implication in Example 5.4.16 (iii) is not.

Let us denote the family of RU-implications as follows:

- $\mathbb{I}_{\mathbb{U}_{\mathbf{I}^*}}$ – the family of all RU-implications generated from uninorms in $\mathcal{U}_{\mathbf{Idem}}$ that are right-continuous and whose associated function g is involutive.

Using the above notations, the presented results can be summarized as follows:

$$\mathbb{I}_{\mathbb{U}_{\mathbf{I}}} \cap \mathbb{I}_{\mathbb{U},\mathbb{N}_{\mathbb{C}}} = \mathbb{I}_{\mathbb{U}_{\mathbf{I}^*}} . \tag{5.17}$$

5.5.4 Intersection between $\mathbb{I}_{\mathbb{U}_{\mathbf{M}}}$ and $\mathbb{I}_{\mathbb{U}_{\mathbf{R}}}$

From Corollary 5.5.2, we know that $\mathbb{I}_{\mathbb{U}_{\mathbf{R}}} \subset \mathbb{I}_{\mathbb{U},\mathbb{N}}$, while from (5.15) we know that $\mathbb{I}_{\mathbb{U},\mathbb{N}} \cap \mathbb{I}_{\mathbb{U}_{\mathbf{M}}} = \emptyset$. Hence

$$\mathbb{I}_{\mathbb{U}_{\mathbf{M}}} \cap \mathbb{I}_{\mathbb{U}_{\mathbf{R}}} = \emptyset . \tag{5.18}$$

5.5.5 Intersection between $\mathbb{I}_{\mathbb{U}_{\mathbf{M}}}$ and $\mathbb{I}_{\mathbb{U}_{\mathbf{I}}}$

From Examples 5.4.8(iii) and 5.4.16(i) with $e = 0.5$, we see that $\mathbb{I}_{\mathbb{U}_{\mathbf{M}}} \cap \mathbb{I}_{\mathbb{U}_{\mathbf{I}}} \neq \emptyset$. In fact, since $\mathcal{U}_{\mathbf{Min}} \cap \mathcal{U}_{\mathbf{Idem}} = \mathcal{U}_{I,G_*}$ (see Remark 5.1.15(ii)), it can be easily seen that

$$\mathbb{I}_{\mathbb{U}_{\mathbf{M}}} \cap \mathbb{I}_{\mathbb{U}_{\mathbf{I}}} = \mathbb{I}_{\mathbb{U}_{\mathbf{I}_*}} , \tag{5.19}$$

where $\mathbb{I}_{\mathbb{U}_{\mathbf{I}_*}}$ denotes the family of RU-implications generated from uninorms in $\mathcal{U}_{I,G_*}$.

5.5.6 Intersection between $\mathbb{I}_{\mathbb{U}_\mathbf{R}}$ and $\mathbb{I}_{\mathbb{U}_\mathbf{I}}$

On the one hand, from Proposition 5.4.13(ii), we see that if $I \in \mathbb{I}_{\mathbb{U}_\mathbf{R}}$, then $I(x,x) = e$, for all $x \in (0,1)$. If $I \in \mathbb{I}_{\mathbb{U}_\mathbf{I}}$, then from Proposition 5.4.22(i) we have $I(x,x) = \max(x, g(x))$ for all $x \in [0,1]$. Now, since g is a decreasing function and $e \in (0,1)$, there exists an $x > e = g(e) \geq g(x)$, i.e., $I(x,x) \neq e$ for all $x \in (0,1)$. Hence

$$\mathbb{I}_{\mathbb{U}_\mathbf{R}} \cap \mathbb{I}_{\mathbb{U}_\mathbf{I}} = \emptyset\,. \tag{5.20}$$

Remark 5.5.7. From the results obtained so far and the previous examples, the following inclusions among the above families can be seen.

(i) The (U,N)-implications $I_{U_\mathbf{P},N_\mathbf{S}}$ and $I_{U_\mathbf{LK},N_\mathbf{S}}$ in Example 5.3.6(i) and (ii) show that

$$\mathbb{I}_{\mathbb{U},\mathbb{N}_\mathbb{C}} \supsetneq \mathbb{I}_{\mathbb{U}_{\mathbf{I}^*}} \cup \mathbb{I}_{\mathbb{U}_\mathbf{R}}\,.$$

(ii) Example 5.4.16(iii) shows that

$$\mathbb{I}_{\mathbb{U}_\mathbf{I}} \supsetneq \mathbb{I}_{\mathbb{U}_{\mathbf{I}^*}} \cup \mathbb{I}_{\mathbb{U}_{\mathbf{I}_*}}\,.$$

(iii) In fact, from Corollary 5.5.2 and the above two inclusions, we have

$$\mathbb{I}_{\mathbb{U}_\mathbf{R}} \subsetneq \mathbb{I}_{\mathbb{U},\mathbb{N}_\mathbb{C}} \subsetneq \mathbb{I}_{\mathbb{U},\mathbb{N}}\,.$$

(iv) The RU-implications $I_{U_\mathbf{LK}}$ and $I_{U_\mathbf{P}}$ from Example 5.4.8(i) and (ii) show that

$$\mathbb{I}_{\mathbb{U}_\mathbf{M}} \supsetneq \mathbb{I}_{\mathbb{U}_{\mathbf{I}_*}}\,.$$

5.6 Bibliographical Remarks

De Baets and Fodor were the first to discuss fuzzy implications from uninorms in [69, 70, 71]. Their study predominantly focused on residual operators from the three classes of uninorms, viz., $\mathcal{U}_{\mathbf{Min}}, \mathcal{U}_{\mathbf{Rep}}$ and the classes of left- and right-continuous idempotent uninorms. It should be noted that when [71] was published only the work of De Baets [67] on left- and right-continuous idempotent uninorms was known.

Once a complete characterization of idempotent uninorms was given by Martin et al. [167], Ruiz and Torrens [215], generalizing some of the results in [71], investigated residual operators from idempotent uninorms in depth.

It should be noted that these are only some nascent efforts and many interesting problems remain. Characterization results in the case of residual operators from uninorms, even in restricted cases, is still unknown.

Once again (U,N)-implications were introduced in passing by De Baets and Fodor in [71]. In fact, they even showed that under certain conditions the residual operators obtained from representable and left- or right-continuous idempotent uninorms coincided with the corresponding (U,N)-implications.

Recently, Jayaram and Baczyński [127] investigated (U,N)-implications relaxing the constraints on the negation N. Along their earlier works on (S,N)-implications they have given a characterization of (U,N)-implications. Needless to state, their characterizations have enabled them to generalize many of the earlier known results, especially on the intersections between the classes of $\mathbb{I}_{\mathbb{U}}$ and $\mathbb{I}_{\mathbb{U},\mathbb{N}}$. Once again, the characterization of (U,N)-implications from non-continuous negations is still unavailable.

It should be noted that (U,N)-implications are closely related to the e-implications investigated in KHALEDI et al. [137], whose representation is yet to be determined.

Generalization of QL-operators to the setting of uninorms has recently been studied by MAS et al. [175]. They considered operations of the form

$$I_{U,V,N}(x,y) = V(N(x),U(x,y)) , \qquad x,y \in [0,1] ,$$

where N is a strong negation and U and V are a conjunctive and a disjunctive uninorm, respectively. As in the setting of t-norms and t-conorms, these operators are not always fuzzy implications. In their study, by the strongness of the negation N considered, their results tend to be quite stringent. For example, V reduces to a t-conorm and hence a nilpotent t-conorm, with the associated strong negation $N_V \leq N$ in the continuous case. Whereas, in the case when N is only a fuzzy negation, it can be shown that more options for V exist. The authors have also investigated the case when $N_V = N$ and show that, among the four classes considered, only when $U \in \mathcal{U}_{\mathbf{Min}}$ we have that $I_{U,V,N} \in \mathcal{FI}$. Hence, a thorough study of such implications is required.

Similar to uninorms, t-operators, usually denoted by F, were proposed by MAS et al. [170]. Like uninorms, these are commutative, associative and increasing binary operations on the unit interval $[0,1]$, but unlike uninorms where the neutral element gets the focus, here the emphasis is on the annihilator $k \in [0,1]$ such that $F(1,0) = F(0,1) = k$. It immediately follows that if $k = 0$, then F is a t-norm, while $k = 1$ implies that F is a t-conorm.

Unfortunately, t-operators do not exactly turn out to be a fertile field for generating fuzzy implications the usual way. For example, consider the generalizations of (S,N)-, QL- and R-implications to the setting of t-operators, with any fuzzy negation N:

(i) Let $I_{F,N}(x,y) = F(N(x),y)$, $x,y \in [0,1]$. Then

$$I_{F,N}(0,0) = F(N(0),0) = F(1,0) = k .$$

Now, if $I_{F,N}$ were to satisfy (I3), i.e., $I_{F,N}(0,0) = 1$, it would fix F to be a t-conorm. Hence $I_{F,N}$ reduces to an (S,N)-implication, $I_{S,N}$.

(ii) Similarly, let

$$I_{F_1,F_2,N}(x,y) = F_1(N(x),F_2(x,y)) , \qquad x,y \in [0,1] ,$$

where F_1, F_2 are any two t-operators with annihilators $k_1, k_2 \in [0,1]$. Then

$$I_{F_1,F_2,N}(0,0) = F_1(N(0),F_2(0,0)) = F_1(1,0) = k_1.$$

Once again (I3) fixes F_1 to be a t-conorm. However,

$$I_{F_1,F_2,N}(1,0) = F_1(N(1),F_2(1,0)) = F_1(0,k_2) = k_2 ,$$

since F_1 is a t-conorm. Now, (I5) insists that $I_{F_1,F_2,N}(1,0) = 0$, which would fix F_2 to be a t-norm, i.e., $I_{F_1,F_2,N}$ reduces to a QL-implication in the setting of t-norms and t-conorms.

(iii) Let us consider now the residual of a t-operator F given as follows:

$$I_F(x,y) = \sup\{t \in [0,1] \mid F(x,t) \leq y\}\,, \qquad x,y \in [0,1]\,.$$

By the definition of a t-operator, we have $F(0,t) = t$, when $t \in [0,k]$ and $F(0,t) = k$, when $t \in [k,1]$. Clearly, this implies that $I_F(0,0) = 0$ and hence I_F cannot be a fuzzy implication.

Recently, there has been a lot of interest on non-commuatative fuzzy conjunctions and disjunctions. One of the earliest studies along these lines was done by FODOR and KERESZTFALVI [104]. Such operators again have proven to be a fertile ground for obtaining fuzzy implications. For example, (S,N)-type implications from copulas/co-copulas were obtained by YAGER [262], while the residuals from semi-copulas, quasi-copulas and copulas were investigated in DURANTE et al. [95], whereas WANG and FANG [253] discuss the residual operations on a complete lattice obtained from left and right uninorms, which are non-commutative.

Part II

Algebraic Study of Fuzzy Implications

Structures are the weapons of the mathematician.
– Nicolas Bourbaki

Let us consider the set of all fuzzy implications, denoted by $\mathcal{FI}$. Since these are basically functions it is both interesting and revealing to study the algebraic structures that can be imposed on $\mathcal{FI}$. This study has been done in Chap. 6 along the following lines:

- Firstly, in Sect. 6.1, we study the lattice structure that exists on $\mathcal{FI}$, by the partial order that is induced on $\mathcal{FI}$ by the natural order in $[0, 1]$.
- Once again, regarding fuzzy implication as functions, we can study their convex combinations and closures. This study is done in Sect. 6.2.
- Let $\varphi \in \Phi$. Now, if $I \in \mathcal{FI}$, then as shown before $I_\varphi \in \mathcal{FI}$. By defining a Φ-conjugacy relation $\sim$ between two fuzzy implications $I, J \in \mathcal{FI}$ as $I \sim J \iff I = (J)_\varphi$ for some $\varphi \in \Phi$, we can study the structure imposed by this relation on $\mathcal{FI}$. Sect. 6.3 explores this approach in detail.
- Since a fuzzy implication is a function from $[0, 1]^2$ to $[0, 1]$, it can be considered as a fuzzy relation on $[0, 1]^2$. Since the composition of fuzzy relations is a fuzzy relation, if $I, J \in \mathcal{FI}$, then it is worthwhile studying whether the composition of two fuzzy implications belong to $\mathcal{FI}$. If so, for which compositions and the subsequent structure imposed on $\mathcal{FI}$. This study is taken up in Sect. 6.4.

It should be emphasized that an interesting fallout of the above study is that each of the above 4 approaches gives us yet other ways of generating fuzzy implications from already existing ones.

The classical implication is involved in many a tautology in that logic. Once again, straightforward generalizations of these tautologies do not hold for all the fuzzy operations involved in them. This necessitates a thorough and deeper study of these tautologies which transform into functional equations involving fuzzy operations. Of all such generalized equivalences four have received maximum attention owing to their applicational value, viz., contrapositive symmetry, the law of importation, distributivity of fuzzy implications over t-norms and t-conorms and T-conditionality. In Chap. 7 we investigate these equivalences for the main families of fuzzy implications introduced in Part 1 of this treatise.

6 Algebraic Structures of Fuzzy Implications

Lattices, like groups, have a fascinating
algebraic theory of their own ...
provide the best algebraic tool for treating combinatorial
and structural problems ...
and play a central role in Universal algebra.
– Garrett Birkhoff (1911-1996)

6.1 Lattice of Fuzzy Implications

As noted in Sect. 1.1, in the family $\mathcal{FI}$ of all fuzzy implications we can consider the partial order induced from the unit interval $[0,1]$. It is interesting and important to note that incomparable pairs of fuzzy implications generate new fuzzy implications by using the standard min (inf) and max (sup) operations. This is another method of generating new fuzzy implications from the given ones.

Theorem 6.1.1. *The family* $(\mathcal{FI}, \leq)$ *is a complete, completely distributive lattice with the lattice operations*

$$(I \vee J)(x,y) := \max(I(x,y), J(x,y)) , \qquad x,y \in [0,1] , \tag{6.1}$$

$$(I \wedge J)(x,y) := \min(I(x,y), J(x,y)) , \qquad x,y \in [0,1] , \tag{6.2}$$

where $I, J \in \mathcal{FI}$.

Proof. Let $I_t \in \mathcal{FI}$ for $t \in \mathcal{T} \neq \emptyset$ and let $y \in [0,1]$ be fixed. Because of (I1) functions $f_{y,t} \colon [0,1] \to [0,1]$ defined by

$$f_{y,t}(x) := I_t(x,y) , \qquad x \in [0,1]$$

are decreasing for all $t \in \mathcal{T}$. Similarly, if $x \in [0,1]$ is fixed, then by (I2), functions $h_{x,t} \colon [0,1] \to [0,1]$ defined by

$$h_{x,t}(y) := I_t(x,y) , \qquad y \in [0,1]$$

are increasing for all $t \in \mathcal{T}$. Thus, by Lemma A.0.1 in Appendix ??, functions $\sup_{t\in\mathcal{T}} I_t$, $\inf_{t\in\mathcal{T}} I_t$ are monotonic. Since (I3), (I4) and (I5) are valid for all functions I_t, we get

$$\sup_{t\in\mathcal{T}} I_t(0,0) = \sup_{t\in\mathcal{T}} I_t(1,1) = \sup_{t\in\mathcal{T}} 1 = 1 ,$$

$$\sup_{t\in\mathcal{T}} I_t(1,0) = \sup_{t\in\mathcal{T}} 0 = 0 .$$

M. Baczyński and B. Jayaram: Fuzzy Implications, STUDFUZZ 231, pp. 183–205, 2008.
springerlink.com

The last equalities are also true for the infimum, so $\sup_{t\in\mathcal{T}} I_t, \inf_{t\in\mathcal{T}} I_t \in \mathcal{FI}$. Therefore, $(\mathcal{FI}, \leq)$ is a complete lattice.

According to [36], Chap. V, family $[0,1]^X$ of all functions $f\colon X \to [0,1]$ forms a completely distributive lattice. Thus $(\mathcal{FI}, \leq)$ is its complete sublattice for $X = [0,1]^2$, and we get that $(\mathcal{FI}, \leq)$ is a completely distributive lattice. □

Since a composition of continuous functions is continuous, we get

Theorem 6.1.2. *The family $(\mathcal{CFI}, \leq)$ is a distributive lattice (sublattice of $(\mathcal{FI}, \leq)$).*

However, the lattice $(\mathcal{CFI}, \leq)$ is not complete. It follows from the well-known fact that certain sequences of continuous functions have limits which are not continuous (cf. Example 6.3.10).

Example 6.1.3. Incomparable pairs of implications from Table 1.3 generate new implications in the distributive lattice of fuzzy implications. Such obtained elements can be combined with other implications, which leads to the following distributive lattice for $I_{\mathbf{LK}}$, $I_{\mathbf{GD}}$, $I_{\mathbf{RC}}$, $I_{\mathbf{KD}}$, $I_{\mathbf{GG}}$, $I_{\mathbf{RS}}$, $I_{\mathbf{WB}}$ and $I_{\mathbf{FD}}$ which, in addition, consists of the following 15 fuzzy implications:

$$H_1 = I_{\mathbf{GG}} \vee I_{\mathbf{RC}}\,, \qquad H_1(x,y) = \begin{cases} 1, & \text{if } x \leq y \\ \max(\frac{y}{x}, 1 - x + xy), & \text{if } x > y \end{cases}$$

$$H_2 = I_{\mathbf{GG}} \wedge I_{\mathbf{RC}}\,, \qquad H_2(x,y) = \begin{cases} 1 - x + xy, & \text{if } x \leq y \\ \min(\frac{y}{x}, 1 - x + xy), & \text{if } x > y \end{cases}$$

$$H_3 = I_{\mathbf{GD}} \vee I_{\mathbf{RC}}\,, \qquad H_3(x,y) = \begin{cases} 1, & \text{if } x \leq y \\ 1 - x + xy, & \text{if } x > y \end{cases}$$

$$H_4 = I_{\mathbf{GD}} \wedge I_{\mathbf{RC}}\,, \qquad H_4(x,y) = \begin{cases} 1 - x + xy, & \text{if } x \leq y \\ y, & \text{if } x > y \end{cases}$$

$$H_5 = I_{\mathbf{GG}} \vee I_{\mathbf{KD}}\,, \qquad H_5(x,y) = \begin{cases} 1, & \text{if } x \leq y \\ \max(\frac{y}{x}, 1 - x), & \text{if } x > y \end{cases}$$

$$H_6 = I_{\mathbf{GG}} \wedge I_{\mathbf{KD}}\,, \qquad H_6(x,y) = \begin{cases} \max(1 - x, y), & \text{if } x \leq y \\ \min(\frac{y}{x}, \max(1 - x, y)), & \text{if } x > y \end{cases}$$

$$H_7 = I_{GD} \wedge I_{\mathbf{KD}}\,, \qquad H_7(x,y) = \begin{cases} \max(1 - x, y), & \text{if } x \leq y \\ y, & \text{if } x > y \end{cases}$$

$$H_8 = H_5 \wedge I_{\mathbf{RC}}\,,$$

$$H_8(x,y) = \begin{cases} 1-x+xy, & \text{if } x \le y \\ \max(1-x, \min(\frac{y}{x}, 1-x+xy)), & \text{if } x > y \end{cases}$$

$$H_9 = I_{\mathbf{FD}} \wedge I_{\mathbf{RC}}\,, \qquad H_9(x,y) = \begin{cases} 1-x+xy, & \text{if } x \le y \\ \max(1-x, y), & \text{if } x > y \end{cases}$$

$$H_{10} = H_3 \wedge I_{\mathbf{GG}}\,, \qquad H_{10}(x,y) = \begin{cases} 1, & \text{if } x \le y \\ \min(\frac{y}{x}, 1-x+xy), & \text{if } x > y \end{cases}$$

$$H_{11} = I_{\mathbf{FD}} \wedge I_{\mathbf{GG}}\,, \qquad H_{11}(x,y) = \begin{cases} 1, & \text{if } x \le y \\ \min(\frac{y}{x}, \max(1-x, y)), & \text{if } x > y \end{cases}$$

$$H_{12} = H_3 \wedge H_5\,,$$

$$H_{12}(x,y) = \begin{cases} 1, & \text{if } x \le y \\ \min(1-x+xy, \max(\frac{y}{x}, 1-x)), & \text{if } x > y \end{cases}$$

$$H_{13} = H_4 \vee H_6\,, \qquad H_{13}(x,y) = \begin{cases} 1-x+xy, & \text{if } x \le y \\ \min(\frac{y}{x}, \max(1-x, y)), & \text{if } x > y \end{cases}$$

$$H_{14} = I_{\mathbf{RC}} \wedge I_{\mathbf{RS}}\,, \qquad H_{14}(x,y) = \begin{cases} 1-x+xy, & \text{if } x \le y \\ 0, & \text{if } x > y \end{cases}$$

$$H_{15} = I_{\mathbf{KD}} \wedge I_{\mathbf{RS}}\,, \qquad H_{15}(x,y) = \begin{cases} \max(1-x, y), & \text{if } x \le y \\ 0, & \text{if } x > y \end{cases}.$$

The lattice of fuzzy implications generated by $I_{\mathbf{LK}}$, $I_{\mathbf{GD}}$, $I_{\mathbf{RC}}$, $I_{\mathbf{KD}}$, $I_{\mathbf{GG}}$, $I_{\mathbf{RS}}$, $I_{\mathbf{WB}}$ and $I_{\mathbf{FD}}$ is presented in Fig. 6.1.

The following result shows that most of the basic properties are preserved by the above lattice operations.

Proposition 6.1.4. *If* $I, J\colon [0,1]^2 \to [0,1]$ *satisfy* (NP) *(*(IP), (OP)*), then* $I \vee J$ *and* $I \wedge J$ *also satisfy* (NP) *(*(IP), (OP)*).*

Proof. Firstly, let us assume that I, J satisfy (NP). Then

$$(I \vee J)(1,y) = \max(I(1,y), J(1,y)) = \max(y,y) = y\,,$$
$$(I \wedge J)(1,y) = \min(I(1,y), J(1,y)) = \min(y,y) = y\,,$$

for all $y \in [0,1]$.

Similarly, if I, J satisfy (IP), then for all $x \in [0,1]$, we have

$$(I \vee J)(x,x) = \max(I(x,x), J(x,x)) = \max(1,1) = 1\,,$$
$$(I \wedge J)(x,x) = \min(I(x,x), J(x,x)) = \min(1,1) = 1\,.$$

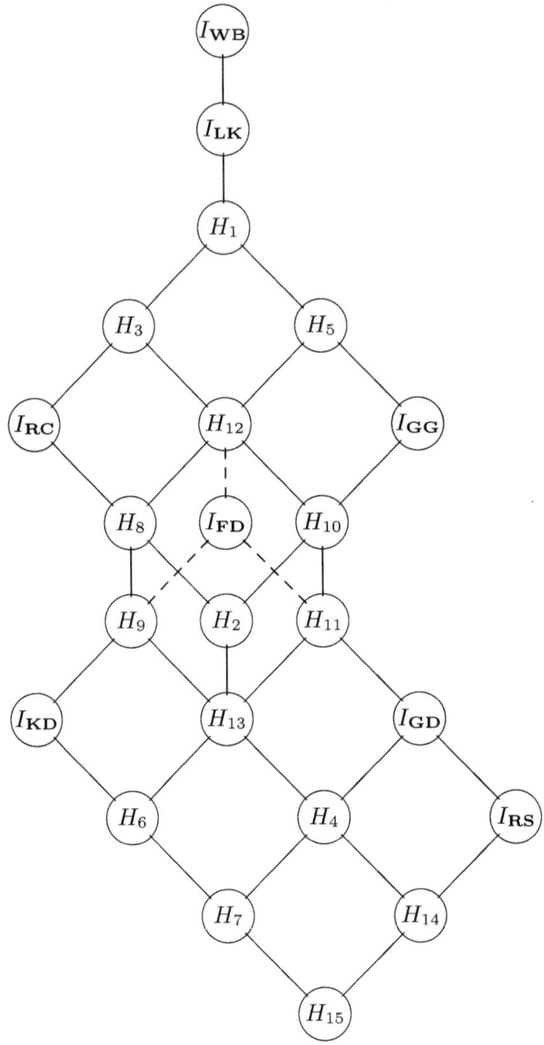

Fig. 6.1. The lattice generated by $I_{\mathbf{LK}}$, $I_{\mathbf{GD}}$, $I_{\mathbf{RC}}$, $I_{\mathbf{KD}}$, $I_{\mathbf{GG}}$, $I_{\mathbf{RS}}$, $I_{\mathbf{WB}}$ and $I_{\mathbf{FD}}$

Finally, assume that I, J satisfy (OP) and take any $x, y \in [0,1]$. We obtain

$$\begin{aligned} x \le y &\Longrightarrow I(x,y) = 1 \text{ and } J(x,y) = 1 \\ &\Longrightarrow \max(I(x,y), J(x,y)) = 1 \text{ and } \min(I(x,y), J(x,y)) = 1 \\ &\Longrightarrow (I \vee J)(x,y) = 1 \text{ and } (I \wedge J)(x,y) = 1 \,. \end{aligned}$$

On the other side, if $(I \vee J)(x,y) = 1$, then $\max(I(x,y), J(x,y)) = 1$, thus $I(x,y) = 1$ or $J(x,y) = 1$, which implies that $x \le y$. A similar proof holds for $I \wedge J$. □

It should be noted that the above fact can be generalized to the infimum and supremum operations.

Remark 6.1.5. One can easily show that (EP) is not generally preserved by the lattice operations. Let us consider the Goguen implication $I_{\mathbf{GG}}$ and the Reichenbach implication $I_{\mathbf{RC}}$, both of which satisfy (EP). Then $I_{\mathbf{GG}} \vee I_{\mathbf{RC}}$ and $I_{\mathbf{GG}} \wedge I_{\mathbf{RC}}$ are, respectively, fuzzy implications H_1 and H_2 given in Example 6.1.3. However, we have

$$H_1(0.7, H_1(0.5, 0.2)) = \frac{6}{7} \neq \frac{22}{25} = H_1(0.5, H_1(0.7, 0.2)) ,$$

and

$$H_2(0.9, H_2(0.5, 0.4)) = \frac{73}{100} \neq \frac{13}{18} = H_2(0.5, H_2(0.9, 0.4)) ,$$

i.e., H_1 and H_2 do not satisfy (EP). In particular, this implies that if I, J are two (S,N)-implications, then $I \vee J$ and $I \wedge J$ are not necessarily (S,N)-implications. The same holds for R-implications generated from left-continuous t-norms, f- and g-implications.

Finally, we discuss the lattice structure of the reciprocals of fuzzy implications.

Proposition 6.1.6. *Let N be a strong negation. If $I, J \in \mathcal{FI}$, then the operation defined by* (1.17) *is order preserving (isotone), i.e.,*

$$I \leq J \Longrightarrow I_N \leq J_N , \tag{6.3}$$

and lattice order preserving, i.e.,

$$(I \vee J)_N = I_N \vee J_N , \qquad (I \wedge J)_N = I_N \wedge J_N . \tag{6.4}$$

Moreover,

$$(\sup_{t \in \mathcal{T}} I_t)_N = \sup_{t \in \mathcal{T}} (I_N)_t , \qquad (\inf_{t \in \mathcal{T}} I_t)_N = \inf_{t \in \mathcal{T}} (I_N)_t , \tag{6.5}$$

for $I_t \in \mathcal{FI}$, $t \in \mathcal{T}$, where $\mathcal{T}$ is any non-empty index set.

Proof. Since (1.17) is a simple substitution, we have (6.3). In particular, we get

$$\sup_{t \in \mathcal{T}} (I_N)_t \leq (\sup_{t \in \mathcal{T}} I_t)_N , \qquad (\inf_{t \in \mathcal{T}} I_t)_N \leq \inf_{t \in \mathcal{T}} (I_N)_t .$$

Moreover, operation (1.17) is involutive i.e., it satisfies (1.18), and therefore it is a bijection. Thus, simultaneously,

$$\sup_{t \in \mathcal{T}} I_t \leq (\sup_{t \in \mathcal{T}} (I_N)_t)_N , \qquad \inf_{t \in \mathcal{T}} ((I_N)_t)_N \leq \inf_{t \in \mathcal{T}} I_t ,$$

which proves (6.5), and (6.4) is a particular case of (6.5). □

This proposition implies that the operation (1.17) preserves the lattice structure in $\mathcal{FI}$. In particular, the reciprocal image of the diagram from Fig. 6.1 has the same form. Now, from Theorem 6.1.1 and Proposition 6.1.6, we have

Theorem 6.1.7. *The set of all contrapositive fuzzy implications is a complete, completely distributive lattice, and the set of all continuous contrapositive fuzzy implications is a distributive lattice.*

6.2 Convex Classes of Fuzzy Implications

In this section, considering fuzzy implications, once again, as functions we study their convex combinations and show that they are closed under such operations.

Lemma 6.2.1. *If $f, g\colon [a,b] \to \mathbb{R}$ are monotonic of the same kind, then for every $\lambda \in [0,1]$ the function $h\colon [a,b] \to \mathbb{R}$ defined as*

$$h = \lambda f + (1-\lambda) g ,$$

is also monotonic of the same kind. Moreover, if f, g are continuous functions, then h is also continuous.

Proof. If f, g are increasing functions, then, for all $\lambda \in [0,1]$, we have $\lambda \geq 0$, $(1-\lambda) \geq 0$, thus

$$h(x) = \lambda f(x) + (1-\lambda) g(x) \leq \lambda f(y) + (1-\lambda) g(y) = h(y) ,$$

for every $x, y \in [a,b]$ such that $x \leq y$. Hence, h is increasing. The proofs for decreasing and constant functions are similar. □

The following result shows that the family of (continuous) fuzzy implications is closed with respect to convex combinations.

Theorem 6.2.2. *$\mathcal{FI}$ and $\mathcal{CFI}$ are convex sets of functions.*

Proof. Let $I, J \in \mathcal{FI}$ and $\lambda \in [0,1]$. From Lemma 6.2.1, the following function $K = \lambda \cdot I + (1-\lambda) \cdot J$ is monotonic. Now,

$$\begin{aligned} K(0,0) &= \lambda \cdot I(0,0) + (1-\lambda) \cdot J(0,0) = \lambda + (1-\lambda) = 1 , \\ K(1,1) &= \lambda \cdot I(1,1) + (1-\lambda) \cdot J(1,1) = \lambda + (1-\lambda) = 1 , \\ K(1,0) &= \lambda \cdot I(1,0) + (1-\lambda) \cdot J(1,0) = 0 + 0 = 0 , \end{aligned}$$

i.e., $K \in \mathcal{FI}$. Of course, if functions I, J are continuous, then the function K is also continuous. □

The above theorem brings a tool for the generation of parameterized families of fuzzy implications.

Example 6.2.3. Let us consider the first segment in the chain (1.2). Observe that it can be parameterized as follows:

$$J^{\lambda} = \lambda I_{\mathbf{GD}} + (1-\lambda) I_{\mathbf{RS}} , \qquad J^{\lambda}(x,y) = \begin{cases} 1, & \text{if } x \leq y , \\ \lambda y, & \text{if } x > y , \end{cases}$$

for $\lambda, x, y \in [0,1]$. If we consider the second segment in the same chain, then we get the following parameterized family of fuzzy implications:

$$K^{\lambda} = \lambda I_{\mathbf{GG}} + (1-\lambda) I_{\mathbf{GD}} , \qquad K^{\lambda}(x,y) = \begin{cases} 1, & \text{if } x \leq y , \\ \lambda \frac{y}{x} + (1-\lambda) y, & \text{if } x > y , \end{cases}$$

for $\lambda, x, y \in [0,1]$. Similarly, any pair of fuzzy implications can be connected by a parameterized segment of fuzzy implications.

As is the case with the lattice operations, convex combination of fuzzy implications preserves most of the basic properties.

Proposition 6.2.4. *If $I, J\colon [0,1]^2 \to [0,1]$ satisfy* (NP) *((*IP*),* (OP)*), then, for any $\lambda \in [0,1]$, the convex combination of I and J also satisfies* (NP) *((*IP*),* (OP)*).*

Proof. Since the case for (NP) and (IP) are straightforward, we show only that convex combination preserves (OP). Assume firstly, that I, J satisfy (OP) and let $\lambda \in [0,1]$ be arbitrary but fixed. If $x, y \in [0,1]$ such that $x \leq y$ then by (OP) we have that

$$\lambda \cdot I(x,y) + (1-\lambda) \cdot J(x,y) = \lambda \cdot 1 + (1-\lambda) \cdot 1 = 1 .$$

On the other hand, if $\lambda \cdot I(x,y) + (1-\lambda) \cdot J(x,y) = 1$, for any $x, y \in [0,1]$, then since $\lambda, 1-\lambda \in [0,1]$ we have that $I(x,y) = J(x,y) = 1$, which by (OP) implies that $x \leq y$. □

Remark 6.2.5. Once again, (EP) is not generally preserved by convex combinations. Let us consider the convex combination K^λ (see Example 6.2.3) of $I_{\mathbf{GG}}$ and $I_{\mathbf{GD}}$, both of which satisfy (EP). Letting $\lambda = 0.25$, $x = 0.2, y = 0.6$ and $z = 0.18$, we have

$$K^\lambda(x, K^\lambda(y,z)) = K^\lambda(0.2, K^\lambda(0.6, 0.18)) = K^\lambda(0.2, 0.21) = 1 ,$$

while

$$K^\lambda(y, K^\lambda(x,z)) = K^\lambda(0.6, K^\lambda(0.2, 0.18)) = K^\lambda(0.6, 0.36) = 0.42 ,$$

i.e., K^λ, with $\lambda = 0.25$, does not satisfy (EP).

In fact, for any $\lambda \in (0,1)$ it can be shown that K^λ does not satisfy (EP). To see this, for any arbitrary $z \in (0,1)$ we can find a $y \in (0,1)$ such that $y > K^\lambda(K^\lambda(y,z), z)$. When this condition is satisfied, simply letting $x = K^\lambda(y,z)$ we have $K^\lambda(x, K^\lambda(y,z)) = 1$ by (OP), while $K^\lambda(y, K^\lambda(x,z)) < 1$, since $y > K^\lambda(x,z) = K^\lambda(K^\lambda(y,z), z)$. In fact, with a little calculation, it can be seen that any y satisfying the following quadratic inequality would suffice: $y^2 - y(z - \lambda z) - \lambda z > 0$. For any given $\lambda \in (0,1)$, by an appropriate choice of z the existence of such a y can be guaranteed.

In particular, this implies that if I, J are two (S,N)-implications, then their convex combinations are not necessarily (S,N)-implications. The same holds for R-implications generated from left-continuous t-norms, f- and g-implications.

Observe now that another way of generating contrapositive fuzzy implications is from convex combinations of given examples of contrapositive implications. Since formula (1.17) leads us to

$$(\lambda I + (1-\lambda) J)_N = \lambda I_N + (1-\lambda) J_N ,$$

for $I, J \in \mathcal{FI}$, $\lambda \in [0,1]$ and a fuzzy negation N, we get

Theorem 6.2.6. *The set of all contrapositive fuzzy implications is convex.*

6.3 Conjugacy Classes of Fuzzy Implications

In the previous chapters we have used the relation of Φ-conjugation in many places. Let us recall that two fuzzy implications I, J are Φ-conjugate, if there exists a $\varphi \in \Phi$ such that $J = I_\varphi$, where

$$I_\varphi(x,y) = \varphi^{-1}(I(\varphi(x), \varphi(y))) , \qquad x, y \in [0,1] . \tag{6.6}$$

Moreover, note that, from Propositions 1.1.7, 1.2.4 and 1.3.6, the standard Φ-conjugation preserves all the properties in Definition A.0.3, continuity and the 4 basic properties defined in Sect. 1.3. In fact, the following result can be proven.

Theorem 6.3.1. *For a bijection $\varphi\colon [0,1] \to [0,1]$ the following statements are equivalent:*

(i) $\varphi \in \Phi$, i.e., φ is an increasing bijection.
(ii) $I_\varphi \in \mathcal{FI}$ for all $I \in \mathcal{FI}$.

Proof. (*i*) $\Longrightarrow$ (*ii*) This implication is another version of Proposition 1.1.7.
(*ii*) $\Longrightarrow$ (*i*) Let us suppose that there exist $x, y \in [0,1]$ such that $x < y$ and $\varphi(x) > \varphi(y)$. Taking the Gödel implication $I_{\mathbf{GD}}$ we see that

$$\begin{aligned}(I_{\mathbf{GD}})_\varphi(x,y) &= \varphi^{-1}(I_{\mathbf{GD}}(\varphi(x), \varphi(y))) = \varphi^{-1}(\varphi(y)) = y ,\\ \varphi(1) &= \varphi((I_{\mathbf{GD}})_\varphi(1,1)) = I_{\mathbf{GD}}(\varphi(1), \varphi(1)) = 1 ,\\ (I_{\mathbf{GD}})_\varphi(x,x) &= \varphi^{-1}(I_{GD}(\varphi(x), \varphi(x))) = \varphi^{-1}(1) = 1 .\end{aligned}$$

Since $(I_{\mathbf{GD}})_\varphi \in \mathcal{FI}$, it satisfies (I2), thus

$$1 = (I_{\mathbf{GD}})_\varphi(x,x) \leq (I_{\mathbf{GD}})_\varphi(x,y) = y ,$$

so $\varphi(y) = \varphi(1) = 1$ contradictory to our assumption $\varphi(y) < \varphi(x) \leq 1$. This shows that φ is an increasing bijection. □

The above result shows why we consider only increasing bijections in the conjugation. In general, we can consider particular families of increasing bijections.

Definition 6.3.2. *Let Ψ denote a certain family of increasing bijections on $[0,1]$.*

(i) We say that $J \in \mathcal{FI}$ is Ψ-conjugate with $I \in \mathcal{FI}$ if $J = I_\varphi$ for some $\varphi \in \Psi$. We will denote this relation by $J \sim_\Psi I$, i.e., $\Psi \subset \Phi$.
(ii) We say that $I \in \mathcal{FI}$ is Ψ-self-conjugate if $I_\varphi = I$ for all $\varphi \in \Psi$. When $\Psi = \Phi$, we say that I is just self-conjugate or invariant.

Theorem 6.3.3. *Let Ψ denote a certain family of increasing bijections on $[0,1]$. Then the following are equivalent:*

(i) $\sim_\Psi$ is an equivalence relation.
(ii) $(\Psi, \circ)$ is a group of increasing bijections.

Proof. (*i*) $\Longrightarrow$ (*ii*) Let Ψ be a fixed family of increasing bijections on $[0,1]$. Assume that the relation $\sim_\Psi$ is an equivalence. Let $I = I_{\mathbf{LK}}$ and $\varphi, \psi \in \Psi$ be arbitrarily fixed. Firstly, letting $J = I_\varphi$ and $K = J_\psi$ we see that $K \sim_\Psi J$ and $J \sim_\Psi I$. Thus, by the transitivity $K \sim_\Psi I$, i.e., $K = I_\chi$, for some $\chi \in \Psi$. Thus $K = J_\psi = (I_\varphi)_\psi = I_{\varphi\circ\psi} = I_\chi$. Since the increasing bijection is uniquely determined for the Łukasiewicz implication (cf. Theorem 2.4.20), we obtain $\varphi \circ \psi = \chi \in \Psi$. Therefore, the composition $\circ$ is an interior operation in Ψ, i.e., $(\Psi, \circ)$ is a semigroup. Next, by the reflexivity, $I \sim_\Psi I$, i.e., $I = I_\varphi$, for some $\varphi \in \Psi$. But $I = I_{\mathrm{id}_{[0,1]}}$, and we get $\mathrm{id}_{[0,1]} = \varphi \in \Psi$, i.e., the semigroup $(\Psi, \circ)$ has the identity element $\mathrm{id}_{[0,1]}$. Finally, putting $J = I_\varphi$, since $J \sim_\Psi I$, by the symmetry also $I \sim_\Psi J$, i.e., $I = J_\psi$, for some $\psi \in \Psi$. Thus,

$$I = J_\psi = (I_\varphi)_\psi = I_{\varphi\circ\psi} = I_{\mathrm{id}_{[0,1]}}.$$

This implies that $\varphi\circ\psi = \mathrm{id}_{[0,1]}$ and, similarly, $\psi\circ\varphi = \mathrm{id}_{[0,1]}$. Thus, $\varphi^{-1} = \psi \in \Psi$, and all elements of semigroup $(\Psi, \circ)$ are invertible. This proves that $(\Psi, \circ)$ is a group.

(*ii*) $\Longrightarrow$ (*ii*) Let us assume that $(\Psi, \circ)$ is a group. Firstly $I_{\mathrm{id}_{[0,1]}} = I$, since $\varphi = \mathrm{id}_{[0,1]} \in \Psi$, so $I \sim_\Psi I$, i.e., the relation $\sim_\Psi$ is reflexive. Next, since $\varphi \in \Psi$ implies $\varphi^{-1} \in \Psi$, we get that $J \sim_\Psi I$ implies $I \sim_\Psi J$, i.e., the relation $\sim_\Psi$ is symmetric. Finally, if $K \sim_\Psi J$ and $J \sim_\Psi I$, then $K = J_\psi$ and $J = I_\varphi$ for some $\varphi, \psi \in \Psi$. Therefore we get $K = J_\psi = (I_\varphi)_\psi = I_{\varphi\circ\psi} = I_\chi$, where $\chi = \varphi \circ \psi \in \Psi$. Thus, $K \sim_\Psi I$, i.e., the relation $\sim_\Psi$ is transitive. This proves that relation $\sim_\Psi$ is an equivalence and finishes the proof. □

The following result shows that Ψ-conjugation preserves the partial order and the lattice order in $\mathcal{FI}$.

Theorem 6.3.4. *If $I, J \in \mathcal{FI}$ and $\varphi \in \Phi$, then*

$$I \le J \Longleftrightarrow I_\varphi \le J_\varphi\,, \tag{6.7}$$

i.e., the conjugation is order preserving (isotone), and

$$(I \vee J)_\varphi = I_\varphi \vee J_\varphi\,, \qquad (I \wedge J)_\varphi = I_\varphi \wedge J_\varphi\,, \tag{6.8}$$

i.e., the conjugation is lattice order preserving. Moreover,

$$(\sup_{t\in\mathcal{T}} I_t)_\varphi = \sup_{t\in\mathcal{T}} (I_t)_\varphi\,, \qquad (\inf_{t\in\mathcal{T}} I_t)_\varphi = \inf_{t\in\mathcal{T}} (I_t)_\varphi\,, \tag{6.9}$$

for $I_t \in \mathcal{FI}$, $t \in \mathcal{T}$, where $\mathcal{T}$ is any non-empty index set.

Proof. Let $x, y \in [0,1]$. If φ is an increasing bijection, then

$$\begin{aligned} I(x,y) \le J(x,y) &\Longleftrightarrow I(\varphi(x),\varphi(y)) \le J(\varphi(x),\varphi(y)) \\ &\Longleftrightarrow \varphi^{-1}(I(\varphi(x),\varphi(y))) \le \varphi^{-1}(J(\varphi(x),\varphi(y))) \\ &\Longleftrightarrow I_\varphi(x,y) \le J_\varphi(x,y)\,, \end{aligned}$$

which proves (6.7).

Let us consider the second equality in (6.9). Using (6.7), we get

$$(\inf_{t\in\mathcal{T}} I_t)_\varphi \leq (I_s)_\varphi\,, \qquad \text{for every } s\in\mathcal{T}\,. \tag{6.10}$$

Thus, we obtain

$$(\inf_{t\in\mathcal{T}} I_t)_\varphi \leq \inf_{t\in\mathcal{T}} (I_t)_\varphi\,. \tag{6.11}$$

Further, from (6.10) and (6.7), we have that

$$\begin{aligned}
&(\inf_{t\in\mathcal{T}}(I_t)_\varphi)_{\varphi^{-1}} \leq I_s\,, \ \text{for every } s\in\mathcal{T}\\
&\Longrightarrow (\inf_{t\in\mathcal{T}}(I_t)_\varphi)_{\varphi^{-1}} \leq \inf_{t\in\mathcal{T}} I_t\\
&\Longrightarrow \inf_{t\in\mathcal{T}}(I_t)_\varphi \leq (\inf_{t\in\mathcal{T}} I_t)_\varphi\,,
\end{aligned}$$

which together with (6.11) gives the second equality in (6.9). The first equality in (6.9) can be proven in a similar way. Finally, (6.8) is a particular case of (6.9), which finishes the proof. □

Corollary 6.3.5. *Let $I, J \in \mathcal{FI}$. If J is invariant, then*

$$I \leq J \Longrightarrow (I_\varphi \leq J,\ \textit{for all } \varphi\in\Phi)\,, \tag{6.12}$$
$$I \geq J \Longrightarrow (I_\varphi \geq J,\ \textit{for all } \varphi\in\Phi)\,. \tag{6.13}$$

Remark 6.3.6. The invariance of the fuzzy implication J in Corollary 6.3.5 is important, since without it, even if J were comparable to another fuzzy implication I, the Φ-conjugate of I need not be comparable to J. To see this, let us consider the Łukasiewicz implication $I_{\mathbf{LK}}$, which is not an invariant fuzzy implication, and the Kleene-Dienes implication $I_{\mathbf{KD}}$. We know, from Example 1.1.5, that $I_{\mathbf{KD}} < I_{\mathbf{LK}}$. Let $\varphi(x) = \sqrt{x}$, for $x\in[0,1]$. Putting $x=\frac{1}{4}$, $y=\frac{1}{16}$ we get

$$(I_{\mathbf{LK}})_\varphi\left(\frac{1}{4},\frac{1}{16}\right) = \left(\frac{3}{4}\right)^2 < \frac{3}{4} = I_{\mathbf{KD}}\left(\frac{1}{4},\frac{1}{16}\right)\,.$$

On the other side,

$$(I_{\mathbf{LK}})_\varphi\left(\frac{1}{16},\frac{1}{4}\right) = 1 > \frac{15}{16} = I_{\mathbf{KD}}\left(\frac{1}{16},\frac{1}{4}\right)\,,$$

which shows that $I_{\mathbf{KD}}$ and $(I_{\mathbf{LK}})_\varphi$ are not comparable.

Theorem 6.3.7. *Let $I, J \in \mathcal{FI}$, $\varphi,\psi\in\Phi$. If I_φ and J_ψ are comparable, then I_χ and J are also comparable, where $\chi = \varphi\circ\psi^{-1}$. In particular,*

$$(J_\psi \leq I_\varphi) \Longleftrightarrow (J \leq I_\chi)\,, \tag{6.14}$$
$$(J_\psi = I_\varphi) \Longleftrightarrow (J = I_\chi)\,. \tag{6.15}$$

Proof. Let $x, y \in [0,1]$ and $u = \psi(x)$, $v = \psi(y)$. Then we get

$$\begin{aligned} J_\psi(x,y) \leq I_\varphi(x,y) &\iff \psi^{-1}(J(\psi(x),\psi(y))) \leq \varphi^{-1}(I(\varphi(x),\varphi(y))) \\ &\iff J(\psi(x),\psi(y)) \leq \psi(\varphi^{-1}(I(\varphi(x),\varphi(y)))) \\ &\iff J(u,v) \leq \psi(\varphi^{-1}(I(\varphi(\psi^{-1}(u)),\varphi(\psi^{-1}(v))))) \\ &\iff J(u,v) \leq \chi^{-1}(I(\chi(u)),\chi(v)) \\ &\iff J(u,v) \leq I_\chi(u,v) \,, \end{aligned}$$

where $\chi = \varphi \circ \psi^{-1}$. This proves (6.14) and implies (6.15). □

From Theorems 6.1.1, 6.3.4 and basic properties of lattices, we get the following results related to invariant (self-conjugate) implications.

Theorem 6.3.8. *The set of all invariant fuzzy implications is a distributive lattice. Moreover, this lattice is finite and consists of the following* 18 *implications* J_1 - J_{18}*:*

$$J_1 = I_1 \,,$$

$$J_2 = I_{\mathbf{WB}} \,,$$

$$J_3(x,y) = \begin{cases} 1, & \textit{if } x = 0 \textit{ or } y > 0 \\ 0, & \textit{otherwise} \end{cases} \,,$$

$$J_4(x,y) = \begin{cases} 1, & \textit{if } x < 1 \textit{ or } y = 1 \\ 0, & \textit{otherwise} \end{cases} \,,$$

$$J_5(x,y) = \begin{cases} 1, & \textit{if } (x < 1 \textit{ and } y > 0) \textit{ or } x = 0 \\ y, & \textit{if } x = 1 \\ 0, & \textit{otherwise} \end{cases} \,,$$

$$J_6(x,y) = \begin{cases} 1, & \textit{if } (x < 1 \textit{ and } y > 0) \textit{ or } x = 0 \textit{ or } y = 1 \\ 0, & \textit{otherwise} \end{cases} \,,$$

$$J_7 = I_{\mathbf{GD}} \,,$$

$$J_8(x,y) = \begin{cases} 1, & \textit{if } x \leq y \\ y, & \textit{if } y < x < 1 \\ 0, & \textit{otherwise} \end{cases} \,,$$

$$J_9(x,y) = \begin{cases} 1, & \textit{if } x < y \textit{ or } x = 0 \\ y, & \textit{otherwise} \end{cases} \,,$$

$$J_{10} = I_{\mathbf{RS}} \,,$$

$$J_{11}(x,y) = \begin{cases} 1, & \textit{if } x < y \textit{ or } x = 0 \textit{ or } y = 1 \\ y, & \textit{if } y \leq x \textit{ and } 0 < x < 1 \\ 0, & \textit{otherwise} \end{cases} \,,$$

$$J_{12}(x,y) = \begin{cases} 1, & \textit{if } x = 0 \\ y, & \textit{otherwise} \end{cases},$$

$$J_{13}(x,y) = \begin{cases} 1, & \textit{if } x < y \textit{ or } x = 0 \\ y, & \textit{if } x = y \textit{ and } x > 0 \\ 0, & \textit{otherwise} \end{cases},$$

$$J_{14}(x,y) = \begin{cases} 1, & \textit{if } x = 0 \textit{ or } y = 1 \\ y, & \textit{if } 0 < x < 1 \\ 0, & \textit{otherwise} \end{cases},$$

$$J_{15}(x,y) = \begin{cases} 1, & \textit{if } x < y \textit{ or } x = 0 \textit{ or } y = 1 \\ 0, & \textit{otherwise} \end{cases},$$

$$J_{16}(x,y) = \begin{cases} 1, & \textit{if } x = 0 \\ y, & \textit{if } 0 < x \le y \\ 0, & \textit{otherwise} \end{cases},$$

$$J_{17}(x,y) = \begin{cases} 1, & \textit{if } x = 0 \textit{ or } y = 1 \\ y, & \textit{if } 0 < x < y < 1 \\ 0, & \textit{otherwise} \end{cases},$$

$$J_{18} = I_0 ,$$

for all $x, y \in [0,1]$.

The lattice of all invariant fuzzy implications was presented by DREWNIAK [80, 81] and is given in Fig. 6.2. The author has also shown that this lattice can be generated by the following 6 fuzzy implications: J_3, J_4, J_{10}, J_{12}, J_{15} and J_{18}.

Theorem 6.3.9. *The greatest lower bound and the least upper bound of any Φ-conjugacy class exist and are included in the family of all invariant fuzzy implications.*

Proof. Let $I \in \mathcal{FI}$, $\psi \in \Phi$, $x, y \in [0,1]$ and

$$J = \sup_{\varphi \in \Phi} I_\varphi , \qquad K = \inf_{\varphi \in \Phi} I_\varphi . \tag{6.16}$$

Firstly, from Theorems 6.1.1 and 6.3.1, we see that $J, K \in \mathcal{FI}$. Since $\psi \in \Phi$ is continuous, we get

$$\begin{aligned} J_\psi(x,y) &= \psi^{-1}(J(\psi(x),\psi(y))) = \psi^{-1}(\sup_{\varphi \in \Phi} I_\varphi(\psi(x),\psi(y))) \\ &= \sup_{\varphi \in \Phi} \psi^{-1}(\varphi^{-1}(I(\varphi(\psi(x)),\varphi(\psi(y))))) \\ &= \sup_{\chi \in \Phi \circ \psi} I_\chi(x,y) = \sup_{\chi \in \Phi} I_\chi(x,y) = J(x,y) , \end{aligned}$$

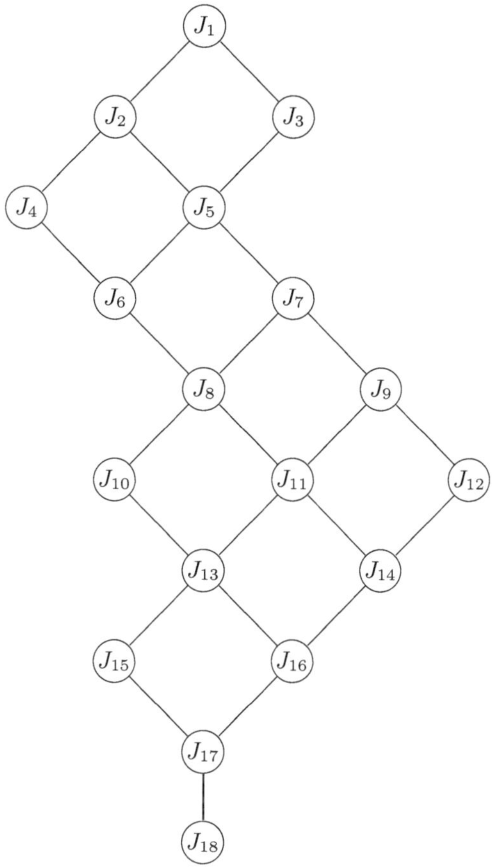

Fig. 6.2. The lattice of invariant fuzzy implications J_1-J_{18}

because $\Phi \circ \psi = \Phi$ in the group $(\Phi, \circ)$. Therefore, $J_\psi = J$ for any $\psi \in \Phi$, which proves that J is an invariant fuzzy implication. Similar arguments prove that the fuzzy implication K is also invariant. □

Example 6.3.10. Let us consider the 9 basic fuzzy implications from Table 1.3. Firstly, observe that the Gödel implication $I_{\mathbf{GD}}$, the Rescher implication $I_{\mathbf{RS}}$ and the Weber implication $I_{\mathbf{WB}}$ are invariant, so

$$\inf_{\varphi\in\Phi}(I_{\mathbf{GD}})_\varphi = \sup_{\varphi\in\Phi}(I_{\mathbf{GD}})_\varphi = I_{\mathbf{GD}} ,$$

$$\inf_{\varphi\in\Phi}(I_{\mathbf{RS}})_\varphi = \sup_{\varphi\in\Phi}(I_{\mathbf{RS}})_\varphi = I_{\mathbf{RS}} ,$$

$$\inf_{\varphi\in\Phi}(I_{\mathbf{WB}})_\varphi = \sup_{\varphi\in\Phi}(I_{\mathbf{WB}})_\varphi = I_{\mathbf{WB}} .$$

For the other 6 basic fuzzy implications, we have

$$\inf_{\varphi\in\Phi}(I_{\mathbf{LK}})_\varphi = I_{\mathbf{GD}} , \qquad \sup_{\varphi\in\Phi}(I_{\mathbf{LK}})_\varphi = I_{\mathbf{WB}} ,$$
$$\inf_{\varphi\in\Phi}(I_{\mathbf{RC}})_\varphi = J_{12} , \qquad \sup_{\varphi\in\Phi}(I_{\mathbf{RC}})_\varphi = I_{\mathbf{WB}} ,$$
$$\inf_{\varphi\in\Phi}(I_{\mathbf{KD}})_\varphi = J_{12} , \qquad \sup_{\varphi\in\Phi}(I_{\mathbf{KD}})_\varphi = I_{\mathbf{WB}} ,$$
$$\inf_{\varphi\in\Phi}(I_{\mathbf{GG}})_\varphi = I_{\mathbf{GD}} , \qquad \sup_{\varphi\in\Phi}(I_{\mathbf{GG}})_\varphi = J_5 ,$$
$$\inf_{\varphi\in\Phi}(I_{\mathbf{YG}})_\varphi = J_{12} , \qquad \sup_{\varphi\in\Phi}(I_{\mathbf{YG}})_\varphi = J_5 ,$$
$$\inf_{\varphi\in\Phi}(I_{\mathbf{FD}})_\varphi = I_{\mathbf{GD}} , \qquad \sup_{\varphi\in\Phi}(I_{\mathbf{FD}})_\varphi = I_{\mathbf{WB}} .$$

The above results can be proven in three steps. Firstly, observe that

$$K \le I \le J , \tag{6.17}$$

for the suitable fuzzy implications I, J, K (e.g. $I_{\mathbf{GD}} \le I_{\mathbf{LK}} \le I_{\mathbf{WB}}$). Next, (6.17) and Corollary 6.3.5 imply that

$$K \le I_\varphi \le J , \tag{6.18}$$

for all $\varphi \in \Phi$. Finally, we choose $\psi_n, \varphi_n \in \Phi$ such that

$$\lim_{n\to\infty} I_{\psi_n} = K , \qquad \lim_{n\to\infty} I_{\varphi_n} = J , \tag{6.19}$$

which will prove our results. We complete this proof by presenting some suitable sequences of bijections that can be considered in (6.19):

$$\psi_n(x) = 1-(1-x)^n , \qquad \varphi_n(x) = x^n , \qquad \text{for } I_{\mathbf{LK}} ,$$
$$\psi_n(x) = \sqrt[n]{x} , \qquad \varphi_n(x) = x^n , \qquad \text{for } I_{\mathbf{RC}}, I_{\mathbf{KD}} , I_{\mathbf{YG}} , I_{\mathbf{FD}} ,$$
$$\psi_n(x) = 1-(1-x)^n , \qquad \varphi_n(x) = 1-\sqrt[n]{1-x} , \qquad \text{for } I_{\mathbf{GG}} ,$$

where $n \in \mathbb{N}$ and $x \in [0,1]$.

Finally, we show that all 9 basic fuzzy implications from Example 1.1.4 are not Φ-conjugate to each other.

Example 6.3.11. (i) Let us consider the Łukasiewicz implication $I_{\mathbf{LK}}$. Since it is the only continuous fuzzy implication from Table 1.3 which satisfies all the 4 basic properties (cf. Table 1.4), we see that it is not Φ-conjugate with any other fuzzy implication from Table 1.3.

(ii) Let us consider the Gödel implication $I_{\mathbf{GD}}$. Since it is an invariant implication, it is not Φ-conjugate with any other fuzzy implication from Table 1.3. The same applies to the Rescher implication $I_{\mathbf{RS}}$ and the Weber implication $I_{\mathbf{WB}}$, since they are also invariant.

(iii) Let us consider the Reichenbach implication $I_{\mathbf{RC}}$. Since it is continuous and satisfies only (NP) and (EP) (cf. Table 1.4), we see that it cannot be Φ-conjugate with $I_{\mathbf{LK}}$, $I_{\mathbf{GD}}$, $I_{\mathbf{GG}}$, $I_{\mathbf{RS}}$, $I_{\mathbf{YG}}$, $I_{\mathbf{WB}}$ and $I_{\mathbf{FD}}$. Finally assume that $I_{\mathbf{RC}}$ is Φ-conjugate with the Kleene-Dienes implication $I_{\mathbf{KD}}$. Since both functions are S-implications, from Theorem 2.4.5 and Table 2.4 we get that the maximum t-conorm $S_{\mathbf{M}}$ is Φ-conjugate with the probabilistic sum t-conorm $S_{\mathbf{P}}$, a contradiction. Thus $I_{\mathbf{RC}}$ is not Φ-conjugate with $I_{\mathbf{KD}}$.

(iv) Let us consider the Kleen-Dienes implication $I_{\mathbf{KD}}$. Since it is continuous and satisfies only (NP) and (EP) (cf. Table 1.4), we see that it cannot be Φ-conjugate with $I_{\mathbf{LK}}$, $I_{\mathbf{GD}}$, $I_{\mathbf{GG}}$, $I_{\mathbf{RS}}$, $I_{\mathbf{YG}}$, $I_{\mathbf{WB}}$ and $I_{\mathbf{FD}}$. From the previous point it is not Φ-conjugate with $I_{\mathbf{RC}}$.

(v) Let us consider the Goguen implication $I_{\mathbf{GG}}$. Since it is not continuous and satisfies all the 4 basic properties (cf. Table 1.4), we see that it cannot be Φ-conjugate with $I_{\mathbf{LK}}$, $I_{\mathbf{RC}}$, $I_{\mathbf{KD}}$, $I_{\mathbf{RS}}$, $I_{\mathbf{YG}}$ and $I_{\mathbf{WB}}$. From point (ii) above it is not Φ-conjugate with $I_{\mathbf{GD}}$. Finally assume that $I_{\mathbf{GG}}$ is Φ-conjugate with the Fodor implication $I_{\mathbf{FD}}$. Since both functions are R-implications, from Theorem 2.5.10 and Table 2.6 we get that the product t-norm $T_{\mathbf{P}}$, which is strict, is Φ-conjugate with the nilpotent minimum t-norm $T_{\mathbf{nM}}$, which is not strict, a contradiction. Thus $I_{\mathbf{GG}}$ is not Φ-conjugate with $I_{\mathbf{FD}}$.

(vi) Let us consider the Yager implication $I_{\mathbf{YG}}$. Since it is the only fuzzy implication in Table 1.4 which is neither continuous nor satisfies (IP), we get that $I_{\mathbf{YG}}$ is not Φ-conjugate with any other fuzzy implication from Table 1.3.

(vii) Taking into consideration all the previous points, we get that $I_{\mathbf{FD}}$ is not Φ-conjugate with any other fuzzy implication from Table 1.3.

6.4 Semigroups of Fuzzy Implications

As noted earlier, fuzzy implications were also introduced as truth space fuzzy relations on $[0,1]$ (see ZADEH [266]). When there is more than one relation composition of such relations is a common operation. Since fuzzy implications are a particular case of fuzzy relations on $[0,1]$, one can compose two fuzzy implications to obtain a fuzzy *relation* - as we will see below it is not always a fuzzy implication. In this section we explore this approach and show that such an investigation eventually leads to an algebraic structure on the family of fuzzy implications $\mathcal{FI}$.

6.4.1 Composition of Fuzzy Implications

Although the sup − min composition of fuzzy relations was the first to be introduced by ZADEH [266], the min can be replaced by any t-norm T. In the following, for notational convenience and enhanced readability, we use the infix notation and hence use $*$ instead of T for a t-norm.

Definition 6.4.1. *Let $I, J \in \mathcal{FI}$ and $*$ be a t-norm. The* $\sup -*$ *composition of I, J is given as follows:*

$$(I \overset{*}{\circ} J)(x,z) = \sup_{y\in[0,1]} \left(I(x,y) * J(y,z)\right) , \qquad x,z \in [0,1] . \tag{6.20}$$

As shown below, the composition $\overset{*}{\circ}$ preserves many properties of a fuzzy implication.

Proposition 6.4.2. *Let $I, J \in \mathcal{FI}$ and $*$ be a t-norm. Then $I \overset{*}{\circ} J$ satisfies* (I1), (I2), (I3), (I4), (LB) *and* (NC).

Proof. Let $x, w, z \in [0,1]$ such that $x \le w$. Then by (I1) of I and the monotonicity of the t-norm $*$ we have the following inequalities, for all $y \in [0,1]$:

$$I(x,y) \ge I(w,y) \Longrightarrow I(x,y) * J(y,z) \ge I(w,y) * J(y,z) ,$$

from which we obtain that

$$\begin{aligned}(I \overset{*}{\circ} J)(x,z) &= \sup_{y\in[0,1]} \left(I(x,y) * J(y,z)\right) \\ &\ge \sup_{y\in[0,1]} \left(I(w,y) * J(y,z)\right) = (I \overset{*}{\circ} J)(w,z) ,\end{aligned}$$

i.e., $I \overset{*}{\circ} J$ satisfies (I1). The fact that $I \overset{*}{\circ} J$ satisfies (I2) can be shown similarly.
Since $I, J \in \mathcal{FI}$ they satisfy (LB), (NC) and we have

$$\begin{aligned}(I \overset{*}{\circ} J)(0,z) &= \sup_{y\in[0,1]} \left(I(0,y) * J(y,z)\right) \\ &= \sup_{y\in[0,1]} \left(1 * J(y,z)\right) = \sup_{y\in[0,1]} J(y,z) = J(0,z) = 1 , \\ (I \overset{*}{\circ} J)(x,1) &= \sup_{y\in[0,1]} \left(I(x,y) * J(y,1)\right) \\ &= \sup_{y\in[0,1]} \left(I(x,y) * 1\right) = \sup_{y\in[0,1]} I(x,y) = I(x,1) = 1 ,\end{aligned}$$

i.e., $I \overset{*}{\circ} J$ satisfies (I3), (I4), (LB) and (NC). □

Remark 6.4.3. It should be noted that, in general, $I \overset{*}{\circ} J$ does not satisfy (I5) and hence (NP). Let us consider the Łukasiewicz and the Kleene-Dienes implications. Then

$$(I_{\mathbf{LK}} \overset{*}{\circ} I_{\mathbf{KD}})(1,0) = \sup_{y\in[0,1]} \left(y * (1-y)\right) .$$

If $*$ is the minimum t-norm $T_{\mathbf{M}}$, then we see that $(I_{\mathbf{LK}} \overset{T_{\mathbf{M}}}{\circ} I_{\mathbf{KD}})(1,0) = 0.5 > 0$, while if $*$ is the Łukasiewicz t-norm $T_{\mathbf{LK}}$, then $(I_{\mathbf{LK}} \overset{T_{\mathbf{LK}}}{\circ} I_{\mathbf{KD}})(1,0) = 0$.

In virtue of Proposition 6.4.2, Remark 6.4.3 and the definition of a fuzzy implication, we have the following result, the proof of which is immediate if one observes that it only associates boundary values of the considered fuzzy implications.

Theorem 6.4.4. *Let $*$ be a t-norm. If $I, J \in \mathcal{FI}$, then*

$$I \overset{*}{\circ} J \in \mathcal{FI} \iff (I \overset{*}{\circ} J)(1,0) = 0 . \tag{6.21}$$

Example 6.4.5. Let $I, J \in \mathcal{FI}$. If for any $t \in (0,1)$ we have $I(1,t) = J(t,0) = 0$, then for any $y \in [0,1]$,

$$\begin{aligned} I(1,y) &= 0 , \qquad \text{for } y \leq t , \\ J(y,0) &= 0 , \qquad \text{for } y \geq t , \end{aligned}$$

and we get

$$(I \overset{*}{\circ} J)(1,0) = \max(\sup_{y\in[0,t]} (0 * J(y,0)), \sup_{y\in[t,1]} (I(1,y) * 0)) = 0 .$$

Therefore $I \overset{*}{\circ} J \in \mathcal{FI}$.

Remark 6.4.6. Note that the composition in (6.20) is not commutative. To see this, let us consider the Gödel implication $I_{\mathbf{GD}}$ and the least fuzzy implication I_0 (see Proposition 1.1.6). Then, for every t-norm $*$, it can be easily verified that

$$I_{\mathbf{GD}} \overset{*}{\circ} I_0 = I_0 ,$$

while

$$(I_0 \overset{*}{\circ} I_{\mathbf{GD}})(x,y) = \begin{cases} 1, & \text{if } x = 0 , \\ y, & \text{otherwise} . \end{cases}$$

Now, we look for subfamilies of $\mathcal{FI}$ closed under the composition (6.20), or equivalently those subfamilies of $\mathcal{FI}$ that satisfy (6.21). Towards this end, we define the following subfamilies.

Definition 6.4.7. *Let $*$ be a t-norm. An implication $I \in \mathcal{FI}$ is called a*

(i) left ideal *if*

$$I \overset{*}{\circ} J \in \mathcal{FI} , \qquad \textit{for all } J \in \mathcal{FI} , \tag{LF}$$

(ii) right ideal *if*

$$J \overset{*}{\circ} I \in \mathcal{FI} , \qquad \textit{for all } J \in \mathcal{FI} , \tag{RF}$$

(iii) central ideal *if it is both a left and right ideal.*

The subfamilies of left, right and central ideals will be denoted by $\mathcal{FI}_{\mathbf{L}}$, $\mathcal{FI}_{\mathbf{R}}$ and $\mathcal{FI}_{\mathbf{C}}$, respectively.

In fact, as the following results show, the above families have a very simple characterization and are independent of a t-norm $*$.

Lemma 6.4.8. *Let $*$ be a t-norm and $I \in \mathcal{FI}$. Consider the following properties:*

$$I(1, y) = 0 , \qquad \text{for all } y \in (0, 1) , \tag{6.22}$$
$$I(y, 0) = 0 , \qquad \text{for all } y \in (0, 1) . \tag{6.23}$$

(i) If I satisfies (6.22) then, for all $z \in [0, 1]$ and for all $J \in \mathcal{FI}$,

$$(I \overset{*}{\circ} J)(1, z) = J(1, z) . \tag{6.24}$$

(ii) If I satisfies (6.23) then, for all $x \in [0, 1]$ and for all $J \in \mathcal{FI}$,

$$(J \overset{*}{\circ} I)(x, 0) = J(x, 0) . \tag{6.25}$$

Proof. (i) From (6.20) and (6.22) we get, for all $z \in [0, 1]$,

$$\begin{aligned}(I \overset{*}{\circ} J)(1, z) &= \max(\sup_{y<1}(I(1, y) * J(y, z)), (I(1, 1) * J(1, z))) \\ &= J(1, z) .\end{aligned}$$

(ii) Once again from (6.20) and (6.23) we get, for all $x \in [0, 1]$,

$$\begin{aligned}(J \overset{*}{\circ} I)(x, 0) &= \max((J(x, 0) * I(0, 0)), \sup_{y>0}(J(x, y) * I(y, 0))) \\ &= J(x, 0) .\end{aligned}$$

□

In fact, (6.22) and (6.23) characterize families of left and right ideals.

Theorem 6.4.9. *The subfamilies of left, right and central ideals are given by*

$$\begin{aligned}\mathcal{FI}_{\mathbf{L}} &= \{I \in \mathcal{FI} \mid I(1, x) = 0, \ \textit{for all } x \in (0, 1)\} , \\ \mathcal{FI}_{\mathbf{R}} &= \{I \in \mathcal{FI} \mid I(x, 0) = 0, \ \textit{for all } x \in (0, 1)\} , \\ \mathcal{FI}_{\mathbf{C}} &= \mathcal{FI}_{\mathbf{L}} \cap \mathcal{FI}_{\mathbf{R}} .\end{aligned}$$

Proof. Let $*$ be a t-norm and $I, J \in \mathcal{FI}$. By Lemma 6.4.8 we see that (6.22) implies $(I \overset{*}{\circ} J)(1, 0) = 0$ and from (6.23) we have $(J \overset{*}{\circ} I)(1, 0) = 0$. Thus, by Theorem 6.4.4, we get (LF), i.e., (6.22) implies $I \in \mathcal{FI}_{\mathbf{L}}$ and (6.23) implies $I \in \mathcal{FI}_{\mathbf{R}}$.

Now, let $I \in \mathcal{FI}$ satisfy (LF). Putting $J = I_{\mathbf{LK}}$ we see that

$$0 = (I \overset{*}{\circ} I_{\mathbf{LK}})(1, 0) = \max(\sup_{y \in [0,1)} (I(1, y) * (1 - y)), 0) ,$$

and therefore we get (6.22). Similarly,

$$0 = (I_{\mathbf{LK}} \overset{*}{\circ} I)(1, 0) = \max(0, \sup_{y \in (0,1]} (y * I(y, 0))) ,$$

implies (6.23). Thus, conditions (6.22) and (6.23) are necessary. The last equality follows from the definition of $\mathcal{FI}_{\mathbf{C}}$. □

Remark 6.4.10. (i) Notice that the set of all right ideals $\mathcal{FI}_{\mathbf{R}}$ is nothing but all those fuzzy implications I whose natural negation is the Gödel negation, i.e., $N_I = N_{\mathbf{D1}}$.
(ii) Once again it is obvious that if I satisfies (NP), then $I \notin \mathcal{FI}_{\mathbf{L}}$.
(iii) Since any $I \in \mathcal{FI}$ satisfies (I3), (I4), from (6.22) and (6.23), we see that if $I \in \mathcal{FI}_{\mathbf{L}} \cup \mathcal{FI}_{\mathbf{R}}$, then I is not continuous.
(iv) With regards the 9 basic fuzzy implications we have the following inclusions, which can be seen from the properties they satisfy (see Table 1.4):

$$\begin{aligned} I_{\mathbf{GD}}, I_{\mathbf{GG}}, I_{\mathbf{YG}} &\in \mathcal{FI}_{\mathbf{R}} \setminus \mathcal{FI}_{\mathbf{L}} , \\ I_{\mathbf{RS}} &\in \mathcal{FI}_{\mathbf{C}} , \\ I_{\mathbf{LK}}, I_{\mathbf{RC}}, I_{\mathbf{KD}}, I_{\mathbf{WB}}, I_{\mathbf{FD}} &\notin \mathcal{FI}_{\mathbf{R}} \cup \mathcal{FI}_{\mathbf{L}} . \end{aligned}$$

(v) On the other hand, see that the following $I \in \mathcal{FI}_{\mathbf{L}} \setminus \mathcal{FI}_{\mathbf{R}}$, thus showing that, indeed, $\mathcal{FI}_{\mathbf{R}}$ and $\mathcal{FI}_{\mathbf{L}}$ are different:

$$I(x, y) = \begin{cases} 0, & \text{if } x = 1 \text{ and } y < 1 , \\ 1 - x, & \text{if } y = 0 , \\ 1, & \text{otherwise} . \end{cases}$$

From Lemma 6.4.8 and Theorem 6.4.9 we have the following result. We only recall that the composition in (6.20) is not commutative (see Remark 6.4.6).

Theorem 6.4.11. *If $*$ is a t-norm, then the following containments exist among the subfamilies of ideals:*

$$\begin{aligned} \mathcal{FI}_{\mathbf{L}} \overset{*}{\circ} \mathcal{FI}_{\mathbf{L}} &\subset \mathcal{FI}_{\mathbf{L}} , \\ \mathcal{FI}_{\mathbf{R}} \overset{*}{\circ} \mathcal{FI} &\subset \mathcal{FI}_{\mathbf{R}} , \\ \mathcal{FI}_{\mathbf{C}} \overset{*}{\circ} \mathcal{FI}_{\mathbf{L}} &\subset \mathcal{FI}_{\mathbf{L}} \cap \mathcal{FI}_{\mathbf{R}} = \mathcal{FI}_{\mathbf{C}} , \\ \mathcal{FI}_{\mathbf{L}} \overset{*}{\circ} \mathcal{FI}_{\mathbf{C}} &\subset \mathcal{FI}_{\mathbf{L}} , \\ \mathcal{FI}_{\mathbf{C}} \overset{*}{\circ} \mathcal{FI}_{\mathbf{R}} &\subset \mathcal{FI}_{\mathbf{R}} , \\ \mathcal{FI}_{\mathbf{R}} \overset{*}{\circ} \mathcal{FI}_{\mathbf{C}} &\subset \mathcal{FI}_{\mathbf{R}} . \end{aligned}$$

6.4.2 Semigroups of Fuzzy Implications

Note that in Sect. 6.1 our main aim was to impose a lattice structure on the family $\mathcal{FI}$ and as a by-product came up with a way of obtaining new fuzzy implications from given fuzzy implications. On the other hand, so far in Sect. 6.4.1 we have explicitly seen how to generate newer fuzzy implications from given fuzzy implications through sup $-*$ composition, by treating fuzzy implications merely as binary fuzzy relations on $[0, 1]$.

In the reminder of this section, we investigate the converse, i.e., does sup $-*$ composition of fuzzy implications give rise to any algebraic structure on the family of fuzzy implications $\mathcal{FI}$? Happily, we can give an affirmative answer to this question, as we show below.

From Theorem 6.4.11, we see that for every t-norm $*$ the subfamilies $\mathcal{FI}_\mathbf{L}$, $\mathcal{FI}_\mathbf{R}$ and $\mathcal{FI}_\mathbf{C}$ are all closed with respect to the composition $\overset{*}{\circ}$ in (6.20). The following result gives a sufficient condition for the composition in (6.20) to be associative.

Theorem 6.4.12. *If the $*$ in the composition in* (6.20) *is a left-continuous t-norm, then $\overset{*}{\circ}$ is associative, i.e., if $I, J, K \in \mathcal{FI}$ then $I \overset{*}{\circ} (J \overset{*}{\circ} K) = (I \overset{*}{\circ} J) \overset{*}{\circ} K$.*

Proof. Let $I, J, K \in \mathcal{FI}$ and let the $*$ in (6.20) be a left-continuous t-norm. For any $x, z \in [0,1]$, we have the following equalities:

$$
\begin{aligned}
(I \overset{*}{\circ} (J \overset{*}{\circ} K))(x,z) &= \sup_{p\in[0,1]} \Big(I(x,p) * (J \overset{*}{\circ} K)(p,z) \Big) \\
&= \sup_{p\in[0,1]} \left(I(x,p) * \big(\sup_{y\in[0,1]} (J(p,y) * K(y,z)) \big) \right) \\
&= \sup_{p,y\in[0,1]} (I(x,p) * (J(p,y) * K(y,z))) \\
&= \sup_{p,y\in[0,1]} ((I(x,p) * J(p,y)) * K(y,z)) \\
&= \sup_{y\in[0,1]} \left(\sup_{p\in[0,1]} ((I(x,p) * J(p,y))) * K(y,z) \right) \\
&= \sup_{y\in[0,1]} \Big((I \overset{*}{\circ} J)(x,y) * K(y,z) \Big) \\
&= ((I \overset{*}{\circ} J) \overset{*}{\circ} K)(x,z) .
\end{aligned}
$$

□

It should be noted that if the t-norm $*$ in the composition in (6.20) is not left-continuous, then the above result does not hold in general (cf. Drewniak and Kula [82], Example 5).

From the monotonicity of supremum and the t-norm $*$, we see that $\overset{*}{\circ}$ is also monotone, i.e., if $I, J, K \in \mathcal{FI}$ such that $I \le K$ then $I \overset{*}{\circ} J \le K \overset{*}{\circ} J$. Recall from Theorem 6.1.1, that $(\mathcal{FI}, \le)$ is a complete, completely distributive lattice. Now, we are ready to state the main result of this section.

Theorem 6.4.13. *If $*$ is a left-continuous t-norm, then the subfamilies of ideals $(\mathcal{FI}_\mathbf{L}, \overset{*}{\circ}, \le)$, $(\mathcal{FI}_\mathbf{R}, \overset{*}{\circ}, \le)$, $(\mathcal{FI}_\mathbf{C}, \overset{*}{\circ}, \le)$ are ordered semigroups of fuzzy implications. Moreover, $(\mathcal{FI}_\mathbf{L}, \le)$, $(\mathcal{FI}_\mathbf{R}, \le)$, $(\mathcal{FI}_\mathbf{C}, \le)$ are also complete, completely distributive lattices.*

Proof. From Theorems 6.4.11 and 6.4.12 we see that $(\mathcal{FI}_\mathbf{L}, \overset{*}{\circ}, \le)$, $(\mathcal{FI}_\mathbf{R}, \overset{*}{\circ}, \le)$, $(\mathcal{FI}_\mathbf{C}, \overset{*}{\circ}, \le)$ are ordered semigroups of fuzzy implications.

From Theorem 6.1.1, $(\mathcal{FI}, \le)$ is a complete, completely distributive lattice. Moreover, (6.22) and (6.23) are preserved by lattice operations. For instance, if

$I(1,x)=0$ and $J(1,x)=0$, then $(I \vee J)(1,x)=0$, $(I \wedge J)(1,x)=0$. Therefore, the subfamilies $(\mathcal{FI}_{\mathbf{L}}, \leq)$, $(\mathcal{FI}_{\mathbf{R}}, \leq)$, $(\mathcal{FI}_{\mathbf{C}}, \leq)$ are sublattices of $(\mathcal{FI}, \leq)$. □

Example 6.4.14. Semigroups from Theorem 6.4.13 are bounded (as complete lattices) and, as can be easily verified, their bounds are as follows:

$$\min \mathcal{FI}_{\mathbf{L}} = \min \mathcal{FI}_{\mathbf{R}} = \min \mathcal{FI}_{\mathbf{C}} = \min \mathcal{FI} = I_0 \,,$$

$$\max \mathcal{FI}_{\mathbf{L}} = I_{\mathbf{LU}} \,, \qquad \max \mathcal{FI}_{\mathbf{R}} = I_{\mathbf{RU}} \,, \qquad \max \mathcal{FI}_{\mathbf{C}} = I_{\mathbf{LU}} \wedge I_{\mathbf{RU}} \,,$$

where, for $x, y \in [0,1]$,

$$I_{\mathbf{LU}}(x,y) = \begin{cases} 1, & \text{if } x < 1 \text{ or } y = 1 \,, \\ 0, & \text{otherwise} \,, \end{cases}$$

$$I_{\mathbf{RU}}(x,y) = \begin{cases} 1, & \text{if } x = 0 \text{ or } y > 0 \,, \\ 0, & \text{otherwise} \,. \end{cases}$$

As mere fuzzy relations, we see that the identity relation is the neutral element of the composition in (6.20), but it is not a fuzzy implication. The family of fuzzy implications $\mathcal{FI}$ has its own neutral element of composition (6.20).

Theorem 6.4.15. *For an arbitrary t-norm $*$, the Rescher implication $I_{\mathbf{RS}}$ is the neutral element of the composition* (6.20) *in $\mathcal{FI}$. Moreover, if the operation $*$ is left-continuous, then the semigroups from Theorem 6.4.13 also have as their neutral element $I_{\mathbf{RS}}$.*

Proof. For arbitrary $I \in \mathcal{FI}$, $x, z \in [0,1]$, we have

$$(I \overset{*}{\circ} I_{\mathbf{RS}})(x,z) = \sup_{y \leq z} I(x,y) = I(x,z) \,,$$

$$(I_{\mathbf{RS}} \overset{*}{\circ} I)(x,z) = \sup_{y \geq x} I(y,z) = I(x,z) \,,$$

which proves that $I \overset{*}{\circ} I_{\mathbf{RS}} = I_{\mathbf{RS}} \overset{*}{\circ} I = I$. The rest is a consequence of Theorem 6.4.13. □

Theorem 6.4.16. *The implication I_0 is the zero element in $\mathcal{FI}_{\mathbf{C}}$, the right zero in $\mathcal{FI}_{\mathbf{R}}$, and the left zero in $\mathcal{FI}_{\mathbf{L}}$.*

Proof. Let $I \in \mathcal{FI}$, $x \in (0,1]$, $z \in [0,1)$. From (6.20) and the definition of I_0 (see Proposition 1.1.6), we have

$$(I_0 \overset{*}{\circ} I)(x,z) = \sup_{y \in [0,1)} (I_0(x,y) * I(y,z)) \vee I_0(x,1) * I(1,z) = I(1,z) \,,$$

$$(I \overset{*}{\circ} I_0)(x,z) = I(x,0) * I_0(0,z) \vee \sup_{y \in (0,1]} (I(x,y) * I_0(y,z)) = I(x,0) \,.$$

If $I \in \mathcal{FI}_{\mathbf{L}}$, then $(I_0 \overset{*}{\circ} I)(x,z) = 0 = I_0(x,z)$. If $I \in \mathcal{FI}_{\mathbf{R}}$, then $(I \overset{*}{\circ} I_0)(x,z) = 0 = I_0(x,z)$. For $x = 0$ or $z = 1$, from (LB) and (NC), we get that $I_0 \overset{*}{\circ} I = I_0$ in $\mathcal{FI}_{\mathbf{L}}$, $I \overset{*}{\circ} I_0 = I_0$ in $\mathcal{FI}_{\mathbf{R}}$, which finishes the proof. □

Finally, we investigate the effect of Φ-conjugacy on the subfamilies of ideals.

Theorem 6.4.17. *The operation* (6.6) *is an interior operation on the subfamilies of ideals, viz.,* $\mathcal{FI}_{\mathbf{L}}, \mathcal{FI}_{\mathbf{R}}$ *and* $\mathcal{FI}_{\mathbf{C}}$.

Proof. If $I \in \mathcal{FI}_{\mathbf{L}}$, $\varphi \in \Phi$, $z \in [0,1)$ then by Theorem 6.4.9 we have

$$I_\varphi(1,z) = \varphi^{-1}(I(\varphi(1),\varphi(z))) = \varphi^{-1}(I(1,\varphi(z))) = \varphi^{-1}(0) = 0\,,$$

i.e., $I_\varphi \in \mathcal{FI}_{\mathbf{L}}$. Similar calculation is valid for an $I \in \mathcal{FI}_{\mathbf{R}}$ and $I \in \mathcal{FI}_{\mathbf{C}}$. □

Corollary 6.4.18. *The subfamilies of ideals* $\mathcal{FI}_{\mathbf{L}}, \mathcal{FI}_{\mathbf{R}}$ *and* $\mathcal{FI}_{\mathbf{C}}$ *with every element* I *contain its* Φ*-conjugacy class too.*

Proposition 6.4.19. *The transformation* (6.6) *is* Φ*-distributive with respect to the composition* (6.20), *i.e., for all* $I, J \in \mathcal{FI}$ *and any* $\varphi \in \Phi$,

$$(I \overset{*}{\circ} J)_\varphi = I_\varphi \overset{*_\varphi}{\circ} J_\varphi\,. \tag{6.26}$$

Proof. Let $I, J \in \mathcal{FI}$, $\varphi \in \Phi$, $x, z \in [0,1]$. Since φ is a bijection on $[0,1]$, for every $y \in [0,1]$ there exists a $t \in [0,1]$ such that $\varphi(t) = y$ and we have

$$\begin{aligned}
(I \overset{*}{\circ} J)_\varphi(x,z) &= \varphi^{-1}((I \overset{*}{\circ} J)(\varphi(x),\varphi(z))) \\
&= \varphi^{-1}\left(\sup_{y\in[0,1]} I(\varphi(x),y) * J(y,\varphi(z))\right) \\
&= \varphi^{-1}\left(\sup_{t\in[0,1]} I(\varphi(x),\varphi(t)) * J(\varphi(t),\varphi(z))\right) \\
&= \sup_{t\in[0,1]} \varphi^{-1}\left(I(\varphi(x),\varphi(t)) * J(\varphi(t),\varphi(z))\right) \\
&= \sup_{t\in[0,1]} \varphi^{-1}\left(\varphi\circ\varphi^{-1}(I(\varphi(x),\varphi(t))) * \varphi\circ\varphi^{-1}(J(\varphi(t),\varphi(z)))\right) \\
&= \sup_{t\in[0,1]} I_\varphi(x,t) *_\varphi J_\varphi(t,z) = (I_\varphi \overset{*_\varphi}{\circ} J_\varphi)(x,z)\,,
\end{aligned}$$

which proves (6.26). □

Recall from Definition 6.3.2(ii) that an $I \in \mathcal{FI}$ is invariant if $I_\varphi = I$ for every $\varphi \in \Phi$. The following results follow immediately from Proposition 6.4.19 and Remark 2.1.4(viii).

Theorem 6.4.20. *If* $*$ *is the minimum t-norm* $T = T_{\mathbf{M}}$ *or the drastic t-norm* $T = T_{\mathbf{D}}$, *then the* $\sup -*$ *composition of invariant fuzzy implications is an invariant function. It is an invariant implication only if* (6.21) *is satisfied.*

Theorem 6.4.21. *Let* $*$ *be a t-norm and* $\varphi \in \Phi$. *If* $I \in \mathcal{FI}$ *is* $*$*-idempotent, i.e.,* $I \overset{*}{\circ} I = I$, *then* I_φ *is* $*_\varphi$*-idempotent.*

Remark 6.4.22. The sup $-*$ composition of conjugate fuzzy implications may not be conjugate with them, i.e., may belong to another conjugacy class. To see this, consider the invariant fuzzy implications J_3 and J_5 from Theorem 6.3.8 and take $* = T_{\mathbf{M}}$. One can easily calculate that $J_3 \overset{T_{\mathbf{M}}}{\circ} J_3 = J_5$ and $J_5 \overset{T_{\mathbf{M}}}{\circ} J_5 = J_3$. As invariant implications, J_3 and J_5 form singleton conjugacy classes.

6.5 Bibliographical Remarks

The lattice operations of 'meet' and 'join' were the first to be employed towards generating new fuzzy implications. BANDLER and KOHOUT [30] obtained fuzzy implications by taking the 'meet' and 'join' of a given fuzzy implication I and its reciprocal I_N, where N was a strong negation. These two techniques, called *upper* and *lower* contrapositivisation, impart contrapositive symmetry to a fuzzy implication I with respect to a strong negation N. These two techniques will be dealt with in detail, presently, in Sect. 7.1.

Although the family of all fuzzy implications $\mathcal{FI}$ is its own convex closure, if we consider some specific families of fuzzy implications, say $\mathbb{I}_{\mathbb{S},\mathbb{N}}$ or $\mathbb{I}_{\mathbb{F}}$, then, as can be seen from Remark 6.2.5, their convex closures are, in general, superset of the original families. Hence we have the following:

Problem 6.5.1. Characterize the convex closures of the following families of fuzzy implications: $\mathbb{I}_{\mathbb{S},\mathbb{N}}, \mathbb{I}_{\mathbb{T}}, \mathbb{I}_{\mathbb{F}}$ and $\mathbb{I}_{\mathbb{G}}$.

However, note that, recently some attempt has been made on exploring the algebraic structures that exist on the family of (S,N)-implications $\mathbb{I}_{\mathbb{S},\mathbb{N}}$ (see JAYARAM and BACZYŃSKI [129]).

Fuzzy implications play a central role in many inference schemes in approximate reasoning (Chap. 8 is entirely dedicated to this topic). The examination of compositions of fuzzy implications is very important in multistage decision making by approximate reasoning. It suffices to consider fuzzy implications as a particular case of fuzzy relations in $[0,1]$. In fact, such an approach was firstly taken by BALDWIN and PILSWORTH [27], where they study the suitability of fuzzy implications, with respect to some 'desirable' properties, in the context of approximate reasoning. They generated new fuzzy implications as a sup $-$ min composition of the given fuzzy implications.

Although all the methods presented in this chapter generate new fuzzy implications from given fuzzy implications, the algebraic structure they impose on $\mathcal{FI}$ has been extensively studied by BACZYŃSKI et al. [11, 13, 14], DREWNIAK [80, 81] and DREWNIAK and SOBERA [83].

7 Fuzzy Implications and Some Functional Equations

Algebra is, properly speaking, the Analysis of equations.
– Joseph Alfred Serret (1819-1885)

Generally speaking functional equations are equations in which the unknowns are functions. In the previous chapters we have seen some functional equations, viz., the exchange property (EP), the contrapositive symmetry (CP) and the like. But as Prof. Aczél writes in his book [1] "merely stating properties (functional equations) satisfied by a function is different from solving and determining all functions that satisfy a given functional equation".

In this chapter we deal with a few functional equations involving fuzzy implications. These equations, once again, arise as the generalizations of the corresponding tautologies in classical logic involving boolean implications.

We study the validity of these functional equations for the different families of fuzzy implications introduced in Chap. 2. Of the many properties or functional equations that involve implication operations, we have chosen four, namely, the contrapositive symmetry, distributivity over t-norms and t-conorms, the law of importation and T-conditionality. This choice is based on the theoretical and applicational value these properties impart to the fuzzy implications possessing them. Finally, a characterization of 8 of the 9 basic fuzzy implications is given based on some functional equations.

7.1 Contrapositive Symmetrization of Fuzzy Implications

In the framework of the classical two-valued logic, contrapositivity of a binary implication is a tautology. In the fuzzy logic, contrapositive symmetry of a fuzzy implication I with respect to strong negation N - CP(N) - plays an important role in the applications of fuzzy implications. Usually, the contrapositive symmetry of a fuzzy implication I is studied with respect to its natural negation N_I. However, not all fuzzy implications satisfy CP(N_I), even if their N_I is strong.

M. Baczyński and B. Jayaram: Fuzzy Implications, STUDFUZZ 231, pp. 207–240, 2008.
springerlink.com

Example 7.1.1. Let us consider the reciprocal Goguen implication (see Example 1.6.3) given by

$$I'_{\mathbf{GG}}(x,y) = \begin{cases} 1, & \text{if } x \leq y , \\ \frac{1-x}{1-y}, & \text{if } x > y , \end{cases} \qquad x,y \in [0,1] .$$

The natural negation of $I'_{\mathbf{GG}}$ is the classical negation $N_{\mathbf{C}}$ which is a strong negation, but $I'_{\mathbf{GG}}$ does not satisfy CP($N_{\mathbf{C}}$). Similarly, the natural negation of the fuzzy implication

$$I_{\mathbf{K}}(x,y) = \sqrt{1-x+xy^2} , \qquad x,y \in [0,1] ,$$

is $N_{I_{\mathbf{K}}}(x) = \sqrt{1-x}$, which is a strict but not strong negation, and hence $I_{\mathbf{K}}$ does not satisfy CP($N_{I_{\mathbf{K}}}$). In fact, since $I_{\mathbf{K}}$ satisfies (NP), by Corollary 1.5.5 we know that $I_{\mathbf{K}}$ does not satisfy (CP) with any fuzzy negation N.

In this section we investigate the different contrapositivisation techniques that have been proposed in the literature. In particular, we study in detail, the upper and lower contrapositivisation techniques proposed in BANDLER and KOHOUT [30] and studied by both FODOR [103] and BALASUBRAMANIAM [21], as also the medium contrapositivisation proposed therein which has a few more advantages than the former techniques.

7.1.1 Upper and Lower Contrapositivisations

BANDLER and KOHOUT [30] have proposed two techniques, viz., upper and lower contrapositivisation, towards imparting contrapositive symmetry to a fuzzy implication I with respect to a strong negation N, whose definitions we give below. Recall that by I_N we denote the N-reciprocal of a fuzzy implication I with respect to a negation N.

Definition 7.1.2. *Let $I \in \mathcal{FI}$ and N be a fuzzy negation. The functions $I_N^u, I_N^l \colon [0,1]^2 \to [0,1]$ defined as*

$$I_N^u(x,y) = \max(I(x,y), I_N(x,y)) = \max(I(x,y), I(N(y),N(x))) ,$$
$$I_N^l(x,y) = \min(I(x,y), I_N(x,y)) = \min(I(x,y), I(N(y),N(x))) ,$$

for all $x,y \in [0,1]$, are called, respectively, the upper and lower contrapositivisations *of I with respect to N.*

Lemma 7.1.3. *If $I \in \mathcal{FI}$ and N is a fuzzy negation, then I_N^u and I_N^l are both fuzzy implications. If, in addition, N is strong, then they satisfy CP(N).*

Proof. From Theorem 1.6.2, the function $I_N \in \mathcal{FI}$. By virtue of Theorem 6.1.1, the family $\mathcal{FI}$ is a lattice and so $I_N^u, I_N^l \in \mathcal{FI}$. Finally, if N is strong, then

$$\begin{aligned} I_N^u(N(y),N(x)) &= \max(I(N(y),N(x)), I(N(N(x)),N(N(y)))) \\ &= \max(I(N(y),N(x)), I(x,y)) \\ &= \max(I(x,y), I(N(y),N(x))) = I_N^u(x,y) , \end{aligned}$$

for all $x, y \in [0,1]$, i.e., I_N^u satisfies CP(N). A similar reasoning can be carried out for I_N^l. □

When a given fuzzy implication I satisfies (CP) with some strong negation N, then of course we do not obtain any new fuzzy implication, since

$$I_N^u = I_N^l = I ,$$

in this case. Therefore, the above methods are mainly important for fuzzy implications which do not satisfy (CP) with any strong negation.

Example 7.1.4. (i) Consider the Gödel implication $I_{\mathbf{GD}}$. Then

$$(I_{\mathbf{GD}})_{N_{\mathbf{C}}}^u(x,y) = I_{\mathbf{FD}}(x,y) ,$$

$$(I_{\mathbf{GD}})_{N_{\mathbf{C}}}^l(x,y) = \begin{cases} 1, & \text{if } x \leq y , \\ \min(1-x, y), & \text{otherwise} , \end{cases}$$

(ii) Consider the Goguen implication $I_{\mathbf{GG}}$. Then

$$(I_{\mathbf{GG}})_{N_{\mathbf{C}}}^u(x,y) = \begin{cases} 1, & \text{if } x \leq y , \\ \max\left(\frac{y}{x}, \frac{1-x}{1-y}\right), & \text{otherwise} , \end{cases}$$

$$(I_{\mathbf{GG}})_{N_{\mathbf{C}}}^l(x,y) = \begin{cases} 1, & \text{if } x \leq y , \\ \min\left(\frac{y}{x}, \frac{1-x}{1-y}\right), & \text{otherwise} , \end{cases}$$

Now we show which basic properties are preserved by the above contrapositivisation techniques.

Proposition 7.1.5. *Let N be a fuzzy negation. If $I \in \mathcal{FI}$ satisfies* (OP)*, then so does I_N^u and I_N^l.*

Proof. Let us assume firstly that $x, y \in [0,1]$ and $x \leq y$. Then $I(x,y) = 1$ from (OP) for the implication I. Moreover, $N(y) \leq N(x)$, thus $I(N(y), N(x)) = 1$. Now,

$$I_N^u(x,y) = \max(I(x,y), I(N(y), N(x))) = \max(1,1) = 1 .$$

Conversely, if $I_N^u(x,y) = 1$ for some $x, y \in [0,1]$, then either $I(x,y) = 1$ or $I(N(y), N(x)) = 1$. On the one hand, if $I(x,y) = 1$, then $x \leq y$ by (OP). On the other hand, if $I(N(y), N(x)) = 1$, then again $N(y) \leq N(x)$ by (OP), so $x \leq y$.

A similar proof can be given for I_N^l. □

Corollary 7.1.6. *Let N be a fuzzy negation. If $I \in \mathcal{FI}$ satisfies* (IP) *then so does I_N^u and I_N^l.*

Example 7.1.7. One can easily show that (EP) is not generally preserved by the above contrapositive techniques.

(i) Let us consider the Goguen implication $I_{\mathbf{GG}}$, which satisfies (EP). Then $(I_{\mathbf{GG}})_{N_{\mathbf{C}}}^u$ does not satisfy (EP).

(ii) Whereas the (S,N)-implication

$$I_{\mathbf{MK}}(x,y) = \max(1-x^2, y) ,$$

satisfies (EP), its lower contrapositivisation with respect $N_{\mathbf{C}}$

$$(I_{\mathbf{MK}})^l_{N_{\mathbf{C}}}(x,y) = \min\left(\max(1-x^2,y), \max(2y-y^2, 1-x)\right) ,$$

does not, since

$$(I_{\mathbf{MK}})^l_{N_{\mathbf{C}}}(0.2, (I_{\mathbf{MK}})^l_{N_{\mathbf{C}}}(0.3, 0.6)) = 0.96 ,$$

while

$$(I_{\mathbf{MK}})^l_{N_{\mathbf{C}}}(0.3, (I_{\mathbf{MK}})^l_{N_{\mathbf{C}}}(0.2, 0.6)) = 0.91 .$$

Example 7.1.8. Similarly, one can easily show that (NP) is not generally preserved by the above contrapositive techniques. Consider the fuzzy implication $I_{\mathbf{K}}$ from Example 7.1.1, which satisfies (NP).

(i) The upper contrapositivisation of $I_{\mathbf{K}}$ with respect to $N_{\mathbf{C}}$ does not satisfy (NP) as shown below:

$$\begin{aligned}(I_{\mathbf{K}})^u_{N_{\mathbf{C}}}(1,y) &= \max(I_{\mathbf{K}}(1,y), I_{\mathbf{K}}(1-y,0)) \\ &= \max(y, \sqrt{1-(1-y)}) = \max(y,\sqrt{y}) = \sqrt{y} \ , y \in [0,1] .\end{aligned}$$

Therefore, from Corollary 1.5.7, its natural negation too is not equal to $N_{\mathbf{C}}$.

(ii) Similarly, as can be easily verified, the lower contrapositivisation of $I_{\mathbf{K}}$ with respect to $\sqrt{1-x^2}$ does not satisfy (NP) either.

The following concept, introduced by BALASUBRAMANIAM [21], gives an equivalent condition under which (NP) is preserved.

Definition 7.1.9. *Let $I \in \mathcal{FI}$ and N be a strong negation. A contrapositivisation technique I^*_N, where $* \in \{u,l\}$, is said to be N-compatible, if $N_{I^*_N} = N$.*

Proposition 7.1.10. *Let $I \in \mathcal{FI}$ and N be a strong negation. Then the following statements are equivalent:*

(i) A contrapositivisation I^*_N, where $* \in \{u,l\}$, is N-compatible.
(ii) I^*_N satisfies (NP).

Proof. (i) $\Longrightarrow$ (ii) Assume that a contrapositivisation technique I^*_N, where $* \in \{u,l\}$, is N-compatible. By Lemma 7.1.3 and Definition 7.1.9, I^*_N satisfies CP($N_{I^*_N}$). Thus, from Lemma 1.5.6(i) we get that I^*_N satisfies (NP).

(ii) $\Longrightarrow$ (i) Assume that I^*_N satisfies (NP). By Lemma 7.1.3 I^*_N satisfies (CP) with N. From Corollary 1.5.8 we obtain $N_{I^*_N} = N$, i.e., I^*_N is N-compatible. □

In other words, N-compatibility of a contrapositivisation technique preserves neutrality.

Proposition 7.1.11. *Let $I \in \mathcal{FI}$ satisfy* (NP) *and N be a strong negation. Then the following statements are equivalent:*

(i) The upper contrapositivisation of I with respect to N is N-compatible.
(ii) $N \geq N_I$.

Proof. By Definition 7.1.2 we have

$$\begin{aligned} I_N^u \text{ is } N\text{-Compatible} &\iff N_{I_N^u}(x) = N(x) \\ &\iff \max(I(x,0), I(1, N(x))) = N(x) \\ &\iff \max(N_I(x), N(x)) = N(x) \\ &\iff N_I(x) \leq N(x)\,, \end{aligned}$$

for all $x \in [0,1]$. □

The following proposition now easily follows.

Proposition 7.1.12. *If $I \in \mathcal{FI}$ satisfies* (NP) *and N is a strong negation such that $N_I \geq N$, then*

(i) $N_{I_N^u} = N_I$,
(ii) $I_N^u(1,y) = N_I(N(y))$, for all $y \in [0,1]$.

Below we present the counterparts of Propositions 7.1.11 and 7.1.12 for the lower contrapositivisation, which can be proven similarly as above.

Proposition 7.1.13. *Let $I \in \mathcal{FI}$ satisfy* (NP) *and N be a strong negation. Then the following statements are equivalent:*

(i) The lower contrapositivisation of I with respect to N is N-compatible.
(ii) $N \leq N_I$.

Proposition 7.1.14. *If $I \in \mathcal{FI}$ satisfies* (NP) *and N is a strong negation such that $N_I \leq N$, then*

(i) $N_{I_N^l} = N_I$,
(ii) $I_N^l(1,y) = N_I(N(y))$, for all $y \in [0,1]$.

Corollary 7.1.15. *Let $I \in \mathcal{FI}$ satisfy* (NP) *and assume that N_I is a strong negation. Then both implications I_N^u and I_N^l are N_I-compatible. In addition, if I satisfies CP(N_I), then $I_{N_I}^U(x,y) = I_{N_I}^L(x,y) = I(x,y)$, for all $x, y \in [0,1]$.*

Remark 7.1.16. Let $I \in \mathcal{FI}$ satisfy (NP).

(i) If N_I is a vanishing negation, then there does not exist any strong negation N such that $N_I \geq N$ and hence the upper contrapositivisation technique is not N-compatible.
(ii) Similarly, if N_I is a filling negation, then there does not exist any strong negation N such that $N_I \leq N$ and hence the lower contrapositivisation technique is not N-compatible.

Thus, only when the natural negation N_I of an implication I is either a non-filling or a non-vanishing negation (or both) one of the above contrapositivisation techniques is N-compatible with respect to a strong negation N.

Now, let us consider $I \in \mathcal{FI}$, whose natural negation N_I is neither non-filling nor non-vanishing as given below. Consider the negation

$$N'(x) = \begin{cases} 1, & \text{if } x \in [0, \alpha] , \\ f(x), & \text{if } x \in (\alpha, \beta) , \\ 0, & \text{if } x \in [\beta, 1] , \end{cases}$$

where $f(x)$ is any decreasing function (possibly discontinuous), $\alpha, \beta \in (0, 1)$. For a t-conorm S, consider the (S,N)-implication obtained from S and N', i.e., let $I_{S,N'}(x, y) = S(N'(x), y)$ for all $x, y \in [0, 1]$. We know that $I_{S,N'} \in \mathcal{FI}$ and the natural negation of $I_{S,N'}$ is N'. Since N' is neither non-filling nor non-vanishing, one cannot find any strong negation N such that either $N \leq N'$ or $N \geq N'$ and thus both the upper and lower contrapositivisation techniques are not N-compatible. This motivates the search for a contrapositivisation technique that is independent of the ordering between N and N_I and still be N-compatible. The following section explores this idea.

7.1.2 Medium Contrapositivisation

As can be seen from Propositions 7.1.11 - 7.1.14, an ordering between the natural negation N_I and the strong negation N is essential for the resulting contrapositivised implication to be N-compatible, when we employ either I_N^u or I_N^l. In this section, we discuss the medium contrapositivisation technique proposed by BALASUBRAMANIAM [21], whose N-Compatibility is independent of the ordering between N and N_I.

Definition 7.1.17. *Let $I \in \mathcal{FI}$ and N be a fuzzy negation. The function $I_N^m \colon [0, 1]^2 \to [0, 1]$ dcfincd as:*

$$I_N^m(x, y) = \min(I(x, y) \vee N(x), I(N(y), N(x)) \vee y) , \qquad x, y \in [0, 1] , \quad (7.1)$$

is called the medium contrapositivisation *of I with respect to N.*

Proposition 7.1.18. *If $I \in \mathcal{FI}$ and N is a fuzzy negation, then $I_N^m \in \mathcal{FI}$. If, in addition, N is strong, then it satisfy CP(N).*

Proof. Once again, by virtue of Theorem 6.1.1 the family $\mathcal{FI}$ is a lattice, so it is enough to show that the following functions $J(x, y) = I(x, y) \vee N(x)$ and $K(x, y) = I(N(y), N(x)) \vee y$ are fuzzy implications. However, the monotonicity of J and K, i.e., (I1) and (I2), are the consequence of the monotonicity of I, N and maximum. Additionally,

$$J(0, 0) = I(0, 0) \vee N(0) = 1 \vee 1 = 1 ,$$
$$J(1, 1) = I(1, 1) \vee N(1) = 1 \vee 0 = 1 ,$$
$$J(1, 0) = I(1, 0) \vee N(1) = 0 \vee 0 = 0 .$$

A similar reasoning can be carried out for the function K. Thus $I_N^m \in \mathcal{FI}$. Finally, if N is strong, then

$$\begin{aligned} I_N^m(N(y), N(x)) &= \min(I(N(y), N(x)) \vee N(N(y)), \\ &\qquad I(N(N(x)), N(N(y))) \vee N(x)) \\ &= \min(I(N(y), N(x)) \vee y, I(x, y) \vee N(x)) \\ &= \min(I(x, y) \vee N(x), I(N(y), N(x)) \vee y) = I_N^m(x, y) , \end{aligned}$$

for all $x, y \in [0, 1]$, i.e., I_N^m satisfies CP(N). □

Proposition 7.1.19. *If $I \in \mathcal{FI}$ satisfies* (NP) *and N is a fuzzy negation, then $N_{I_N^m} = N$ and I_N^m satisfies* (NP).

Proof. By the definition of I_N^m and (NP) for I we have

$$\begin{aligned} N_{I_N^m}(x) &= \min(I(x, 0) \vee N(x), I(1, N(x)) \vee 0) \\ &= \min(N_I(x) \vee N(x), N(x)) = N(x) , \end{aligned}$$

for any $x \in [0, 1]$. Further, for any $y \in [0, 1]$, we have

$$\begin{aligned} I_N^m(1, y) &= \min(I(1, y) \vee N(1), I(N(y), 0) \vee y) \\ &= \min(y \vee 0, N_I(N(y)) \vee y) \\ &= \min(y, N_I(N(y)) \vee y) = y . \end{aligned}$$

□

Corollary 7.1.20. *If $I \in \mathcal{FI}$ satisfies* (NP) *and N is a strong negation then the medium contrapositivisation of I with respect to N is N-Compatible.*

Similarly to the upper and lower contrapositivisations, the medium contrapositivisation too preserves (IP) and (OP), as the following result shows.

Proposition 7.1.21. *Let N be a fuzzy negation. If I satisfies* (OP), *then so does I_N^m.*

Proof. Let $x \leq y$. Then $I(x, y) = 1$, by (OP). Moreover, $N(y) \leq N(x)$ and $I(N(y), N(x)) = 1$. Now,

$$\begin{aligned} I_N^m(x, y) &= \min(I(x, y) \vee N(x), I(N(y), N(x)) \vee y) \\ &= \min(1 \vee N(x), 1 \vee y) \\ &= \min(1, 1) = 1 . \end{aligned}$$

Conversely, if $I_N^m(x, y) = 1$, then by definition, $I(x, y) \vee N(x) = 1$ and $I(N(y), N(x)) \vee y = 1$. Now, if $I(x, y) \vee N(x) = 1$, then either $I(x, y) = 1$ or $N(x) = 1$. $I(x, y) = 1$, by (OP), implies that $x \leq y$, or equivalently, $N(y) \leq N(x)$ and hence $I(N(y), N(x)) = 1$. If $N(x) = 1$, since N is strict we have $x = 0$ in which case $x \leq y$ for all $y \in [0, 1]$. Similarly, if $y = 1$, then obviously $x \leq y$ for all $x \in [0, 1]$. Thus $I_N^m(x, y) = 1$ implies $x \leq y$, i.e, $I_N^m(x, y)$ satisfies (OP) if I does. □

Corollary 7.1.22. *Let N be a fuzzy negation. If $I \in \mathcal{FI}$ satisfies* (IP), *then so does I_N^m.*

With regards to upper and lower contrapositivisation, when I satisfies (CP) with respect to the strong N, we have that $I_N^u = I_N^l = I$. Whereas, in the case of medium contrapositivisation we have that

$$\begin{aligned} I_N^m(x,y) &= \min(I(x,y) \vee N(x), I(N(y),N(x)) \vee y) \\ &= \min(I(x,y) \vee N(x), I(x,y) \vee y) \\ &= I(x,y) \vee (N(x) \wedge y) , \end{aligned}$$

which is not always equal to the original fuzzy implication I. To see this, consider the fuzzy implication $I_{U_{\mathbf{P}},N_{\mathbf{C}}}$ given in Example 5.3.6(ii). Clearly, $I_{U_{\mathbf{P}},N_{\mathbf{C}}}$ satisfies CP($N_{\mathbf{C}}$). However,

$$I_{U_{\mathbf{P}},N_{\mathbf{C}}}(0.6, 0.4) = 0.32 < 0.4 = \min(1 - 0.6, 0.4) .$$

Hence, $(I_{U_{\mathbf{P}},N_{\mathbf{C}}})_{N_{\mathbf{C}}}^m \neq I_{U_{\mathbf{P}},N_{\mathbf{C}}}$. For the Kleene-Dienes implication $I_{\mathbf{KD}}$, which satisfies CP($N_{\mathbf{C}}$), as can be verified, $(I_{\mathbf{KD}})_{N_{\mathbf{C}}}^m = I_{\mathbf{KD}}$. Thus even when I satisfies CP(N) the medium contrapositivisation of I with respect to N can be a different implication and allows us to construct newer fuzzy implications that satisfy CP(N).

Moreover, the above process does not continue indefinitely. In fact, in the case when I satisfies CP(N) the medium contrapositivisation of I_N^m with respect to the same N is

$$\begin{aligned} (I_N^m)_N^m(x,y) &= \min(I_N^m(x,y) \vee N(x), I_N^m(N(y),N(x)) \vee y) \\ &= (I(x,y) \vee (N(x) \wedge y)) \vee (N(x) \wedge y) \\ &= (I(x,y) \vee (N(x) \wedge y)) = I_N^m(x,y) , \end{aligned}$$

for all $x, y \in [0,1]$, i.e., $(I_N^m)_N^m = I_N^m$.

7.2 Distributivity of Fuzzy Implications

In classical logic, distributivity of a binary operator over another determines the underlying structure of the algebra imposed by these operators. In fuzzy logic too, we find many works discussing the distributivity of t-norms over t-conorms (see Bertoluzza [32], Bertoluzza and Doldi [33], Klement et al. [146], Carbonell et al. [49]), of uninorms over t-operators (Mas et al. [171], Ruiz and Torrens [214]) and the like.

Following are the four basic distributive equations involving an implication:

$$(p \vee q) \rightarrow r \equiv (p \rightarrow r) \wedge (q \rightarrow r) , \tag{7.2}$$

$$(p \wedge q) \rightarrow r \equiv (p \rightarrow r) \vee (q \rightarrow r) , \tag{7.3}$$

$$r \rightarrow (s \wedge t) \equiv (r \rightarrow s) \wedge (r \rightarrow t) , \tag{7.4}$$

$$r \rightarrow (s \vee t) \equiv (r \rightarrow s) \vee (r \rightarrow t) . \tag{7.5}$$

Each of the above equivalences is a tautology in classical logic. The generalizations of these in fuzzy logic leads to the distributivity of fuzzy implications over t-norms and t-conorms, as given below:

$$I(S(x,y),z) = T(I(x,z), I(y,z)) , \tag{7.6}$$
$$I(T(x,y),z) = S(I(x,z), I(y,z)) , \tag{7.7}$$
$$I(x, T_1(y,z)) = T_2(I(x,y), I(x,z)) , \tag{7.8}$$
$$I(x, S_1(y,z)) = S_2(I(x,y), I(x,z)) , \tag{7.9}$$

for $x, y, z \in [0,1]$, where I is a fuzzy implication, T, T_1, T_2 are t-norms and S, S_1, S_2 are t-conorms.

In this section, we study the distributivity of fuzzy implications over t-norms and t-conorms, by studying conditions under which the different families of fuzzy implications, viz., (S,N)-implications, R-implications, QL-implications and f- and g-generated implications, satisfy equations (7.6) - (7.9).

7.2.1 On the Equation $I(S(x,y),z) = T(I(x,z), I(y,z))$

Proposition 7.2.1. *Let a function $I\colon [0,1]^2 \to [0,1]$ satisfy* (NP)*, T be a t-norm and S a t-conorm. If the triple (I, T, S) satisfies* (7.6) *for all $x, y, z \in [0,1]$, then $T = T_{\mathbf{M}}$.*

Proof. Putting $x = y = 1$ in (7.6) we get $I(S(1,1),z) = T(I(1,z), I(1,z))$, which implies $z = T(z,z)$, for all $z \in [0,1]$. Hence $T = \min$, the only idempotent t-norm. □

Proposition 7.2.2. *Let a function $I\colon [0,1]^2 \to [0,1]$ satisfy* (NP) *and $I(\,\cdot\,, z_0)$ be one-to-one for some $z_0 \in [0,1)$. If the triple (I, T, S), where T is a t-norm and S a t-conorm, satisfies* (7.6) *for all $x, y, z \in [0,1]$, then $S = S_{\mathbf{M}}$.*

Proof. Proposition 7.2.1 implies that I reduces (7.6) to

$$I(S(x,y),z) = \min(I(x,z), I(y,z)) .$$

Letting $x = y$ in the above equation with this fixed $z_0 \in [0,1)$, we have, $I(S(x,x),z_0) = \min(I(x,z_0), I(x,z_0))$, i.e., $I(S(x,x),z_0) = I(x,z_0)$, which implies $S(x,x) = x$, for all $x \in [0,1]$, since I is one-to-one in the first variable for z_0. Hence $S = S_{\mathbf{M}}$, the only idempotent t-conorm. □

The following result can be easily obtained.

Proposition 7.2.3. *For a function $I\colon [0,1]^2 \to [0,1]$ the following statements are equivalent:*

(i) I is decreasing in the first variable, i.e., I satisfies (I1).
(ii) I satisfies $I(\max(x,y),z) = \min(I(x,z), I(y,z))$ for all $x, y, z \in [0,1]$, i.e., I satisfies (7.2).

Corollary 7.2.4. *A fuzzy implication I that satisfies* (NP) *and is one-to-one in the first variable, when the second variable is in* $[0,1)$, *reduces* (7.6) *to* (7.2) *and satisfies* (7.2).

In the rest of this section, we discuss (7.6) for the different families of fuzzy implications, viz., (S,N)-, R-, QL-, f- and g-implications.

From Propositions 7.2.2 and 7.2.3 we have the following result.

Theorem 7.2.5. *Let T be a t-norm and S a t-conorm. Further, let I be any one of the following families of fuzzy implication:*

(a) an (S,N)-implication $I_{S',N'}$ obtained from a t-conorm S' and a strict negation N', or
(b) a QL-implication obtained from a triple $I_{T^\bullet,S^\bullet,N^\bullet}$, where $N^\bullet$ is a strict negation.

Then the following statements are equivalent:

(i) The triple (I,T,S) satisfies (7.6) *for all $x,y,z \in [0,1]$.*
(ii) $S = S_{\mathbf{M}}$ and $T = T_{\mathbf{M}}$.

Theorem 7.2.6. *Let I_{T^*} be an R-implication generated from a left-continuous t-norm T^*, T a t-norm and S a t-conorm. Then the following statements are equivalent:*

(i) The triple (I_{T^*},T,S) satisfies (7.6) *for all $x,y,z \in [0,1]$.*
(ii) $S = S_{\mathbf{M}}$ and $T = T_{\mathbf{M}}$.

Proof. (i) $\Longrightarrow$ (ii) From Theorem 2.5.4, we know that I_{T^*} satisfies (NP) and hence $T = T_{\mathbf{M}}$ in (7.6). Now, letting $x = y$ in (7.6) we have

$$I_{T^*}(S(x,x),z) = \min(I_{T^*}(x,z), I_{T^*}(x,z)) = I_{T^*}(x,z)\,. \tag{7.10}$$

Since S is a t-conorm, $S(x,x) \geq x$. If $S(x_0,x_0) > x_0$ for some $x_0 \in [0,1]$, then there exists a $t_0 \in [0,1]$ such that $S(x_0,x_0) = s_0 > t_0 > x_0$. Thus from (7.10) we have,

$$I_{T^*}(S(x_0,x_0),t_0) = I_{T^*}(x_0,t_0)\,,$$

so

$$I_{T^*}(s_0,t_0) = I_{T^*}(x_0,t_0)\,.$$

By Theorem 2.5.7, I_{T^*} satisfies (OP), which implies $I_{T^*}(x_0,t_0) = 1$ and therefore, $I_{T^*}(s_0,t_0) = 1$, which is not true since this would imply $s_0 \leq t_0$, a contradiction. Therefore, $s_0 = x_0$, i.e., $S(x,x) = x$, for all $x \in [0,1]$, hence $S = S_{\mathbf{M}}$. Thus (7.6) reduces to (7.2), when I is an R-implication obtained from a left-continuous t-norm.

(ii) $\Longrightarrow$ (i) Obvious from Proposition 7.2.3. □

From Theorem 3.1.7, Lemma 3.1.8 and Corollary 7.2.4 we have

Theorem 7.2.7. *Let I_f be an f-generated implication, T a t-norm and S a t-conorm. Then the following statements are equivalent:*

(i) The triple (I_f, T, S) satisfies (7.6) *for all $x, y, z \in [0,1]$.*
(ii) $S = S_{\mathbf{M}}$ and $T = T_{\mathbf{M}}$.

Since $\mathbb{I}_{\mathbb{F},\infty} = \mathbb{I}_{\mathbb{G},\infty}$ we have that Theorem 7.2.7 holds also for a g-generated implication obtained from a g-generator such that $g(1) = \infty$. In fact, the following result shows that this is true even if $g(1) < \infty$.

Theorem 7.2.8. *Let I_g be a g-generated implication with $g(1) < \infty$, T a t-norm and S a t-conorm. Then the following statements are equivalent:*

(i) The triple (I_g, T, S) satisfies (7.6) *for all $x, y, z \in [0,1]$.*
(ii) $S = S_{\mathbf{M}}$ and $T = T_{\mathbf{M}}$.

Proof. *(i)* $\Longrightarrow$ *(i)* Let I_g satisfy (7.6). From Theorem 3.2.8, we know that I_g satisfies (NP) and hence $T = T_{\mathbf{M}}$ in (7.6). If $S \neq S_{\mathbf{M}}$, then there exists an $x \in (0,1)$ such that $S(x,x) > x$. Now, $x \cdot g(1) < S(x,x) \cdot g(1) \leq g(1)$ and by the continuity of g there exists a $z \in (0,1)$ such that

$$g(z) < x \cdot g(1) < S(x,x) \cdot g(1)\,,$$

and hence

$$\frac{g(z)}{S(x,x)} < \frac{g(z)}{x} < g(1)\,.$$

By the strict monotonicity of g, we have

$$g^{-1}\left(\frac{g(z)}{x}\right) > g^{-1}\left(\frac{g(z)}{S(x,x)}\right)\,,$$

thus $I_g(x,z) > I_g(S(x,x),z)$. For the above $x, z \in (0,1)$, we see that I_g does not satisfy (7.6), a contradiction. Hence $S = S_{\mathbf{M}}$.

(ii) $\Longrightarrow$ *(i)* Obvious by Proposition 7.2.3. □

7.2.2 On the Equation $I(T(x,y),z) = S(I(x,z), I(y,z))$

Once again the following propositions are easy to obtain along the above lines.

Proposition 7.2.9. *Let a function $I\colon [0,1]^2 \to [0,1]$ satisfy* (NP), *T be a t-norm and S a t-conorm. If the triple (I, T, S) satisfies* (7.7) *for all $x, y, z \in [0,1]$, then $S = S_{\mathbf{M}}$.*

Proposition 7.2.10. *Let a function $I\colon [0,1]^2 \to [0,1]$ satisfy* (NP) *and $I(\,\cdot\,, z_0)$ be one-to-one for some $z_0 \in [0,1)$. If the triple (I, T, S), where T is a t-norm and S a t-conorm, satisfies* (7.7) *for all $x, y, z \in [0,1]$, then $T = T_{\mathbf{M}}$.*

Corollary 7.2.11. *If $I\colon [0,1]^2 \to [0,1]$ is any function that satisfies* (NP) *such that $I(\,\cdot\,, 0)$ is one-one then $T = T_{\mathbf{M}}$ in* (7.7).

Proposition 7.2.12. *For a function $I\colon [0,1]^2 \to [0,1]$ the following statements are equivalent:*

(i) I is decreasing in the first variable, i.e., I satisfies (I1).
(ii) I satisfies $I(\min(x,y),z) = \max(I(x,z),I(y,z))$ for all $x,y,z \in [0,1]$, i.e., I satisfies (7.3).

Theorem 7.2.13. *A fuzzy implication I that satisfies* (NP) *and is one-to-one in the first variable, when the second variable is in $[0,1)$, reduces* (7.7) *to* (7.3) *and satisfies* (7.3).

Once again, as shown above, the following result can be proven:

Theorem 7.2.14. *Let T be a t-norm and S a t-conorm. Further, let I be any one of the following families of fuzzy implication:*

(a) an (S,N)-implication $I_{S',N'}$ obtained from a t-conorm S' and a strict negation N', or
(b) a QL-implication obtained from a triple $I_{T^\bullet,S^\bullet,N^\bullet}$, where $N^\bullet$ is a strict negation, or
(c) an R-implication I_{T^*} obtained from a left continuous t-norm T^*, or
(d) an f-generated implication I_f, or
(e) a g-generated implication I_g.

Then the following statements are equivalent:

(i) The triple (I,T,S) satisfies (7.7) *for all $x,y,z \in [0,1]$.*
(ii) $S = S_{\mathbf{M}}$ and $T = T_{\mathbf{M}}$.

7.2.3 On the Equation $I(x,T_1(y,z)) = T_2(I(x,y),I(x,z))$

The first two results can be easily proven.

Proposition 7.2.15. *For a function $I\colon [0,1]^2 \to [0,1]$ the following statements are equivalent:*

(i) I is increasing in the second variable, i.e., I satisfies (I2).
(ii) I satisfies $I(x,\min(y,z)) = \min(I(x,y),I(x,z))$ for all $x,y,z \in [0,1]$, i.e., I satisfies (7.4).

Proposition 7.2.16. *Let T_1,T_2 be t-norms. If $I\colon [0,1]^2 \to [0,1]$ is any function that satisfies* (NP), *then $T_1 = T_2$ in* (7.8).

In this case, (7.8) becomes

$$I(x,T(y,z)) = T(I(x,y),I(x,z))\,, \qquad x,y,z \in [0,1]\,. \tag{7.11}$$

Proposition 7.2.17. *Let I be a fuzzy implication such that its natural negation N_I is continuous. If I satisfies* (7.11) *with a t-norm T, then $T = T_{\mathbf{M}}$.*

Proof. Letting $y = z = 0$ we have, $I(x, 0) = I(x, T(0, 0)) = T(I(x, 0), I(x, 0))$, i.e., $N_I(x) = T(N_I(x), N_I(x))$ for all $x \in [0, 1]$. Since N_I is continuous, it is onto on $[0, 1]$, i.e., for every $x \in [0, 1]$ there exists a $y \in [0, 1]$ such that $N_I(x) = y$. Hence we have that $y = T(y, y)$, for all $y \in [0, 1]$, which implies $T = T_{\mathbf{M}}$. □

Proposition 7.2.18. *Let I be a fuzzy implication that satisfies* (NP), *is continuous except at the point* $(0, 0)$ *and its natural negation* $N_I = N_{\mathbf{D1}}$. *If I satisfies* (7.11) *with a t-norm T, then T is positive.*

Proof. Assuming to the contrary, let T not be positive. Then there exists some $y \in (0, 1)$ such that $T(y, y) = 0$. Let $y^* = \sup\{t \in (0, 1) \mid T(t, t) = 0\} > 0$. Let us fix $y' < y^*$. Observe that $T(y', y') = 0$. From (NP) and (I1) we get

$$I(1, y') = y' < I(0, y') = 1 .$$

Thus, by the continuity of I except at $(0, 0)$, we have that there exists an $x_0 \in (0, 1)$ such that $I(x_0, y') = y_0 > y^* > y'$.

Now, letting $x = x_0$ and $y = z = y'$ in (7.11) we have

$$I(x_0, T(y', y')) = I(x_0, 0) = N_I(x_0) = N_{\mathbf{D1}}(x_0) = 0 ,$$

whereas

$$T(I(x_0, y'), I(x_0, y')) = T(y_0, y_0) .$$

If I satisfies (7.11), this implies that $T(y_0, y_0) = 0$, contradicting the fact that y^* is the supremum. Hence there does not exist any $y \in (0, 1)$ such that $T(y, y) = 0$, i.e., T is a positive t-norm. □

The following result follows from Proposition 7.2.17:

Theorem 7.2.19. *Let T_1, T_2 be t-norms. Further, let I be any one of the following families of fuzzy implication:*

(a) an (S,N)-implication $I_{S',N'}$ obtained from a t-conorm S' and a continuous negation N', or
(b) a QL-implication obtained from a triple $I_{T^\bullet,S^\bullet,N^\bullet}$, where $N^\bullet$ is a continuous negation.

Then the following statements are equivalent:

(i) The triple (I, T_1, T_2) satisfies (7.8) *for all $x, y, z \in [0, 1]$.*
(ii) $T_1 = T_2 = T_{\mathbf{M}}$.

Although in the case of R-implications the above result still holds, since not all R-implications have their natural negations to be continuous, we present a different proof in the following result.

Theorem 7.2.20. *Let I_{T^*} be an R-implicaion generated from left continuous t-norm T^* and T_1, T_2 be t-norms. Then the following statements are equivalent:*

(i) The triple (I_{T^*}, T_1, T_2) satisfies (7.8) *for all $x, y, z \in [0, 1]$.*
(ii) $T_1 = T_2 = T_{\mathbf{M}}$.

Proof. Firstly, see that from Theorem 2.5.4 we know that I_{T^*} satisfies (NP) and hence it suffices to consider (7.11).

(i) $\Longrightarrow$ *(ii)* Assume that $T \neq T_{\mathbf{M}}$ in (7.11). Then there exists a $t \in (0,1)$ such that $T(t,t) < t$. Let r be so chosen that $T(t,t) < r < t$. Now, with $x = r$ and $y = z = t$ in (7.11) we have,

$$I_{T^*}(r, T(t,t)) = T(I_{T^*}(r,t), I_{T^*}(r,t)) ,$$

but $T(I_{T^*}(r,t), I_{T^*}(r,t)) = T(1,1) = 1$, by (OP) of I_{T^*}. This implies that $I_{T^*}(r, T(t,t)) = 1$, which again by (OP) implies $r \leq T(t,t)$, a contradiction to our assumption. Hence $T_1 = T_2 = T = T_{\mathbf{M}}$ in (7.8).

(ii) $\Longrightarrow$ *(i)* This follows from Proposition 7.2.15. □

In the Yager's class of fuzzy implications, firstly, see that if the f-generator is such that $f(0) < \infty$, from Theorem 3.1.7, Propositions 3.1.6(ii) and 7.2.17 we have the following result.

Theorem 7.2.21. *Let f be an f-generator such that $f(0) < \infty$ and T_1, T_2 be t-norms. Then the following statements are equivalent:*

(i) The triple (I_f, T_1, T_2) satisfies (7.8) for all $x, y, z \in [0,1]$.
(ii) $T_1 = T_2 = T_{\mathbf{M}}$.

From Theorem 3.1.7 and Proposition 7.2.18 we know that in the case of an I_f obtained from an f-generator with $f(0) = \infty$, if I_f satisfies (7.11) with a t-norm T, then T is positive. The following result shows that there are t-norms other than $T_{\mathbf{M}}$ for which such an I_f satisfies (7.11).

Theorem 7.2.22. *Let f be an f-generator such that $f(0) = \infty$. The f-implication I_f satisfies (7.11) with a t-norm T if either $T = T_{\mathbf{M}}$, or T is the strict t-norm obtained using f as the additive generator, i.e., $T(x,y) = f^{(-1)}(f(x) + f(y))$ for all $x, y \in [0,1]$*

Proof. Firstly, observe that since I_f is a fuzzy implication, it satisfies (7.11) when $T = T_1 = T_2 = T_{\mathbf{M}}$. So let us assume that T is the t-norm obtained using f as the additive generator, i.e., $T(x,y) = f^{(-1)}(f(x) + f(y))$. Since $f(0) = \infty$ we know that $f^{(-1)} = f^{-1}$ and we have

$$\begin{aligned}
I_f(r, T(s,t)) &= f^{-1}\left(r \cdot f(T(s,t))\right) \\
&= f^{-1}\left(r \cdot f \circ f^{-1}(f(s) + f(t))\right) \\
&= f^{-1}\left(r \cdot (f(s) + f(t))\right) \\
&= f^{-1}\left(r \cdot f(s) + r \cdot f(t)\right) \\
&= f^{-1}\left(f \circ f^{-1}(r \cdot f(s)) + f \circ f^{-1}(r \cdot f(t))\right) \\
&= f^{-1}\left(f(I_f(r,s)) + f(I_f(r,t))\right) \\
&= T(I_f(r,s), I_f(r,t)) .
\end{aligned}$$

□

If I_g is a g-implication, then from Proposition 3.2.7 we see that $N_{I_g} = N_{\mathbf{D1}}$, while from Theorem 3.2.8 we also know I_g satisfies (NP) and is continuous except at $(0,0)$. Hence, once again, from Proposition 7.2.18 we know that if I_g satisfies (7.11) with a t-norm T, then T is positive.

From $\mathbb{I}_{\mathbb{F},\infty} = \mathbb{I}_{\mathbb{G},\infty}$, we see that Theorem 7.2.22 is also valid for an I_g obtained from a g-generator with $g(1) = \infty$. For an I_g obtained from a g-generator where $g(1) < \infty$, we have the following result.

Theorem 7.2.23. *Let I_g be obtained from a g-generator where $g(1) < \infty$ and T_1, T_2 be t-norms. Then the following statements are equivalent:*

(i) The triple (I_g, T_1, T_2) satisfies (7.8) *for all $x, y, z \in [0,1]$.*
(ii) $T_1 = T_2 = T_{\mathbf{M}}$.

Proof. (i) $\Longrightarrow$ (ii) Let I_g be obtained from a g-generator where $g(1) < \infty$. From Theorem 3.2.8, it satisfies (NP), so $T_1 = T_2 = T$. If $T \neq T_{\mathbf{M}}$, then there exists a $y_0 \in (0,1)$ such that $T(y_0, y_0) < y_0$ and by the strict monotonicity of g there exists a $y' \in (0,1)$ such that $g(T(y_0, y_0)) < y' < g(y_0)$. Moreover, since $g(1) < \infty$, let $x_0 = \dfrac{y'}{g(1)}$. Now we have the following inequalities:

$$\begin{aligned} T(y_0, y_0) < y_0 &\Longrightarrow g(T(y_0, y_0)) < y' < g(y_0) \\ &\Longrightarrow g(T(y_0, y_0)) < x_0 \cdot g(1) < g(y_0) \\ &\Longrightarrow \frac{g(T(y_0, y_0))}{x_0} < g(1) < \frac{g(y_0)}{x_0} \,. \end{aligned}$$

Notice that,

$$I_g(x_0, T(y_0, y_0)) = g^{(-1)}\left(\frac{g(T(y_0, y_0))}{x_0}\right) < 1 \,,$$

while

$$I_g(x_0, y_0) = g^{(-1)}\left(\frac{g(y_0)}{x_0}\right) = 1 \,.$$

Now letting $x = x_0$ and $y = y_0$ it can be easily seen that I_g does not satisfy (7.11). Hence $T = T_{\mathbf{M}}$.

(ii) $\Longrightarrow$ (i) Obvious from Proposition 7.2.15. □

7.2.4 On the Equation $I(x, S_1(y, z)) = S_2(I(x, y), I(x, z))$

Once again the following propositions are easy to obtain along the above lines.

Proposition 7.2.24. *Let S_1, S_2 be t-conorms. If $I\colon [0,1]^2 \to [0,1]$ is any function that satisfies* (NP), *then $S_1 = S_2$ in* (7.9).

In this case, (7.9) becomes

$$I(x, S(y, z)) = S(I(x, y), I(x, z)) \,, \qquad x, y, z \in [0,1] \,. \tag{7.12}$$

Proposition 7.2.25. *Let $I\colon [0,1]^2 \to [0,1]$ be any function that satisfies* (I1) *and let $I(\,\cdot\,, 0)$ be continuous. If I satisfies* (7.12) *with a t-conorm S, then $S = S_{\mathbf{M}}$.*

Proposition 7.2.26. *For a function $I\colon [0,1]^2 \to [0,1]$ the following statements are equivalent:*

(i) I is increasing in the second variable, i.e., I satisfies (I2).
(ii) I satisfies $I(x, \max(y,z)) = \max(I(x,y), I(x,z))$ for all $x, y, z \in [0,1]$, i.e., I satisfies (7.5).

The following result immediately follows from Proposition 7.2.25:

Theorem 7.2.27. *Let S_1, S_2 be t-conorms and let I be any one of the following families of fuzzy implication:*

(a) an (S,N)-implication $I_{S',N}$ obtained from a t-conorm S' and a continuous negation N, or
(b) a QL-implication obtained from a triple $I_{T^\bullet,S^\bullet,N^\bullet}$, where $N^\bullet$ is a continuous negation, or
(c) an R-implication I_{T^*} obtained from a left continuous t-norm T^*, whose natural negation $N_{I_{T^*}}$ is continuous (and hence strong), or
(d) an f-implication I_f obtained from an f-generator with $f(0) < \infty$.

Then the following statements are equivalent:

(i) The triple (I, S_1, S_2) satisfies (7.9) *for all $x, y, z \in [0,1]$.*
(ii) $S_1 = S_2 = S_{\mathbf{M}}$.

Lemma 7.2.28. *Let $I \in \mathcal{FI}$ satisfy* (OP). *If I satisfies* (7.12) *with a t-conorm S, then S is either idempotent or S is not a positive t-conorm.*

Proof. If S is idempotent, then we know $S = S_{\mathbf{M}}$ and that any fuzzy implication I satisfies (7.12) with $S_{\mathbf{M}}$.

On the other hand, let $S \neq S_{\mathbf{M}}$. Then there exists a $y \in (0,1)$ such that $S(y,y) > y$. Now, since I satisfies (7.12), for any $x \in (y, S(y,y))$ we have $I(x,y) = y_0 < 1$ by (OP) and

$$1 = I(x, S(y,y)) = S(I(x,y), I(x,y)) = S(y_0, y_0)\,,$$

and hence S is non-positive. □

From Lemma 7.2.28 it suffices to consider non-positive t-conorms in the analysis of (7.12) with I being an R-implication. The following result presented by Baczyński and Jayaram [18] gives an equivalence condition in the case S is a nilpotent t-conorm. We omit the proof of this result, since it requires many new facts and suggest, instead, the above work for the full proof.

Theorem 7.2.29. *For a nilpotent t-conorm S and an R-implication I_T generated from a strict t-norm T the following statements are equivalent:*

(i) The pair (I_T, S) satisfies (7.12) for all $x, y, z \in [0,1]$.
(ii) S is Φ-conjugate with the Łukasiewicz t-conorm $S_{\mathbf{LK}}$ and I_T is Φ-conjugate with the Goguen implication $I_{\mathbf{GG}}$ with the same increasing bijection, i.e., there exists $\varphi \in \Phi$, which is uniquely determined, such that

$$S(x,y) = (S_{\mathbf{LK}})_\varphi(x,y) = \varphi^{-1}(\min(\varphi(x) + \varphi(y), 1)) ,$$

and

$$I_T(x,y) = (I_{\mathbf{GG}})_\varphi(x,y) = \begin{cases} 1, & \text{if } x \leq y , \\ \varphi^{-1}\left(\frac{\varphi(y)}{\varphi(x)}\right), & \text{if } x > y , \end{cases}$$

for all $x, y \in [0,1]$.

Note that, in the case of an R-implication I_T obtained from a nilpotent t-norm T, the natural negation $N_T = N_{I_T}$ is strong and this case is already covered in Theorem 7.2.27(c).
Along the lines of Theorem 7.2.22, the following result shows that in the case of a g-generated implication obtained from a g-generator such that $g(1) = \infty$, there are t-conorms other than $S_{\mathbf{M}}$ for which I_g satisfies (7.12). Once again, since $\mathbb{I}_{\mathbb{F},\infty} = \mathbb{I}_{\mathbb{G},\infty}$ we see that Theorem 7.2.30 is also valid for an I_f obtained from an f-generator with $f(0) = \infty$, whose proof can be obtained along the lines as that of Theorem 7.2.22.

Theorem 7.2.30. *Let I_g be obtained from a g-generator with $g(1) = \infty$. I_g satisfies (7.12) if either $S = S_{\mathbf{M}}$, or S is the t-conorm obtained using g as the additive generator, i.e., $S(x,y) = g^{-1}(g(x) + g(y))$.*

7.3 The Law of Importation

The equation

$$(p \wedge q) \rightarrow r \equiv (p \rightarrow (q \rightarrow r)) ,$$

known as the *law of importation*, is a tautology in classical logic. The general form of the above equivalence is given by

$$I(T(x,y), z) = I(x, I(y,z)) , \qquad x, y, z \in [0,1] , \tag{LI}$$

where I is a fuzzy implication and T is a t-norm.

Remark 7.3.1. It can be immediately seen that if a fuzzy implication I satisfies (LI) with respect to any t-norm T, by the commutativity of the t-norm T, we have that I satisfies the exchange principle (EP). On the other hand, a fuzzy implication I may satisfy (EP) without satisfying (LI) with respect to any t-norm T. To see this, consider the fuzzy implication given in Example 5.3.6(iii), which can also be written in the following form:

$$I_{U_{\mathbf{M}},N_{\mathbf{C}}}(x,y) = \begin{cases} \min(1-x,y), & \text{if } \max(1-x,y) \leq 0.5 \,, \\ \max(1-x,y), & \text{otherwise} \,. \end{cases}$$

Putting $x = 0.7, y = 1, z = 0.4$ in (LI) we get, for any t-norm T,

$$\begin{aligned} I(T(0.7,1),0.4) &= I(0.7,0.4) = \min(1-0.7,0.4) = 0.3 \,, \\ I(0.7,I(1,0.4)) &= I(0.7,0) = 0 \neq 0.3 \,. \end{aligned}$$

Hence I does not satisfy (LI) with any t-norm T.

In the following subsections we discuss the validity of (LI) for the usual families of fuzzy implications.

7.3.1 (S,N)-Implications and the Law of Importation

Theorem 7.3.2. *Let T be a t-norm and $I_{S,N}$ an (S,N)-implication obtained from a strict negation N. Then the following statements are equivalent:*

(i) The couple of functions $I_{S,N}$ and T satisfies (LI) *for all $x, y, z \in [0,1]$.*
(ii) T is the N-dual of the t-conorm S.

Proof. *(i)* $\Longrightarrow$ *(ii)* Let an (S,N)-implication $I_{S,N}$ generated from a strict negation satisfy (LI) with a t-norm T. Then, for all $x, y \in [0,1]$ we have

$$I_{S,N}(T(x,y),z) = I_{S,N}(x, I_{S,N}(y,z)) \,.$$

Putting $z = 0$ above and using the formula for (S,N)-implication we get

$$N(T(x,y)) = S(N(x), N(y)) \,.$$

Since N is a strict negation, we get

$$T(x,y) = N^{-1}(S(N(x),N(y))) \,,$$

i.e., T is the N-dual t-norm of S.

(ii) $\Longrightarrow$ *(i)* If T is the N-dual of S, then by the associativity (S2) of S we get

$$\begin{aligned} I_{S,N}(T(x,y),z) &= S(N(T(x,y)),z) = S(N(N^{-1}(S(N(x),N(y)))),z) \\ &= S(S(N(x),N(y)),z) = S(N(x),S(N(y),z)) \\ &= I_{S,N}(x, I_{S,N}(y,z)) \end{aligned}$$

for all $x, y, z \in [0,1]$. □

Example 7.3.3. In the case the negation N is not strict, then the t-norm T with which $I_{S,N}$ satisfies (LI) need not be unique.

(i) Consider the least (S,N)-implication $I_{\mathbf{D}}$ obtained from the Gödel negation $N_{\mathbf{D1}}$, which is not continuous (see Table 2.4). Let T be any positive t-norm. Now, if $x = 0$ or $y = 0$, then

$$I_{\mathbf{D}}(T(x,y),z) = I_{\mathbf{D}}(0,z) = 1 = I_{\mathbf{D}}(x, I_{\mathbf{D}}(y,z))$$

while if $\min(x,y) > 0$, then again

$$I_{\mathbf{D}}(T(x,y),z) = z = I_{\mathbf{D}}(x, I_{\mathbf{D}}(y,z)),$$

for any $z \in [0,1]$. Hence $I_{\mathbf{D}}$ satisfies (LI) with any positive t-norm T.

(ii) On the other hand, the Weber implication $I_{\mathbf{WB}}$, which is the greatest (S,N)-implication obtained from the non-continuous negation $N_{\mathbf{D2}}$ (see Table 2.4), satisfies (LI) with any t-norm T. To see this, let $x, z \in [0,1]$. If $y = 1$, then

$$I_{\mathbf{WB}}(T(x,y),z) = I_{\mathbf{WB}}(x,z) = I_{\mathbf{WB}}(x, I_{\mathbf{WB}}(y,z)) \ .$$

Now, let $y \in [0,1)$. Since $T(x,y) < 1$, we have $I_{\mathbf{WB}}(T(x,y),z) = 1$, and so is $I_{\mathbf{WB}}(x, I_{\mathbf{WB}}(y,z)) = I_{\mathbf{WB}}(x,1) = 1$.

Corollary 7.3.4. *Let $\varphi \in \Phi$. The Φ-conjugate $(I_{S,N})_\varphi$ of an (S,N)-implication $I_{S,N}$, where N is a strict negation, satisfies* (LI) *if and only if T is the N_φ-dual of S_φ.*

7.3.2 R-Implications and the Law of Importation

Theorem 7.3.5. *Let T be a t-norm and I_{T^*} an R-implication obtained from a left-continuous t-norm T^*. Then the following statements are equivalent:*

(i) The couple of functions I_{T^*} and T satisfies (LI) *for all $x, y, z \in [0,1]$.*
(ii) $T = T^*$.

Proof. (i) $\Longrightarrow$ (ii) Let I_{T^*} satisfy (LI) for some t-norm T. Let $x, y \in [0,1]$ be arbitrary but fixed. Now putting x, y and $z = T^*(x,y)$ into (LI) and using (EP), we get

$$\begin{aligned} I_{T^*}(T(x,y), T^*(x,y)) &= I_{T^*}(x, I_{T^*}(y, T^*(x,y))) \\ &= I_{T^*}(x, I_{T^*}(y, T^*(y,x))) \ . \end{aligned}$$

Observe that from Proposition 2.5.2, T^* and I_{T^*} satisfy (RP), so in particular, since $T^*(x,y) \le T^*(x,y)$, we get $I_{T^*}(y, T^*(y,x)) \ge x$. Therefore, from the above equality, (I2) and (IP), we get

$$I_{T^*}(T(x,y), T^*(x,y)) \ge I_{T^*}(x,x) = 1 \ .$$

Now, from (OP), we obtain

$$T(x,y) \le T^*(x,y) \ .$$

On the other hand, for the above fixed x, y, let $z = T(x, y)$. By (IP), we get

$$I_{T^*}(T(x,y), T(x,y)) = 1 .$$

Therefore, putting the above fixed x, y and z into (LI), we have

$$1 = I_{T^*}(x, I_{T^*}(y, T(x,y))) .$$

By (OP) we get $x \leq I_{T^*}(y, T(y,x))$, so by (RP) we obtain

$$T^*(x,y) \leq T(x,y) .$$

Since x, y are arbitrary, from the above inequalities we have that $T = T^*$.

(ii) $\Longrightarrow$ *(i)* If $T = T^*$, the fact that I_{T^*} and T satisfy (LI) has already been shown in Theorem 2.5.7. □

In the case the t-norm T^* considered is not left-continuous, the corresponding I_{T^*} may not satisfy (LI) for any T, or in the case it does satisfy (LI), the t-norm T may or may not be unique.

Example 7.3.6. (i) Consider the fuzzy implications $I_{\mathbf{TB}^*}$ and $I_{\mathbf{TB}}$, given in Example 2.5.6, which are obtained from the non-left-continuous t-norms $T_{\mathbf{B}^*}$ and $T_{\mathbf{B}}$, respectively. Since $I_{\mathbf{TB}^*}, I_{\mathbf{TB}}$ do not satisfy (EP), from Remark 7.3.1, we see that they do not satisfy (LI) with respect to any t-norm T, either.

(ii) On the other hand, if we consider the non-left-continuous nilpotent minimum t-norm $T_{\mathbf{nM}^*}$, then the R-implication generated from $T_{\mathbf{nM}^*}$ is the Fodor implication $I_{\mathbf{FD}}$, which satisfies both (EP) and also (LI) only with the left-continuous nilpotent minimum t-norm $T_{\mathbf{nM}}$.

(iii) In fact, as was already shown in Example 7.3.3(ii), the Weber implication $I_{\mathbf{WB}}$, which is also an R-implication obtained from the drastic t-norm $T_{\mathbf{D}}$, satisfies (LI) with any t-norm T.

From Theorems 7.3.2, 7.3.5 and Propositions 2.4.5, 2.5.10, we have:

Corollary 7.3.7. *Let $\varphi \in \Phi$. The Φ-conjugate $(I_{T^*})_\varphi$ of an R-implication I_{T^*} obtained from a left-continuous t-norm T^* satisfies* (LI) *if and only if* $T = (T^*)_\varphi$.

7.3.3 QL-Implications and the Law of Importation

From Proposition 2.6.7 and Example 7.3.3(ii) we have:

Proposition 7.3.8. *Let the QL-operation I_{T^*,S^*,N^*} obtained from a positive t-conorm S^*, a t-norm T^* and a negation N^* be a fuzzy implication. Then I_{T^*,S^*,N^*} satisfies* (LI) *with any t-norm T.*

Proposition 7.3.9. *Let T be a t-norm and let the QL-operation I_{T^*,S^*,N^*} obtained from a t-conorm S^*, a t-norm T^* and a strict negation N^* be a fuzzy implication. Then the following statements are equivalent:*

(i) *The couple of functions* I_{T^*,S^*,N^*} *and* T *satisfies* (LI) *for all* $x, y, z \in [0,1]$.
(ii) I_{T^*,S^*,N^*} *is an (S,N)-implication generated from* N *and a t-conorm* $S(x,y) = S^*(x, T^*((N^*)^{-1}(x), y))$ *and* T *is the* N^* *dual of* S.

Proof. (i) $\Longrightarrow$ (ii) Let I_{T^*,S^*,N^*} satisfy (LI) with the t-norm T. From Remark 7.3.1, it immediately follows that I_{T^*,S^*,N^*} satisfies (EP). From Theorem 2.6.19 and Remark 2.6.20(i), we have that I_{T^*,S^*,N^*} is an (S,N)-implication generated from N^* and $S(x,y) = S^*(x, T^*(N^*(x), y))$. Hence, from Theorem 7.3.2, we have that if I_{T^*,S^*,N^*} satisfies (LI), then T is the N^* dual of S.

(ii) $\Longrightarrow$ (i) Let us assume that I_{T^*,S^*,N^*} is an (S,N)-implication generated from N and a t-conorm $S(x,y) = S^*(x, T^*((N^*)^{-1}(x), y))$. If T is the N^* dual of S, then again from Theorem 7.3.2, we have that I_{T^*,S^*,N^*} satisfies (LI) with this T. □

7.3.4 f- and g-Implications and the Law of Importation

Theorem 7.3.10. *Let* T *be a t-norm and* I_f *an* f*-generated implication. Then the following statements are equivalent:*

(i) *The couple of functions* I_f *and* T *satisfies* (LI) *for all* $x, y, z \in [0,1]$.
(ii) $T = T_{\mathbf{P}}$.

Proof. (i) $\Longrightarrow$ (ii) Let I_f satisfy (LI) with a t-norm T. Then

$$\begin{aligned} I_f(T(x,y), z) &= I_f(x, I_f(y,z)) \\ &\Longrightarrow f^{-1}(T(x,y) \cdot f(z)) = f^{-1}(x \cdot f \circ f^{-1}(y \cdot f(z))) \\ &\Longrightarrow T(x,y) \cdot f(z) = x \cdot y \cdot f(z)\,, \end{aligned}$$

for all $x, y, z \in [0,1]$. Now, if $z \in (0,1)$, then, since f is strictly decreasing, $0 < f(z) < \infty$ and hence $T(x,y) = xy$ for all $x, y \in [0,1]$, i.e., $T = T_{\mathbf{P}}$.

(ii) $\Longrightarrow$ (i) If T is the product t-norm $T_{\mathbf{P}}$, then

$$\begin{aligned} I_f(x, I_f(y,z)) &= f^{-1}(x \cdot f(I_f(y,z))) \\ &= f^{-1}(x \cdot f \circ f^{-1}(y \cdot f(z))) \\ &= f^{-1}(x \cdot (y \cdot f(z))) \\ &= I_f(T_{\mathbf{P}}(x,y), z)\,, \end{aligned}$$

for any $x, y, z \in [0,1]$. □

In fact, the above result holds even for a g-generated implication I_g. Towards proving it, we firstly prove the following useful result.

Lemma 7.3.11. *If a* g*-implication* I_g *satisfies* (LI) *with a t-norm* T*, then* T *is positive.*

Proof. On the contrary, let $T(x_0, y_0) = 0$ for some $x_0, y_0 \in (0,1)$. Thus, by (LB), $I_g(T(x_0,y_0), z) = I_g(0,z) = 1$ for any $z \in [0,1]$. Since, I_g satisfies (LI) with respect to T we have $1 = I_g(x_0, I_g(y_0, z))$. In particular, if $z = 0$, then from Lemma 3.2.7 we get $I_g(y_0, z) = 0$ and $I_g(x_0, I_g(y_0,z)) = I_g(x_0, 0) = 0$, which implies $1 = 0$, a contradiction. Hence T is positive. □

Theorem 7.3.12. *Let T be a t-norm and I_g a g-generated implication. Then the following statements are equivalent:*

(i) The couple of functions I_g and T satisfies (LI) *for all $x, y, z \in [0,1]$.*
(ii) $T = T_{\mathbf{P}}$.

Proof. Firstly, see that $\mathbb{I}_{\mathbb{F},\infty} = \mathbb{I}_{\mathbb{G},\infty}$ from Proposition 4.4.1(iii). Thus, by Theorem 7.3.10, we know that the result holds for all I_g obtained from a g-generator with $g(1) = \infty$. Hence in the following we only consider the case when $g(1) < \infty$.

(*i*) $\Longrightarrow$ (*ii*) Let I_g be a g-generated implication such that $g(1) < \infty$ and T be a t-norm for which I_g satisfies the law of importation. Note that from Lemma 7.3.11 T is positive. On the one hand, let us suppose that for some $x, y \in (0,1)$ we have $T(x,y) > x \cdot y$ and let $z = g^{-1}(x \cdot y \cdot g(1))$. Then we have

$$\frac{g(z)}{y} \leq \frac{g(z)}{T(x,y)} < \frac{g(z)}{x \cdot y} = g(1)\,,$$

and so,

$$\begin{aligned} I_g(x, I_g(y,z)) &= g^{(-1)}\left(\frac{1}{x} \cdot g \circ g^{(-1)}\left(\frac{g(z)}{y}\right)\right) \\ &= g^{(-1)}\left(\frac{1}{x} \cdot g \circ g^{-1}\left(\frac{g(z)}{y}\right)\right) \\ &= g^{(-1)}\left(\frac{g(z)}{x \cdot y}\right) = 1\,, \end{aligned}$$

whereas

$$I_g(T(x,y), z) = g^{(-1)}\left(\frac{g(z)}{T(x,y)}\right) < 1\,,$$

a contradiction.

On the other hand, if for some $x, y \in (0,1)$ we have $T(x,y) < x \cdot y < y$, then let $z = g^{-1}(T(x,y) \cdot g(1))$. Since T is positive, $T(x,y) \neq 0$ and

$$\frac{g(z)}{y} < \frac{g(z)}{x \cdot y} < \frac{g(z)}{T(x,y)} = g(1)\,,$$

and so,

$$I_g(x, I_g(y,z)) = g^{(-1)}\left(\frac{g(z)}{x \cdot y}\right) < 1\,,$$

whereas

$$I_g(T(x,y), z) = g^{(-1)}\left(\frac{g(z)}{T(x,y)}\right) = 1\,,$$

a contradiction. Hence necessarily $T(x,y) = x \cdot y$ for all $x, y \in (0,1)$, i.e., $T = T_{\mathbf{P}}$.

(*ii*) $\Longrightarrow$ (*i*) This is a straightforward verification. □

Example 7.3.13. In Table 7.1 we list the 9 basic fuzzy implications along with the t-norms for which they satisfy the law of importation (LI). From Table 1.4, we see that the Rescher implication $I_{\mathbf{RS}}$ does not satisfy (EP) and hence by Remark 7.3.1 it does not satisfy (LI) with respect to any t-norm T.

Table 7.1. Basic fuzzy implications and the law of importation

Implication	t-norm
$I_{\mathbf{LK}}$	$T_{\mathbf{LK}}$
$I_{\mathbf{GD}}$	$T_{\mathbf{M}}$
$I_{\mathbf{RC}}$	$T_{\mathbf{P}}$
$I_{\mathbf{KD}}$	$T_{\mathbf{M}}$
$I_{\mathbf{GG}}$	$T_{\mathbf{P}}$
$I_{\mathbf{RS}}$	$\times$
$I_{\mathbf{YG}}$	$T_{\mathbf{P}}$
$I_{\mathbf{WB}}$	any T
$I_{\mathbf{FD}}$	$T_{\mathbf{nM}}$

7.4 Fuzzy Implications and T-Conditionality

An implication operator plays an important role in the deductive process of a logic which is usually realized by some rules of inference. Modus ponens is one such rule of inference, wherein given two classical logic propositions $A \to B$ and A we infer B. A similar rule of inference in the case of dealing with fuzzy propositions is called the *generalized modus ponens* (GMP) wherein given two fuzzy propositions $A \to B$ and A' we infer B'. The highlight of this inference is even if $A' \neq A$ we still will be able to infer a reasonable conclusion B'. One of the conditions that any inference scheme employed to realize GMP is expected to satisfy is that the GMP should coincide with MP in the case $A' = A$, i.e., B' should be B.

In this section we will consider the following functional inequality, which as will be seen later in Chap. 8, will enable one of the important schemes that implements GMP, viz., the compositional rule of inference (CRI) Zadeh [266], to have the above property (see Alsina and Trillas [5]). GMP and CRI will be dealt with in detail in Sects. 8.3.1 and 8.3.2.

Definition 7.4.1. *Let $I \in \mathcal{FI}$ and T be any t-norm. Then I is called an* MP-fuzzy implication *for T, if the pair (I, T) satisfies the following condition:*

$$T(x, I(x, y)) \leq y\,, \qquad x, y \in [0, 1]\,. \tag{TC}$$

It should be noted that investigations of (TC), called also as *T-conditional* in the literature, have been done only for the three main families of fuzzy implications, viz., (S,N)-, R- and QL-implications, and that too under some restricted conditions, typically continuity of the underlying operators is assumed. In this section, we gather all such results available in the literature so far.

Firstly, we present some general facts. Since $T(x, y) \leq x$ for any $x, y \in [0, 1]$, we see that for a given fuzzy implication I, if $x \leq y$, then $T(x, I(x, y)) \leq x \leq y$ for any t-norm T. Hence, it actually suffices to consider only the cases when $x > y$. The following gives a necessary condition on I when $x > y$.

Proposition 7.4.2. *If $I \in \mathcal{FI}$ is such that there exist $x, y \in (0, 1)$ such that $x > y$ and $I(x, y) = 1$, then I does not satisfy* (TC) *with any t-norm T.*

Proof. Let there exist $x_0, y_0 \in (0, 1)$ such that $x_0 > y_0$ and $I(x_0, y_0) = 1$. Then we have $T(x_0, I(x_0, y_0)) = T(x_0, 1) = x_0 > y_0$ for any t-norm T. Hence I does not satisfy (TC) with any t-norm T. □

If a fuzzy implication I satisfies (TC) with a t-norm T, there can exist other fuzzy implications and t-norms that satisfy (TC) with T or I, respectively, depending on their relative ordering as can be seen from the following result (see also Trillas et al. [234]).

Proposition 7.4.3. *Let an $I \in \mathcal{FI}$ and a t-norm T satisfy* (TC).

(i) $N_I \leq N_T$, the natural negation of T.
(ii) If $N_I \neq N_{\mathbf{D1}}$, the least fuzzy negation, then T has zero-divisors.
(iii) If N_I is a non-vanishing negation and T is continuous, then T is Φ-conjugate with $T_{\mathbf{LK}}$.
(iv) If J is a fuzzy implication such that $J < I$, i.e., $J(x, y) \leq I(x, y)$ for all $x, y \in [0, 1]$, then J satisfies (TC) *with T.*
(v) If T' is a t-norm such that $T' \leq T$, i.e., $T'(x, y) \leq T(x, y)$ for all $x, y \in [0, 1]$, then I satisfies (TC) *with T'.*

Proof. (i) Letting $y = 0$ in (TC) we have $T(x, I(x, 0)) \leq 0$, so $T(x, N_I(x)) = 0$ for all $x \in [0, 1]$ and hence $N_I \leq N_T$, the natural negation of T.
(ii) It is immediate from (i) above that if the natural negation of the fuzzy implication I is such that there exists some $x \in (0, 1)$ such that $N_I(x) > 0$, then the T in (TC) is necessarily not positive, i.e., T has zero-divisors.
(iii) Let N_I be a non-vanishing negation, i.e., for every $x \in [0, 1)$ we have that $N_I(x) > 0$. This implies that every $x \in [0, 1)$ is a zero-divisor of T and T being continuous is Φ-conjugate with $T_{\mathbf{LK}}$, the only continuous t-norm whose set of zero-divisors is equal to $(0, 1)$.
(iv) This fact is obvious from the monotonicity of the t-norm T, since for all $x, y \in [0, 1]$ we have $T(x, J(x, y)) \leq T(x, I(x, y)) \leq y$.
(v) Once again, $T'(x, I(x, y)) \leq T(x, I(x, y)) \leq y$, for all $x, y \in [0, 1]$. □

7.4.1 (S,N)-Implications and T-Conditionality

Let us first consider the least (S,N)-implication obtained when $N = N_{\mathbf{D1}}$, viz., $I_{\mathbf{D}}$. It is easy to see that $I_{\mathbf{D}}$ satisfies (TC) with any t-norm T. In fact, if $x = 0$, then it is obvious. If $x > 0$, then $T(x, I_{\mathbf{D}}(x, y)) = T(x, y) \leq y$, for any $y \in [0, 1]$.

Since $N_{I_{S,N}} = N$, from Proposition 7.4.3 we see that an (S,N)-implication obtained from an $N \neq N_{\mathbf{D1}}$ does not satisfy (TC) with a positive t-norm T. From Proposition 7.4.2 the following result can be proven.

Theorem 7.4.4. *Let an (S,N)-implication $I_{S,N}$ satisfy* (TC) *with a t-norm T and let N_S^d be the set of discontinuity points of the natural negation of S. Then*

(i) $N_S \geq N$ on $[0, 1] \backslash N_S^d$,
(ii) $N_S < N$ on N_S^d.

Proof. Let $I_{S,N}$ be the (S,N)-implication obtained from a t-conorm S whose natural negation is N_S. Let N_S^d be the set of all points at which the natural negation of S is discontinuous. Let us assume that there exists an $x \in (0, 1)$ such that $N_S(x) < N(x)$.

(i) If $x \in [0, 1] \backslash N_S^d$, then there exists a $y' \in (N_S(x), N(x))$ such that there exists a y such that $y' = N_S(y) > N_S(x)$. Obviously, $x > y$. Now,

$$I_{S,N}(x, y) = S(N(x), y) \geq S(N_S(y), y) = 1 \ .$$

From Proposition 7.4.2 we know then that I cannot satisfy (TC) with any t-norm T, a contradiction.

(ii) This is immediate from point (i) above. □

Corollary 7.4.5. *If an (S,N)-implication $I_{S,N}$ satisfies* (TC) *with a t-norm T and N_S is continuous then $N_S \geq N$.*

The following result was proven by TRILLAS et al. [234] when the considered N is strong.

Theorem 7.4.6. *Let $I_{S,N}$ be an (S,N)-implication obtained from a strong negation N and T be a continuous t-norm. Then the following statements are equivalent:*

(i) $I_{S,N}$ and T satisfy (TC).
(ii) There exists $\varphi \in \Phi$ such that $T = (T_{\mathbf{LK}})_\varphi$, $N \leq N_\varphi$ and

$$S(x, y) \leq (S_{\mathbf{LK}})_\varphi(N_\varphi \circ N(x), y) \ , \qquad x, y \in [0, 1] \ .$$

Proof. *(i)* $\Longrightarrow$ *(ii)* From Proposition 7.4.3(iii) we know that there exists $\varphi \in \Phi$ such that $T = (T_{\mathbf{LK}})_\varphi$. From Proposition 7.4.3(i) we get $N_{I_{S,N}} = N \leq N_T = N_\varphi$. Since N is strong, we get

$$\begin{aligned} T(N(x), I_{S,N}(N(x), y)) &= (T_{\mathbf{LK}})_\varphi(N(x), S(x, y)) \\ &= \varphi^{-1}(\max(0, \varphi(N(x)) + \varphi(S(x, y)) - 1)) \ , \end{aligned}$$

for all $x, y \in [0,1]$. However, $I_{S,N}$ and T satisfy (TC), so we obtain

$$\max(0, \varphi(N(x)) + \varphi(S(x,y)) - 1) \leq \varphi(y) ,$$

i.e., either,

$$\varphi(N(x)) + \varphi(S(x,y)) - 1 \leq 0 \Longrightarrow S(x,y) \leq N_\varphi \circ N(x) ,$$

or,

$$\begin{aligned}
&0 \leq \varphi(N(x)) + \varphi(S(x,y)) - 1 \leq \varphi(y) \\
&\quad \Longrightarrow \varphi(S(x,y)) \leq \min(1, 1 - \varphi(N(x)) + \varphi(y)) \\
&\quad \Longrightarrow S(x,y) \leq \varphi^{-1}(\min(1, \varphi \circ \varphi^{-1}(1 - \varphi(N(x))) + \varphi(y))) \\
&\quad \Longrightarrow S(x,y) \leq (S_{\mathbf{LK}})_\varphi(N_\varphi \circ N(x), y) .
\end{aligned}$$

(*ii*) $\Longrightarrow$ (*i*) This can be easily verified by straightforward substitution. □

Remark 7.4.7. Under the conditions of Theorem 7.4.6, we see from the above proof that $N_\varphi \geq N$. Hence $x \leq (N_\varphi \circ N)(x)$ for any $x \in [0,1]$ and $S_{\mathbf{LK}_\varphi}(x,y) \leq S_{\mathbf{LK}_\varphi}((N_\varphi \circ N)(x), y)$ for all $x, y \in [0,1]$. Thus any t-conorm S such that $S \leq S_{\mathbf{LK}_\varphi}$ also would satisfy the conditions under the assumptions of Theorem 7.4.6.

7.4.2 R-Implications and *T*-Conditionality

In the case of R-implications obtained from left-continuous t-norms we have the following sufficiency condition:

Theorem 7.4.8. *Let T^* be a left-continuous t-norm and I_{T^*} the R-implication obtained from it. Then I_{T^*} satisfies* (TC) *with T^*.*

Proof. Since T^* is left-continuous, from Proposition 2.5.2 we know that the pair I_{T^*} satisfies the residuation principle (RP) with T^* and since $I_{T^*}(x,y) \geq I_{T^*}(x,y)$ the result is immediate. □

Remark 7.4.9. Note that, even when T^* is a left-continuous t-norm, $T = T^*$ may not be the only t-norm for which I_{T^*} satisfies (TC).

(i) Consider the t-norm $T_{\mathbf{M}}$ which is a continuous, and hence a left-continuous, t-norm whose residuum is the Gödel implication $I_{\mathbf{GD}}$. It is easy to see that $I_{\mathbf{GD}}$ satisfies (TC) with any t-norm T. In fact, if $x \leq y$, then $T(x, I_{\mathbf{GD}}(x,y)) = T(x,1) \leq y$, while in the case $x > y$ we again have $T(x, I_{\mathbf{GD}}(x,y)) = T(x,y) < y$.

(ii) We know that if T^*, T_1^* are two left-continuous t-norms such that $T_1^* \leq T^*$, then $I_{T_1^*} \geq I_{T^*}$ and hence from Proposition 7.4.3(v) we see that I_{T^*} satisfies (TC) also with T_1^*.

However, in the case T^* is not a left-continuous t-norm, the obtained I_{T^*} may not satisfy (TC) with any t-norm T. To see this, let us recall that an I_{T^*} obtained from any t-norm T^* satisfies

$$x \le y \Longrightarrow I_{T^*}(x,y) = 1 , \qquad x,y \in [0,1] .$$

Whereas, if T^* is not a left-continuous t-norm, then we know it may not satisfy (OP), i.e., there may exist $x > y$ such that $I_{T^*}(x,y) = 1$ and hence from Proposition 7.4.2 we know I_{T^*} may not satisfy (TC) with any t-norm T. For example, consider the R-implications $I_{\mathbf{WB}}$ and $I_{\mathbf{TB}}$ obtained from the non left-continuous t-norms $T_{\mathbf{D}}$ and $T_{\mathbf{B}}$. It is easy to see that both $I_{\mathbf{WB}}$ and $I_{\mathbf{TB}}$ do not satisfy (TC) with any T.

7.4.3 QL-Implications and T-Conditionality

From Remark 2.6.6(i) we know that the natural negation of a QL-implication is never equal to $N_{\mathbf{D1}}$ and hence no QL-implication satisfies (TC) with a t-norm T that is strictly monotone. Further, we know that

$$I_{T,S,N}(x,y) = S(N(x), T(x,y)) \le S(N(x), y) = I_{S,N}(x,y) ,$$

for any $x, y \in [0,1]$. Hence from Proposition 7.4.3(iv) we see that $I_{T,S,N}$ satisfies (TC) also with the same t-norm T that the corresponding $I_{S,N}$ satisfies (TC).

7.5 Characterization through Functional Equations

While in the previous sections we have characterized fuzzy implications satisfying certain functional equations, in this section we look at the functional equations that characterize some of the 9 basic fuzzy implications given in Table 1.3 (included below in Table 7.2 for ready reference).

From Theorem 2.5.33 we have the following characterization for the Łukasiewicz implication $I_{\mathbf{LK}}$.

Theorem 7.5.1. *For a function $I\colon [0,1]^2 \to [0,1]$ the following statements are equivalent:*

(i) I is Φ-conjugate with the Łukasiewicz implication $I_{\mathbf{LK}}$, i.e., there exists $\varphi \in \Phi$, which is uniquely determined, such that I has the form

$$I(x,y) = (I_{\mathbf{LK}})_{\varphi}(x,y) = \varphi^{-1}(\min(1 - \varphi(x) + \varphi(y), 1)) , \qquad x,y \in [0,1] .$$

(ii) I is continuous and satisfies both (EP) *and* (OP).

From Proposition 2.5.11 we see the following:

Table 7.2. The 9 basic fuzzy implications

Name	Formula
Łukasiewicz	$I_{\mathbf{LK}}(x,y) = \min(1, 1-x+y)$
Gödel	$I_{\mathbf{GD}}(x,y) = \begin{cases} 1, & \text{if } x \leq y \\ y, & \text{if } x > y \end{cases}$
Reichenbach	$I_{\mathbf{RC}}(x,y) = 1 - x + xy$
Kleene-Dienes	$I_{\mathbf{KD}}(x,y) = \max(1-x, y)$
Goguen	$I_{\mathbf{GG}}(x,y) = \begin{cases} 1, & \text{if } x \leq y \\ \frac{y}{x}, & \text{if } x > y \end{cases}$
Rescher	$I_{\mathbf{RS}}(x,y) = \begin{cases} 1, & \text{if } x \leq y \\ 0, & \text{if } x > y \end{cases}$
Yager	$I_{\mathbf{YG}}(x,y) = \begin{cases} 1, & \text{if } x = 0 \text{ and } y = 0 \\ y^x, & \text{if } x > 0 \text{ or } y > 0 \end{cases}$
Weber	$I_{\mathbf{WB}}(x,y) = \begin{cases} 1, & \text{if } x < 1 \\ y, & \text{if } x = 1 \end{cases}$
Fodor	$I_{\mathbf{FD}}(x,y) = \begin{cases} 1, & \text{if } x \leq y \\ \max(1-x, y), & \text{if } x > y \end{cases}$

Theorem 7.5.2. *For a function $I\colon [0,1]^2 \to [0,1]$ the following statements are equivalent:*

(i) I is the Gödel implication $I_{\mathbf{GD}}$.
(ii) $I_\varphi = I$, for all $\varphi \in \Phi$ and the T_I obtained from I by (2.19) *is an idempotent t-norm.*

Characterization of the Kleene-Dienes implication was firstly obtained by Baczyński [6].

Theorem 7.5.3. *For a function $I\colon [0,1]^2 \to [0,1]$ the following statements are equivalent:*

(i) I is Φ-conjugate with the Kleene-Dienes implication $I_{\mathbf{KD}}$, i.e., there exists $\varphi \in \Phi$, such that I has the form

$$I(x,y) = (I_{\mathbf{KD}})_\varphi(x,y) = \max(\varphi^{-1}(1-\varphi(x)), y)\,, \qquad x,y \in [0,1]\,.$$

(ii) $I \in \mathcal{CFI}$ satisfies (NP)*, N_I is strong, and $I(N_I(x), x) = x$ for all $x \in [0,1]$.*

Proof. *(i)* $\Longrightarrow$ *(ii)* The proof in this direction is straightforward.

(ii) $\Longrightarrow$ *(i)* Let $I \in \mathcal{CFI}$ and define a function $S\colon [0,1]^2 \to [0,1]$ by

$$S(x,y) = I(N_I(x), y)\,, \qquad x,y \in [0,1]\,. \tag{7.13}$$

From Proposition 2.4.6 we see that S satisfies (S3). Moreover, for any $x \in [0,1]$, we have

$$S(0,x) = I(I(0,0),x) = I(1,x) = x \,,$$
$$S(x,0) = I(I(x,0),0) = x \,,$$
$$S(x,x) = I(I(x,0),x) = x \,,$$

i.e., S satisfies (S4) and S is idempotent. Hence, from Remark 2.2.5(ii) we obtain that $S = S_{\mathbf{M}}$. Next, we also have by the strongness of N_I

$$S(N_I(x),y) = I(N_I(N_I(x)),y) = I(x,y) \,.$$

Therefore,

$$I(x,y) = \max(N_I(x),y) \,, \qquad x,y \in [0,1] \,. \tag{7.14}$$

Thus, Theorem 1.4.13 implies that there exists $\varphi \in \Phi$, such that

$$N_I(x) = \varphi^{-1}(1-\varphi(x)) \,, \qquad x \in [0,1] \,. \tag{7.15}$$

From (7.14) and (7.15), we obtain that

$$I(x,y) = \max(\varphi^{-1}(1-\varphi(x)),y) \,, \qquad x,y \in [0,1] \,, \tag{7.16}$$

i.e., $I = (I_{\mathbf{KD}})_\varphi$. □

It should be pointed out that the bijection φ of the unit interval in Theorem 7.5.3 is not determined uniquely, owing to the fact that the isomorphism between any two strong negations is not unique (see BACZYŃSKI and DREWNIAK [11], Corollary 4).

The characterization of the Goguen implication has been obtained by BACZYŃSKI [9] and is based on the properties of R-implications.

Theorem 7.5.4. *For a function $I\colon [0,1]^2 \to [0,1]$ the following statements are equivalent:*

(i) I is Φ-conjugate with the Goguen implication $I_{\mathbf{GG}}$, i.e., there exists $\varphi \in \Phi$, which is uniquely determined up to a positive constant exponent, such that I has the form as given earlier in (2.34)*, viz.,*

$$I(x,y) = (I_{\mathbf{GG}})_\varphi(x,y) = \varphi^{-1}\left(\min\left(1, \frac{\varphi(y)}{\varphi(x)}\right)\right) \,, \qquad x,y \in [0,1] \,.$$

(ii) I is continuous except at the point $(0,0)$*, satisfies* (I2), (EP), (OP) *and* $N_I = N_{\mathbf{D1}}$.

Proof. (i) $\Longrightarrow$ (ii) Let I be conjugate with the Goguen implication $I_{\mathbf{GG}}$, i.e., there exists an increasing bijection φ, such that I has the form (2.34). Since $I_{\mathbf{GG}}$ is the R-implication obtained from the product t-norm $T_{\mathbf{P}}$, which is continuous, and hence left-continuous, Theorem 2.5.10 implies that I is also an R-implication obtained from $(T_{\mathbf{P}})_\varphi$, which is again continuous. Now, by Theorem 2.5.17 we observe that I is a fuzzy implication which satisfies (I2), (EP), (OP). It is easy to check that I is continuous except at the point $(0,0)$ and that $N_I = N_{\mathbf{D1}}$.

(ii) $\Longrightarrow$ *(i)* Let I satisfy the required properties. Then, by Lemma 2.5.20 and Theorem 2.1.8, the function T defined by (2.20) is a t-norm and is conjugate with the product t-norm $T_{\mathbf{P}}$, i.e., $T(x,y) = \varphi^{-1}(\varphi(x)\varphi(y))$ for some increasing bijection $\varphi\colon [0,1] \to [0,1]$. Since I is still right continuous with respect to the second variable, Theorem 2.5.14 implies $I = I_T = I_{(T_{\mathbf{P}})_\varphi}$. Now, it immediately follows from Lemma 2.5.22 that I has the form (2.34). □

The characterization of the Rescher implication is trivial.

Theorem 7.5.5. *For a function $I\colon [0,1]^2 \to [0,1]$ the following statements are equivalent:*

(i) I is the Rescher implication $I_{\mathbf{RS}}$.
(ii) I satisfies (OP) *and the following property*

$$x > y \Longleftrightarrow I(x,y) = 0 \,.$$

The characterization of the Yager implication is based on the law of importation (LI) and the distributive equation (7.11). For a proof, the readers are referred to Baczyński [7].

Theorem 7.5.6. *For a function $I\colon [0,1]^2 \to [0,1]$ the following statements are equivalent:*

(i) I is Φ-conjugate with the Yager fuzzy implication, i.e., there exists $\varphi \in \Phi$, such that I has the form

$$I(x,y) = (I_{\mathbf{YG}})_\varphi(x,y) = \begin{cases} \varphi^{-1}\left(\varphi(y)^{\varphi(x)}\right), & \text{if } x > 0 \text{ or } y > 0\,, \\ 1, & \text{if } x = 0 \text{ and } y = 0\,. \end{cases}$$

(ii) $I \in \mathcal{FI}$ is continuous except the point $(0,0)$ and it satisfies the law of importation

$$I(T(x,y),z) = I(x, I(y,z))\,, \qquad x,y,z \in [0,1]\,,$$

the distributive equation

$$I(x, T(y,z)) = T(I(x,y), I(x,z))\,, \qquad x,y,z \in [0,1]\,,$$

and the following functional equation

$$I\left(\varphi^{-1}(0.5), T(y,y)\right) = y\,, \qquad y \in [0,1]\,,$$

with a strict t-norm T generated from an increasing bijection $\varphi \in \Phi$.

The following characterization of the Fodor implication is from Pei [205].

Theorem 7.5.7. *For a function $I\colon [0,1]^2 \to [0,1]$ the following statements are equivalent:*

(i) I is the Fodor implication $I_{\mathbf{FD}}$.
(ii) I satisfies (EP), (OP), *CP($N_{\mathbf{C}}$) and for any $x, y, z \in [0,1]$,*

$$\max(I(x,y), I(I(x,y), \max(N_{\mathbf{C}}(x), y))) = 1 \,. \tag{7.17}$$

Proof. (*i*) $\Longrightarrow$ (*ii*) The proof in this direction is straightforward.

(*ii*) $\Longrightarrow$ (*i*) Let $x \leq y$. Then by (OP) we have that $I(x,y) = 1$. We only need to show that whenever $x > y$, $I(x,y) = \max(1-x, y)$. Now, for any $x, y \in [0,1]$ by (EP) we have the following:

$$I(y, I(x,y)) = I(x, I(y,y)) = I(x,1) = 1 \,,$$
$$I(N_{\mathbf{C}}(x), I(x,y)) = I(N_{\mathbf{C}}(x), I(N_{\mathbf{C}}(y), N_{\mathbf{C}}(x))) = 1 \,.$$

Thus, from (OP), we get $y \leq I(x,y)$ and $N_{\mathbf{C}}(x) \leq I(x,y)$, from whence we obtain that $\max(N_{\mathbf{C}}(x), y) \leq I(x,y)$.

Now, if $x > y$ by (OP) we have that $I(x,y) < 1$ and hence from (7.17) we have that $I(I(x,y), \max(N_{\mathbf{C}}(x), y)) = 1$, which again by (OP) implies max $(N_{\mathbf{C}}(x), y) \geq I(x,y)$. Obviously, $I(x,y) = \max(1-x, y)$, whenever $x > y$, i.e., $I = I_{\mathbf{FD}}$. □

Finally, the characterization of the Weber implication is the following.

Theorem 7.5.8. *For a function $I\colon [0,1]^2 \to [0,1]$ the following statements are equivalent:*

(i) I is the Weber implication $I_{\mathbf{WB}}$.
(ii) There exists a function $F\colon [0,1]^2 \to [0,1]$ such that
(F1) $F(x,1) = 0$ for all $x \in [0,1)$,
(F2) $F(1,x) = x$ for all $x \in [0,1]$,
and I is the residual of F.

Proof. (*i*) $\Longrightarrow$ (*ii*) Consider the function F obtained from $I_{\mathbf{WB}}$ using (2.19) as follows:

$$F(x,y) = \inf\{t \in [0,1] \mid I_{\mathbf{WB}}(x,t) \geq y\} \,, \qquad x, y \in [0,1] \,. \tag{7.18}$$

It can be easily checked that F satisfies $(F1)$, $(F2)$ and that I is the residual of F.

(*ii*) $\Longrightarrow$ (*i*) Once again, it is a straightforward verification that if F is any function that satisfies $(F1)$, $(F2)$, then $I_{\mathbf{WB}}$ can be obtained as the residual of F. □

7.6 Bibliographical Remarks

Bandler and Kohout [30] were the first to study the contrapositivisation of fuzzy implications, while exploring the semantics of fuzzy implications. It was in this work the two contrapositivisation techniques, called Upper and Lower Contrapositivisation, were proposed. The lower contrapositivisation of the Goguen implication $I_{\mathbf{GG}}$ can also be found in the work of WANGMING [254] while discussing the goodness of the operations used in CRI.

The first major work to explore contrapositivisation in detail, in its own right, was that of FODOR [103], where he discusses the contrapositive symmetry of fuzzy implications for the three main families, viz., S-, R- and QL-implications. It bears repetition that it was during this study that Fodor discovered the nilpotent minimum t-norm $T_{\mathbf{nM}}$, which is perhaps the first left-continuous but non-continuous t-norm known in the literature. If T is a left-continuous t-norm and N a strong negation, then defining a binary operation $*_T$ on $[0,1]$ by

$$x *_T y = \min\{T(x,y), N(I_T(y, N(x)))\} , \tag{7.19}$$

Fodor showed that $*_T$ is almost a t-norm, and that $*_T$ indeed becomes a t-norm if $T(x,y) \leq N[I_T(y,N(x))]$ for $y > N(x)$, in which case it is given by

$$x *_T y = \begin{cases} T(x,y), & \text{if } y > N(x) , \\ 0, & \text{if } y \leq N(x) . \end{cases} \tag{7.20}$$

One can easily recognize that if $T = T_{\mathbf{M}}$ in (7.20), then we obtain the nilpotent minimum t-norm $T_{\mathbf{nM}}$. In fact, $*_T$ in addition to having very attractive properties has also opened up avenues for many subsequent research works. For example, JENEI [131] has investigated into which t-norms can be employed in (7.20) to obtain newer t-norms. He terms such transformations of t-norms as N-annihilation and has obtained many new families of left-continuous t-norms with strong associated negations. On the other hand, CIGNOLI et al. [54] have considered (7.20) with N being a non involutive negation, and have characterized continuous t-norms T such that (7.20) is a t-norm with its associated negation coinciding with N.

However, Fodor considered only the upper contrapositivisation of the R-implication I_T, with T a strict t-norm, citing the lack of neutrality of the lower contrapositivisation of I_T as one of the reasons for not considering it. Later on BALASUBRAMANIAM [21] investigated this further and proposed conditions under which both the contrapositivization techniques give rise to fuzzy implications that satisfy the neutrality property. Moreover, taking cue from $*_T$ of Fodor, by defining an appropriate binary operator, he has shown that the lower contrapositivisation of a fuzzy implication can also be seen as the residuation of a binary conjunction. Again, $*_t$ is not a t-norm in general.

Distributivity of fuzzy implications over t-norms and t-conorms came to critical attention - see COMBS [57, 58], DICK and KANDEL [76], MENDEL and LIANG [179] - after the work by COMBS and ANDREWS [56], wherein they attempted to exploit the distributivity (7.3) towards eliminating combinatorial rule explosion in fuzzy systems (see also BALASUBRAMANIAM and RAO [23, 24]). Conditions under which the general form of (7.3), viz., (7.7), hold for R- and S-implications first appeared in a paper by TRILLAS and ALSINA [233]. Following this, a similar study was undertaken in BALASUBRAMANIAM and RAO [25] for the remaining distributive equations (7.6) - (7.9) for the R- and S-implications. Moreover, (7.8) has been discussed in BACZYŃSKI [7, 8] under the assumptions that $T = T_1 = T_2$ is a strict t-norm and the implication I is continuous except at the point $(0, 0)$.

Just as we have considered different t-norms and t-conorms in the above distributive equations, one can also consider different fuzzy implications I in the same equation. For example, the most general version of (7.8) would be

$$I_1(x, T_1(y, z)) = T_2(I_2(x, y), I_3(x, z)) .$$

BACZYŃSKI [10] has shown that in all the above equations (7.6) - (7.9) it suffices to consider only a single fuzzy implication I.

The equations (7.7) - (7.9) continue to play an important role in lossless rule reduction in Fuzzy Rule Based Systems BALASUBRAMANIAM and RAO [23, 24, 25], SOKHANSANJ and FITCH [228], WEINSCHENK [256].

RUIZ and TORRENS [216, 217] have investigated the distributivity of the families of (U,N)-implications obtained from strong negations N and that of RU-implications. Quite obviously, their results subsume many of the results presented in Sect. 7.2.

In the framework of fuzzy logic, the law of importation has mostly been studied in conjunction with another property or functional equation, usually a distributive equation. In A-implications defined by TÜRKSEN et al. [244], (LI) with T as the product t-norm $T_{\mathbf{P}}$, was taken as one of the axioms. BACZYŃSKI [7] has studied the law of importation in conjunction with (7.8) and has given a characterization. BOUCHON-MEUNIER and KREINOVICH [42] have characterized fuzzy implications that have the law of importation as one of the axioms along with (7.8). They have considered $T_{\mathbf{M}}$ for the t-norm T and claim that Mamdani's choice of implication 'min' is "not so strange after all". In JAYARAM [126] the law of importation for the above families of fuzzy implications is investigated with the t-norm T being substituted both by a uninorm U and a t-operator F.

TRILLAS and his group have worked quite extensively on T-Conditionality of fuzzy implications, especially of the three early families. Some of their earliest works on this topic are [240] and [241]. For a more recent work, see [237]. In fact, in these works, (TC) is dealt with under more lenient terms, viz., given a fuzzy implication I they determine a binary operation M such that (I, M) satisfy (TC) and a few other properties. The operation M is not required to be either commutative or associative, and hence need not be a t-norm. Subsequently, in [5, 234, 235], they went on to consider (TC) for an (I, T) pair, with T a t-norm. As noted in [240] and DUBOIS and PRADE [88] the R-implication obtained from a t-norm T has an interesting connection to (TC) and can be interpreted as follows: "Given a t-norm T and two statements A, B with truth-values a, b, respectively, $I_T(A, B)$ gives the largest truth-value such that B can be inferred by means of T from both A and $I_T(A, B)$."

Of course, as noted earlier, the T-conditionality of Yager's families of fuzzy implications f- and g-implications is yet to be studied - a study which can lend more justification to their use in approximate reasoning.

Other than the above four tautologies involving fuzzy implication operators there have been a few others that have been considered in the literature.

IGEL and TEMME [123] have investigated the classical hypothetical syllogism given by the equation

$$(p \longrightarrow q) \wedge (q \longrightarrow r) \Longrightarrow p \longrightarrow r \, .$$

They analyze the validity of the chaining syllogism in fuzzy systems, i.e., whether two fuzzy rules IF F THEN G and IF G THEN H imply the rule IF F THEN H and give conditions under which this basic deduction scheme holds.

BUSTINCE et al. [45] studied the following property of a fuzzy implication:

$$I(x, N(x)) = N(x) \, , \qquad x \in [0,1] \, ,$$

where N is a strong negation. Their motivation for investigating the above equation stems from the fact that a fuzzy implication satisfying the above property with respect to a strong negation N when used in the *inclusion grade indicator* of DE BAETS and KERRE [68] satisfies the axioms of SINHA and DOUGHERTY [225], which are some of the desirable properties of indicators for fuzzified set inclusion.

HÁJEK and KOHOUT [121] call a fuzzy implication *special* if it has the following property: For any $x, y \in [0,1]$ and every $\varepsilon > 0$ such that $x + \varepsilon, y + \varepsilon \leq 1$,

$$I(x, y) \leq I(x + \varepsilon, y + \varepsilon) \, .$$

The work finds motivation in the fact that each special implicational quantifier determines a special implication connective and, conversely, each special implication connective is given by a special implicational quantifier. Implicational quantifiers play a role in obtaining extremal values of some statistics on contingency tables with fixed marginals. Recently, SAINIO et al. [219] have shown that an R-implication from a left-continuous t-norm T is special if and only if T is indeed continuous.

A characterization for the family of f- and g-implications is yet to be proposed. However, based on some recent results of BACZYŃSKI and JAYARAM [18] (see also [19]), a not-so-elegant but still a useful characterization can be obtained. Similarly, a characterization of the Reichenbach implication $I_{\mathbf{RC}}$ is still unknown. Although, what can be considered as a partial result on this topic appears in BUSTINCE et al. [45] (see Theorem 8 and Corollary 5), a complete characterization is difficult to obtain owing to the fact that the isomorphism between any two strong negations, or that between any two strict t-norms, is not unique (see FODOR and ROUBENS [105], Proposition 1.1).

Part III

Applicational Study of Fuzzy Implications

There is no branch of mathematics,
however abstract, which may not some day
be applied to phenomena of the real world.
– Nikolai Lobachevsky (1792-1856)

In Parts I and II the theoretical aspects of fuzzy implications, viz., analytical and algebraic, were discussed. In this part, we discuss their applicational value. One of the best known application areas of fuzzy logic is *approximate reasoning* (AR), wherein from imprecise inputs and fuzzy premises or rules we obtain, often, imprecise conclusions. AR with fuzzy sets encompasses a wide variety of inference schemes and have been readily embraced in many fields, especially among others, decision making, expert systems and control. Fuzzy implications play a vital role in many of these inference mechanisms, the exposition of which forms the main motivation for this Part III.

The approach taken while employing the different AR methods can be broadly classified into two types, namely, fuzzy reasoning methods that are used for function approximation, as in fuzzy control, and those that are used, say in decision making, where only a heuristic measure of the quality of the inferred output is either known, available or expected. In this treatise, since our primary agenda is to illustrate the role of fuzzy implications in AR, we restrict our studies to inference schemes in AR that use fuzzy sets, or equivalently, possibility theory.

Part III, consisting only of Chap. 8, highlights the role played by fuzzy implications with their myriad properties in the different inference schemes in AR, all of which are different realizations of the *generalized modus ponens* (GMP). In particular, we concentrate on the *compositional rule of inference* (CRI) of Zadeh [266] and those inference schemes that can be grouped as *similarity based reasoning* (SBR). Firstly, we give a general overview of the structure and inference in the above schemata. Following this, we investigate the role played by the different properties of a fuzzy implication employed in these reasoning procedures, especially, in the interpretability and correctness or reasonableness of the inferred outputs and in the efficiency of the inference procedure, thus partially vindicating our choice of functional equations (properties) considered in Chap. 7.

8 Fuzzy Implications in Approximate Reasoning

I have hardly ever known a mathematician
who was capable of reasoning.
– Plato (ca 429-347 BC)

Boolean implications are employed in inference schemas like *modus ponens*, *modus tollens*, etc., where the reasoning is done with statements or propositions whose truth-values are two-valued. Fuzzy implications play a similar role in the generalizations of the above inference schemas, where reasoning is done with fuzzy statements whose truth-values lie in $[0, 1]$ instead of $\{0, 1\}$.

One of the best known application areas of fuzzy logic is *approximate reasoning* (AR) (Driankov et al. [84]), wherein from imprecise inputs and fuzzy premises or rules we obtain, often, imprecise conclusions. Approximate reasoning with fuzzy sets encompasses a wide variety of inference schemes and have been readily applied in many fields, among others, decision making, expert systems and fuzzy control.

An inference scheme proposed under AR is validated or assessed mainly based on the reasonableness of inference (effectiveness) and the complexity of the algorithm (efficiency), other than its intuitive elegance. For example, they are used in fuzzy control primarily to approximate a function, which usually describes the system under consideration. On the other hand, in the areas of decision making and expert systems, AR techniques are employed for their inferential capabilities that conform to the basic rules of *generalized modus ponens* (GMP) as envisaged in fuzzy logic.

In this chapter our emphasis will be on the role played by fuzzy implications in the inference schemes from the point of view of both their effectiveness and efficiency.

Firstly, an introduction to approximate reasoning is given in Sect. 8.1, following which Sect. 8.2 discusses the basic elements that are employed in approximate reasoning, namely, possibility distributions, fuzzy statements, fuzzy IF-THEN rules. Following this, in Sect. 8.3, we discuss the two important inference schemes in AR employing fuzzy sets, viz., *compositional rule of inference* (CRI) of Zadeh [266] and *similarity based reasoning* (SBR).

Similarly, there are two main ways in which the goodness of inference of an AR scheme is usually evaluated - with respect to what are now termed as 'GMP rules' and with respect to its functional approximation capabilities. Fuzzy reasoning methods that are used for function approximation are different from

M. Baczyński and B. Jayaram: Fuzzy Implications, STUDFUZZ 231, pp. 245–278, 2008.
springerlink.com

those that are used in approximate reasoning (the generalized modus ponens extended to several fuzzy rules). The influence of fuzzy implications in this context is investigated in Sect. 8.4.

Inferencing schemes in AR can be resource consuming - both memory and time (see, for example, CORNELIS et al. [62], MARTIN-CLOUAIRE [168], DEMIRLI and TÜRKSEN [73]). Moreover, an increase in the number of rules only exacerbates the problem. As the number of input variables and/or input fuzzy sets increases, there is a combinatorial explosion of rules in multiple fuzzy rule based systems. Many techniques, mainly, by way of transformation of the rule structure or the underlying inference scheme itself, have been proposed towards efficient inferencing in AR. Sect. 8.5 discusses the properties that fuzzy implications employed in these schemes should possess for these techniques to be applicable.

This chapter, especially, highlights the roles of the different functional equations considered in Chap. 7.

8.1 Approximate Reasoning

8.1.1 Classical Implication in Inference Schemas

Implication in classical logic is employed to relate propositional formulae. For example, if A, B are any classical logic propositional formulae, then $A \longrightarrow B$ is called a *conditional statement* and is interpreted as 'A implies B'. Here A is called the antecedent and B the consequent of the conditional statement.

The classical implication operator plays a central role in the inference schemas like *modus ponens*, *modus tollens*, *hypothetical syllogism*, etc. If we know that the proposition A and the conditional $A \longrightarrow B$ are true, then from modus ponens we can conclude that B is true. Similarly, if $\neg B$ and $A \longrightarrow B$ are true, then from modus tollens we have $\neg A$ is true, which is nothing but the contrapositive form of modus ponens. For example, let $A =$ 'It is raining' and $B =$ 'Take the umbrella'. Then $A \longrightarrow B$ states that

IF it is raining THEN take the umbrella.

Now, if we know that 'It is raining', i.e., A is true, then using modus ponens we can deduce that B is true, i.e., we need to take the umbrella.

8.1.2 Fuzzy Implication in Inference Schemas

Similar to implication in classical logic, fuzzy implication is employed to relate fuzzy propositional formulae in fuzzy logic. For example, if A, B are any fuzzy logic propositional formulae, then $A \longrightarrow B$ is called a *fuzzy conditional statement* or more commonly as a fuzzy IF-THEN rule and is again interpreted as 'A implies B'. Here A is called the fuzzy antecedent and B the fuzzy consequent of the fuzzy IF-THEN rule.

Quite expectedly, fuzzy implication operators play a pivotal role in the inference schemas in fuzzy logic like fuzzy extensions of modus ponens, modus tollens, hypothetical syllogism, etc. But there is a major difference in the classical modus ponens and its fuzzy extension, termed generalized modus ponens, which we illustrate with the help of the following example.

Now, let A = 'Weather is cold', B = 'Dress warmly' and $A \longrightarrow B$. Now, if it is only known that 'Weather is quite chilly' what possible conclusion can we infer? After all, our common usage of the language suggests that 'quite chilly' though is a form of being 'cold', it is not exactly equal to the proposition A, i.e., if A' = 'Weather is chilly', then $A \approx A'$ but $A \not\equiv A'$.

8.1.3 Approximate Reasoning

The idea of linguistic fuzzy models imitating the human way of thinking was proposed by Zadeh in his pioneering work ZADEH [266]. The term *approximate reasoning* refers to methods and methodologies that enable reasoning with imprecise inputs to obtain meaningful outputs (see DRIANKOV et al. [84]). Inference in approximate reasoning is in sharp contrast to inference in classical logic - in the former the consequence of a given set of fuzzy propositions depends in an essential way on the meaning attached to these fuzzy propositions. Thus, inference in approximate reasoning is computation with fuzzy sets that represent the meaning of a certain set of fuzzy propositions.

Crucial to these methods is the presence and availability of a knowledge base, usually in the form of conditional statements. One way to realize approximate reasoning when the knowledge base consists of fuzzy IF-THEN rules is by employing the fuzzy logic inference schemes like generalized modus ponens (GMP), generalized modus tollens (GMT), hypothetical syllogism, etc. Although GMP is a general inference schema various implementations and methods have been proposed to realize it. Since the knowledge base consists of fuzzy conditionals, fuzzy implications play an important role in all the different realizations of GMP.

8.2 Fuzzy IF-THEN Rules

8.2.1 Possibility Distribution

Let $\mathcal{U} \neq \emptyset$ be a set and let x be any variable that can assume a value in $\mathcal{U}$. A possibility distribution for x is a mapping π_x from $\mathcal{U}$ to $[0, 1]$. In the absence of absolute information that $x = u_0 \in \mathcal{U}$, $\pi_x(u)$ gives the degree to which it is possible for x to assume the value $u \in \mathcal{U}$. The higher the value of $\pi_x(u)$ the more possible it is for x to assume the value u. It is possible that $\pi_x(u) = 1$, for some $u = u_0 \in \mathcal{U}$ and still $\pi_x(u) > 0$ for all other $u \neq u_0 \in \mathcal{U}$. BEZDEK et al. [34] view π_x as an elastic constrainment of the possible values x can assume. In the case $\pi_x(u) = 1$, for some $u = u_0 \in \mathcal{U}$ and $\pi_x(u) = 0$ for all other $u \neq u_0 \in \mathcal{U}$, we have precise information about the value of x.

8.2.2 Fuzzy Statements

An expression of the form "$\tilde{x}$ is A", where A is a fuzzy set on an appropriate domain $\mathcal{U}$, with reference to the context, is termed as a *fuzzy statement*. The above fuzzy statement can be viewed in two different ways according to BEZDEK et al. [34][1]:

- Let it be given that "$\tilde{x}$ is A" and also that $\tilde{x}$ assumes the precise value, let us say, $\tilde{x} = u$, where $u \in \mathcal{U}$, the domain of A. Then the truth value of the above fuzzy statement is obtained as follows:

$$t(\text{"}\tilde{x}\ \ is\ \ A\text{"}) = A(u)\ ,$$

  i.e., the truth value of the above fuzzy statement, given that $\tilde{x}$ is precisely known, is equal to the degree to which u - the value $\tilde{x}$ assumes - is itself compatible with the fuzzy set A. Thus greater the membership degree of u in the concept A is, the higher the truth value of the fuzzy statement.
- While in the above case a fuzzy statement was looked upon as a fuzzy proposition to be evaluated based on some precise information, it can also be used to express something precise when the only information regarding the variable x is imprecise. In other words, given that "$\tilde{x}$ is A" it can be determined to what extent $\tilde{x} = u_0 \in \mathcal{U}$ is possible, which is simply given as the possibility distribution for $\tilde{x}$ on $\mathcal{U}$ constructed using A, i.e., $\pi_{\tilde{x}}(u) = A(u)$, for all $u \in \mathcal{U}$.

Consider the statement 'John is tall' and that $\tilde{x}$ - the height of John - is precisely given to be $5'10'' \in \mathcal{U}$. This example falls under the first setting for the following reasons. Now, $A(5'10'')$ gives the membership degree of $5'10''$ in the concept $A =$ Tall, which can be interpreted as how much John belongs to the concept of being tall, or equivalently, how much 'John is tall' is true, which is nothing but the truth-value t('John is tall').

Conversely, given that 'John is tall' what can we say about the height of John, $\tilde{x}$, being equal to $5'10''$? This is given by the possibility distribution for $\tilde{x}$ on $\mathcal{U}$ constructed using the concept (fuzzy set) $A =$ Tall, i.e., $\pi_{\tilde{x}}(5'10'') = A(5'10'')$.

In other words, the membership degree $A(u)$ while it gives the degree of truth of the fuzzy statement in the former case, it is interpreted as the degree of possibility of $\tilde{x} = u$ in the latter.

8.2.3 Fuzzy IF-THEN Rules

A fuzzy statement of the type discussed above "$\tilde{x}$ is A" can be interpreted in yet another way - as a linguistic statement. Let $A\colon \mathcal{U} \to [0,1]$ be a fuzzy set on a suitable domain $\mathcal{U}$. Then A represents a concept and hence can be thought of as a *linguistic value.* A *linguistic variable* is a symbol $\tilde{x}$ that can assume or be assigned a linguistic value. Then a linguistic statement "$\tilde{x}$ is A" is interpreted as the linguistic variable $\tilde{x}$ taking the linguistic value A.

For example, let $\mathcal{U}$ denote the set of all values in degrees centigrade. If the linguistic variable $\tilde{x}$ denotes 'Temperature', then it can assume the following linguistic values A, viz., *high, more or less high, medium, cool, very cold, etc.* Each

[1] The above two interpretations bear a close resemblance to the 'adjunctive' and 'connective' interpretations as given in SMETS and MAGREZ [226], p. 331, though they are originally given for a binary operator.

concept, say high temperature, represented by a fuzzy set A is again context-dependent. For example, high temperature (fever) for a human being is different from the high temperature in a blast furnace, and accordingly the domain of the linguistic value is selected.

A fuzzy IF-THEN rule is of the form

$$\text{IF } \tilde{x} \text{ is } A \text{ THEN } \tilde{y} \text{ is } B \ ,$$

where $\tilde{x}, \tilde{y}$ are linguistic variables and A, B are linguistic expressions/values assumed by the linguistic variables and are defined over suitable universes of discourse $\mathcal{U}, \mathcal{V}$. For example,

$$\text{IF } \tilde{x} \text{ (volume) is } A \text{ (high) THEN } \tilde{y} \text{ (pressure) is } B \text{ (low)} \ .$$

8.3 Inference Schemes in Approximate Reasoning

8.3.1 Generalized Modus Ponens (GMP)

Modus ponens, modus tollens and hypothetical syllogism are some of the rules of deduction employed in classical two-valued logic. For example, given a propositional formula p and the conditional statement $p \longrightarrow q$, modus ponens allows us to infer the propositional formula q. On the other hand, if it is known that the propositional formula q is not true, or equivalently $\neg q$, and the conditional statement $p \longrightarrow q$ is valid, modus tollens allows us to infer that the propositional formula p is also not-true, i.e., $\neg p$.

Inference schemes in AR deal with fuzzy propositions, as explained in Sect. 8.1, and hence the classical modus ponens has been extended to the context of fuzzy logic under the inference pattern called *generalized modus ponens* (henceforth GMP).

Given the fuzzy rule (see Sect. 8.2) of the form

$$\text{IF } \tilde{x} \text{ is } A \text{ THEN } \tilde{y} \text{ is } B \ ,$$

and the fact

$$\tilde{x} \text{ is } A' \ ,$$

GMP allows the following conclusion to be drawn

$$\tilde{y} \text{ is } B' \ ,$$

where A, A' (A' not necessarily identical to A) and B, B' are fuzzy sets on non-empty sets X, Y, respectively, i.e., $A, A' \in \mathcal{F}(X)$ and $B, B' \in \mathcal{F}(Y)$.

Of all the various inference schemes that realize GMP in approximate reasoning, two of them have been prevalent in the literature, viz., reasoning methods based on the

(i) combination-projection principle, of which the *compositional rule of inference* (CRI) of Zadeh [266] is a good example;

(ii) similarity between inputs and antecedents and the subsequent modification of the consequent, usually known as *similarity based reasoning* (SBR), also called as *plausible reasoning* by DUBOIS and PRADE [89], of which *compatibility modification inference* (CMI) of CROSS and SUDKAMP [65] and the *approximate analogical reasoning scheme* (AARS) of TÜRKSEN and ZHONG [245] are some representative samples.

8.3.2 Compositional Rule of Inference (CRI)

We begin this section with the following definitions, with the assumption that $X = \{x_1, x_2, \ldots, x_n\}$ is a finite set.

Definition 8.3.1. *A fuzzy set $A\colon X \to [0,1]$, is said to be a 'fuzzy singleton', if there exists an $x_0 \in X$ such that A has the following representation:*

$$A(x) = \begin{cases} 1, & \text{if } x = x_0 \, , \\ 0, & \text{if } x \neq x_0 \, . \end{cases} \tag{8.1}$$

We say A attains normality at $x_0 \in X$.

Recall that, for a non-empty set X, $\mathcal{F}(X)$ denotes the fuzzy power set of X, i.e., $\mathcal{F}(X) = \{A | A\colon X \to [0,1]\}$.

Definition 8.3.2. *Let X be a non-empty set of finite cardinality, i.e., card $X < \infty$. Let $A \in \mathcal{F}(X)$ and F be any associative binary operation on $[0,1]$, then $F_{x\in X}(A(x)) = F_{i=1}^{n}(A(x_i)) = F(A(x_1), A(x_2), \ldots, A(x_n))$.*

Definition 8.3.3. *If $A, B \in \mathcal{F}(X)$ and F is any binary operation on $[0,1]$, then $F(A,B)$ is a fuzzy set on X, i.e., $F(A,B) \in \mathcal{F}(X)$ and is defined as $F(A,B)(x) = F(A(x), B(x))$, for all $x \in X$.*

Definition 8.3.4. *Let X, Y be finite, nonempty sets, $A \in \mathcal{F}(X)$ and $B \in \mathcal{F}(Y)$. The* Cartesian product *of the fuzzy sets A and B with respect to a t-norm T, denoted as $T(A,B)$, is the fuzzy set on $X \times Y$ defined as follows:*

$$T(A,B)(x,y) = T(A(x), B(y)) \, , \qquad (x,y) \in X \times Y \, .$$

The compositional rule of inference (CRI) of ZADEH [266] is one of the earliest implementations of GMP. Here, a fuzzy IF-THEN rule of the form:

$$\text{IF } \widetilde{x} \text{ is } A \text{ THEN } \widetilde{y} \text{ is } B, \tag{8.2}$$

is represented by a fuzzy relation $R(x,y)\colon X \times Y \to [0,1]$ as follows:

$$R(x,y) = I(A(x), B(y)) \, , \qquad (x,y) \in X \times Y \, , \tag{8.3}$$

where I is usually a fuzzy implication (or any other binary operation over the unit interval, e.g., a t-norm) and A, B are fuzzy sets on their respective domains

X, Y. In this section, unless otherwise explicitly stated $x \in X$, $y \in Y$ and $z \in Z$. Then given a fact $\widetilde{x}$ is A', the inferred output B' is obtained as sup $-T$ composition (cf. (6.20)) of $A'(x)$ and $R(x, y)$, i.e.,

$$B'(y) = A'(x) \overset{T}{\circ} R(x,y) = \sup_{x \in X} T(A'(x), R(x,y)) \qquad (8.4)$$
$$= \sup_{x \in X} T(A'(x), I(A(x), B(y))) \ ,$$

where T can be any t-norm (see WANGMING [254]). Observe that here we use the prefix notation instead of the infix notation used in Section 6.4.

In the case when the input A' is a fuzzy singleton (8.1) attaining normality at an $x_0 \in X$, then

$$B'(y) = A'(x) \overset{T}{\circ} R(x,y) = \sup_{x \in X} T(A'(x), R(x,y))$$
$$= T(A'(x_0), R(x_0, y)) = T(1, R(x_0, y)) = R(x_0, y) \ . \qquad (8.5)$$

In the case of Multi-Input Single-Output (MISO) fuzzy rule given by

$$\text{IF } \widetilde{x} \text{ is } A \text{ AND } \widetilde{y} \text{ is } B \text{ THEN } \widetilde{z} \text{ is } C \ , \qquad (8.6)$$

the relation R is given by

$$R(x, y; z) = I(A(x) \odot B(y), C(z)) \ , \qquad (8.7)$$

where $\odot$, also called the *antecedent combiner*, is usually a t-norm. In fact, if $A \in \mathcal{F}(X)$ and $B \in \mathcal{F}(Y)$, for some non-empty sets X, Y, then we have $A(x) \odot B(y) = T(A, B)(x, y)$, the Cartesian product of the fuzzy sets A, B with respect to the t-norm T. We will exploit the above notational equivalence in the sequel for enhanced readability.

Consider a single MISO rule of the type (8.6), denoted by $(A, B) \longrightarrow C$, for simplicity. Then given a multiple-input (A', B') the inferred output C', taking the sup $-T$ composition, is given by

$$C' = (A', B') \overset{T}{\circ} ((A, B) \longrightarrow C) \ , \qquad (8.8)$$

i.e.,

$$C'(z) = \sup_{(x,y) \in X \times Y} T(A'(x) \odot B'(y), I(A(x) \odot B(y), C(z))) \ . \qquad (8.9)$$

As in the SISO case, when the inputs are fuzzy singletons A' and B' attaining normality at the points $x_0 \in X$, $y_0 \in Y$, respectively, we have the form similar to (8.5), i.e.,

$$C'(z) = (A(x_0) \odot B(y_0)) \longrightarrow C(z) = R(x_0, y_0; z) \ . \qquad (8.10)$$

Example 8.3.5. Let

$$A = [0.9\ 0.8\ 0.7\ 0.7]\,,$$
$$B = [1\ 0.6\ 0.8]\,,$$
$$C = [0.1\ 0.1\ 0.2]$$

denote fuzzy sets defined on, respectively, the following classical sets

$$X = \{x_1, x_2, x_3, x_4\}\,,$$
$$Y = \{y_1, y_2, y_3\}\,,$$
$$Z = \{z_1, z_2, z_3\}\,.$$

Let the antecedent combiner $\odot$ be the product t-norm $T_{\mathbf{P}}(x,y) = xy$. Then taking the Cartesian product of A and B with respect to $T_{\mathbf{P}}$, we have

$$A \odot B = T_{\mathbf{P}}(A,B) = \begin{pmatrix} 0.9 & 0.54 & 0.72 \\ 0.8 & 0.48 & 0.64 \\ 0.7 & 0.42 & 0.56 \\ 0.7 & 0.42 & 0.56 \end{pmatrix}.$$

Let I be the Reichenbach implication $I_{\mathbf{RC}}(x,y) = 1 - x + xy$. Now, from (8.7), with $R(z_i) = T_{\mathbf{P}}(A,B) \longrightarrow z_i$, for $i = 1,2,3$, we have:

$$R(z_1) = R(z_2) = \begin{pmatrix} 0.19 & 0.514 & 0.352 \\ 0.28 & 0.568 & 0.424 \\ 0.37 & 0.622 & 0.496 \\ 0.37 & 0.622 & 0.496 \end{pmatrix},$$

$$R(z_3) = \begin{pmatrix} 0.28 & 0.568 & 0.424 \\ 0.36 & 0.616 & 0.488 \\ 0.44 & 0.664 & 0.552 \\ 0.44 & 0.664 & 0.552 \end{pmatrix}.$$

Let $A' = [0\ 0\ 1\ 0], B' = [0\ 1\ 0]$ be the given fuzzy singleton inputs. Then

$$A' \odot B' = T_{\mathbf{P}}(A',B') = \begin{pmatrix} 0 & 0 & 0 \\ 0 & 0 & 0 \\ 0 & 1 & 0 \\ 0 & 0 & 0 \end{pmatrix}. \tag{8.11}$$

Taking the sup $-T_{\mathbf{M}}$ composition, we finally have

$$C' = T_{\mathbf{P}}(A',B') \overset{T_{\mathbf{M}}}{\circ} I_{\mathbf{RC}}(T_{\mathbf{P}}(A,B),C) = [0.622\ 0.622\ 0.664]\,. \tag{8.12}$$

8.3.3 Inference in CRI with Multiple Rules

Given a fuzzy rule base of m rules, of the form

$$R_i : \text{ IF } \widetilde{x} \text{ is } A_i \text{ THEN } \widetilde{y} \text{ is } B_i \ , \qquad i = 1, 2, \ldots, m \ , \tag{8.13}$$

where $A_i \in \mathcal{F}(X)$, $B_i \in \mathcal{F}(Y)$ for $i = 1, 2, \ldots, m$, the corresponding fuzzy relations R_i are given as in (8.3). Then given a fact $\widetilde{x}$ is A', there are two ways in which one can aggregate over the rules and infer the output fuzzy set B'.

In the case of Multi-Input Single-Output (MISO) rules of the form given below, with n input domains X_j,

$$\begin{aligned} R_i : \text{ IF } \widetilde{x}_1 \text{ is } A_{i1} \text{ AND } \widetilde{x}_2 \text{ is } A_{i2} \text{ AND } \ldots \text{ AND } \widetilde{x}_n \text{ is } A_{in} \\ \text{THEN } \widetilde{y} \text{ is } B_i \ , \qquad i = 1, 2, \ldots, m \ , \end{aligned} \tag{8.14}$$

where $A_{ij} \in \mathcal{F}(X_j)$, $B_i \in \mathcal{F}(Y)$ for $i = 1, 2, \ldots, m$; $j = 1, \ldots, n$, the corresponding fuzzy relations R_i, $i = 1, 2, \ldots, m$, are given as in (8.7), using the Cartesian product of fuzzy sets A_{ij}, for $j = 1, \ldots, n$.

Remark 8.3.6. Since both in the case of MISO and SISO rules the input fuzzy set(s) can be seen to be a fuzzy set on either a single domain or a Cartesian product of the domains, in the sequel we consider only the SISO case. The MISO case will only be considered wherever the t-norm employed to obtain the Cartesian product of the antecedents plays a role.

First Infer Then Aggregate - FITA

In FITA, all the m relations R_i are preserved. Given an input $\widetilde{x}$ is A', we first compose A' with each of the relations R_i to get B_i's, i.e., infer B_i' individually from R_i:

$$B_i'(y) = A'(x) \overset{T}{\circ} R_i(x, y) = A'(x) \overset{T}{\circ} I(A_i(x), B_i(y)) \ . \tag{8.15}$$

Then the overall output fuzzy set $B'(y)$ is obtained by the aggregating the individually inferred B_i's, i.e., from

$$\begin{aligned} U_{ind} : B'(y) &= \widehat{T}(B_1'(y), B_2'(y), \ldots, B_m'(y)) \\ &= \widehat{T}_{i=1}^{m}(B_i'(y)) = T_{i=1}^{m}(A'(x) \overset{T}{\circ} R_i(x, y)) \\ &= \widehat{T}_{i=1}^{m}(\sup_{x \in X} T(A'(x), R_i(x, y))) \\ &= \widehat{T}_{i=1}^{m}(\sup_{x \in X} T(A'(x), I(A_i(x), B_i(y)))) \ , \end{aligned} \tag{8.16}$$

where $\widehat{T}$ is once again a t-norm and I is a fuzzy implication.

First Aggregate Then Infer - FATI

In FATI, we first aggregate all the rules (or relations) R_i into a single 'aggregated' relation R,

$$R(x,y) = \widehat{T}_{i=1}^{m}(R_i(x,y)) = \widehat{T}_{i=1}^{m}(I(A_i(x), B_i(y)) , \tag{8.17}$$

where $\widehat{T}$ is any t-norm. Given an input $\widetilde{x}$ is A', we then infer B' by,

$$\begin{aligned} U_{comb} : B'(y) &= A'(x) \overset{T}{\circ} R(x,y) = A'(x) \overset{T}{\circ} \widehat{T}_{i=1}^{m}(R_i(x,y)) \\ &= \sup_{x\in X} T\big(A'(x), \widehat{T}_{i=1}^{m}(R_i(x,y))\big) \\ &= \sup_{x\in X} T\big(A'(x), \widehat{T}_{i=1}^{m}(I(A_i(x), B_i(y)))\big) . \end{aligned} \tag{8.18}$$

Remark 8.3.7. (i) One can also use any t-conorm $\widehat{S}$ for the aggregation instead of a $\widehat{T}$ in (8.15). However, in the case of implicative rules - when the antecedents of the rules and its consequents are related by a fuzzy implication - i.e., when $R = I$, we know that the rules are viewed as fuzzy constraints and hence the more rules the more constraints to be satisfied and hence based on the principle of minimum specificity, the rules are aggregated conjunctively. Hence in the sequel we only consider a t-norm $\widehat{T}$ for aggregating over the rules. For more details, see, e.g., Kiszka et al. [139], Dubois et al. [91], Di Nola et al. [79] and Mantaras [166].

(ii) Moreover, when the fuzzy rule base consists of implication-based rules, i.e., $R = I$, it is well known that (see Dubois et al. [91] or Klir and Yuan [147]) FATI gives more specific result than that obtained by inferring rule by rule as in FITA. In fact, the result of FITA is a superset of that obtained by FATI, i.e.,

$$A'(x) \overset{T}{\circ} \Big(\bigcap R_i(x,y)\Big) \subseteq \bigcap \Big(A'(x) \overset{T}{\circ} R_i(x,y)\Big) .$$

8.3.4 Similarity Based Reasoning (SBR)

Consider once again the fuzzy IF-THEN rule (8.2). Let the given input be $\tilde{x}$ is A'. Inference in *similarity based reasoning* (SBR) schemes in AR is based on the calculation of a measure of compatibility or similarity $M(A, A')$ of the input A' to the antecedent A of the rule, and the use of a modification function J to modify the consequent B, according to the value of $M(A, A')$.

Some of the well known examples of SBR are *compatibility modification inference* (CMI) in Cross and Sudkamp [65], *approximate analogical reasoning scheme* (AARS) in Türksen and Zhong [245] and *consequent dilation rule* (CDR) in Morsi and Fahmy [187], Magrez and Smets [162], Chen [52], etc. In this section, we detail the typical inferencing mechanism in SBR, both in the case of SISO and MISO fuzzy rule bases.

Matching Function M

Definition 8.3.8. *Let X be a non-empty set and $\mathcal{F}(X)$ the fuzzy power set of X. Any function $M\colon \mathcal{F}(X) \times \mathcal{F}(X) \to [0,1]$ is said to be a* matching function.

Let X be a non-empty set and $A, A' \in \mathcal{F}(X)$. A matching function M operates on A, A' to give a real number in $[0,1]$, which can be interpreted in many ways, for instance, as a degree of similarity between A and A' or as a subsethood measure of A' in A. In general, an $s = M(A, A') \in [0,1]$ gives a measure of the compatibility of A' to A.

Example 8.3.9. Let X be a non-empty set and $A, A' \in \mathcal{F}(X)$. Below we list a few of the matching functions employed in the literature.

- Zadeh's max-min:

$$M_{\mathbf{Z}}(A, A') = \max_{x \in X} \min(A(x), A'(x)) .$$

- Magrez - Smets' measure [162]:

$$M_{\mathbf{M}}(A, A') = \max_{x \in X} \min(N(A(x)), A'(x)) ,$$

  where N is a fuzzy negation.
- Measure of subsethood [187]:

$$M_{\mathbf{S}}(A, A') = \min_{x \in X} I(A'(x), A(x)) ,$$

  where I is a fuzzy implication.
- Scalar product [52]:

$$M_{\mathbf{C}}(A, A') = \frac{A \cdot A'}{\max(A \cdot A, A' \cdot A')} ,$$

  where the domain X is discretized into n points, i.e., $X = \{x_1, x_2, \ldots, x_n\}$ and hence $A, A' \in [0,1]^n$ with '$\cdot$' as the scalar product of the 'vectors' A, A'.
- Disconsistency measure [245]:

$$M_{\mathbf{Tk}}(A, A') = \sqrt{\frac{\sum_{i=1}^{n} (A(x_i) - A'(x_i))^2}{n}} ,$$

  where once again the domain X is discretized into n points.

The following are some of the desirable properties of a matching function M. If $A_1, A_2, A \in \mathcal{F}(X)$, then

$$A_1 \subseteq A_2 \Longrightarrow M(A_1, A) \geq M(A_2, A) , \tag{MF1}$$

$$M(A, \overline{A}) = M(\overline{A}, A) = 0 , \tag{MF2}$$

$$M(A, A) = 1 , \tag{MF3}$$

$$M(A_1, A_2) = M(A_2, A_1) . \tag{MF4}$$

Property (MF1) states that if A_1 is more specific than A_2 then it should reflect higher on its similarity value with respect to any A. Properties (MF2), (MF3) and (MF4) are all self-explanatory.

Modification Function J

Let us again consider

$$\text{IF } \tilde{x} \text{ is } A \text{ THEN } \tilde{y} \text{ is } B ,$$

to be the given SISO fuzzy IF-THEN rule and $\tilde{x}$ *is* A' the observed fuzzy input. Let $s = M(A, A') \in [0, 1]$ be a measure of the compatibility of A' to A.

Let Y be a non-empty set and $B \in \mathcal{F}(Y)$. The modification function J is again a function from $[0, 1]^2$ to $[0, 1]$ and produces a modification $B' \in \mathcal{F}(Y)$ based on s and B, i.e., the consequence in SBR, using the modification function J, is given by

$$B'(y) = J(s, B(y)) = J(M(A, A'), B(y)) , \qquad y \in Y . \tag{8.19}$$

In approximate analogical reasoning scheme (AARS) the following modification operators have been proposed (cf. [245]):

(i) *More or Less*:

$$J_{\mathbf{ML}}(s, B) = B'(x) = \min\left(1, \frac{B(x)}{s}\right) ,$$

(ii) *Membership Value Reduction*:

$$J_{\mathbf{MVR}}(s, B) = B'(x) = B(x) \cdot s ,$$

for any $x \in X$. Once again in $J_{\mathbf{MVR}}$ the multiplication '·' can be generalized to any t-norm T. In CMI [65] and CDR [187] J is taken to be a fuzzy implication. Note that J need not be either commutative or associative.

Aggregation Function G

In the case of multiple rules of the form given in (8.13) we infer the final output by aggregating over the rules, using an associative aggregation operation $G\colon [0, 1]^2 \to [0, 1]$:

$$B'(y) = G_{i=1}^{m} \left(J\left(M(A_i, A'), B_i(y) \right) \right) , \qquad y \in Y . \tag{8.20}$$

Usually, G is either a t-norm, t-conorm or a uninorm.

SBR Inference for MISO Fuzzy Rule Base

So far, we have seen the inference process in SBR scheme for the case of (multiple) SISO rules. On the other hand, if we consider a lone Multi Input Single Output (MISO) fuzzy IF-THEN rule of the form (8.14) (where $m = 1$, of course), then given the input that ($\widetilde{x}_1$ is A'_1; ... ; $\widetilde{x}_n$ is A'_n), the input along every dimension, i.e., every $A'_j \in \mathcal{F}(X_j)$ is matched to the corresponding antecedent $A_j \in \mathcal{F}(X_j)$ to produce the similarity value s_j, i.e., $s_j = M(A_j, A'_j)$, $j = 1, \ldots, n$.

Now, the consequent is modified, based not on the individual s_i values, but on the aggregated value of these n similarity values s_i. To this end the following operator is employed.

Table 8.1. Some SBR inference schemes along with their inference operations, where T is any t-norm, S is any t-conorm, I is any fuzzy implication and Avg. is the averaging operation

SBR scheme	G	J	K	M
CMI [65]	T	I	T	$M_{\mathbf{Z}}$
AARS [245]	$S_{\mathbf{M}}$	$J_{\mathbf{MVR}}, J_{\mathbf{ML}}$	Avg.	$M_{\mathbf{Tk}}, M_{\mathbf{Z}}$
CDR [187]	$T_{\mathbf{M}}$	I	–	$M_{\mathbf{S}}$

Combiner Function K

Let $K\colon [0,1]^2 \to [0,1]$ be an associative and commutative function that combines the matching degrees of A_j to A'_j, for all $j = 1, \ldots, n$. We refer to K as the *combiner* in the sequel. Now, the consequence of an individual MISO rule is given by

$$\begin{aligned} B'(y) &= J(K(M(A_1, A'_1), \ldots, M(A_n, A'_n)), B(y)) \,, \\ &= J\left(K(s_1, \ldots, s_n), B(y)\right) \,, \qquad y \in Y \,. \end{aligned} \tag{8.21}$$

Once again, typically, K is a t-norm or a t-conorm or a uninorm.

In the case of multiple MISO rules of the form (8.14) (where $m > 1$), given the input that ($\widetilde{x}_1$ is A'_1; ... ; $\widetilde{x}_n$ is A'_n), we infer the final output by aggregating over the rules,

$$B'(y) = G_{i=1}^{m}\left(J\left(K_{j=1}^{n}(M(A_{ij}, A'_j)), B(y)\right)\right) \,, \qquad y \in Y \,. \tag{8.22}$$

8.4 Effectiveness of Inference Schemes in AR

Inferencing in AR, in essence, is to obtain fuzzy conclusions from imprecise inputs and fuzzy IF-THEN rules. Moreover, the inference schemes in AR often have to deal with a set of fuzzy IF-THEN rules *simultaneously*. This inherent fuzziness involved and the parallelism in the inference in determining the final conclusion have necessitated validation of the 'goodness' of the obtained conclusions.

Since AR is predominantly used in fuzzy control (FC), expert systems (ES) and decision support systems (DSS), the measures proposed so far can broadly be categorized as follows:

(i) **Objective:** In FC, the set of fuzzy IF-THEN rules can be seen as embedding the knowledge required to control the system under consideration. Hence the relationship between the inputs and the outputs obtained can be seen as interpolating an unknown system function through a fuzzy mapping (see GOTTWALD [114]). This obviously leads to a function approximation problem. Hence the inference mechanism is judged on how well it can approximate a given continuous function.

(ii) **Heuristic:** On the other hand, unlike in FC, in the domain of ES and DSS there are no *apriori* functions to approximate and hence objective evaluation is difficult, if not impossible. Still, based on some sound principles of inference in logic many works have proposed (see FUKAMI et al. [108, 109], BALDWIN and PILSWORTH [27], MIZUMOTO [186], DUBOIS and PRADE [92], KLIR and YUAN [147]) sets of axioms that an inference scheme is expected to obey and these in turn become a measure in determining the effectiveness of the inference. One such set of axioms is what has now come to be termed as 'GMP Rules'.

8.4.1 GMP Rules and AR

The following axioms were proposed to measure the goodness of an inference scheme that can be considered as a realisation of GMP. It should be emphasized that the conformance of an inference scheme to these axioms depend on the underlying fuzzy logic operators employed therein.

Let X, Y be some non-empty sets, $A, A', A'' \in \mathcal{F}(X)$ and $B, B', B'' \in \mathcal{F}(Y)$. Let us again consider

$$\text{IF } \tilde{x} \text{ is } A \text{ THEN } \tilde{y} \text{ is } B\ ,$$

to be the given SISO fuzzy IF-THEN rule. Assume also that whenever $\tilde{x}$ is $A'(A''$, respectively) is the observed fuzzy input, $\tilde{y}$ is $B'(B''$, respectively) is the obtained fuzzy output. Let us also define a special fuzzy set $V \in \mathcal{F}(Y)$ such that $V(y) = 1$ for all $y \in Y$. By $\overline{A}$, we denote the fuzzy-set complement of A with respect to a negation N and is given as $\overline{A}(x) = N(A(x))$, for all $x \in X$.

The commonly accepted axioms (see, for example, MAGREZ and SMETS [162], DEMIRLI and TÜRKSEN [73]) are as follows:

$$B' \supseteq B\ , \tag{GMP1}$$

$$A' \subseteq A'' \Longrightarrow B' \subseteq B''\ , \tag{GMP2}$$

$$A' = \overline{A} \Longrightarrow B' = V\ , \tag{GMP3}$$

$$A' = A \Longrightarrow B' = B\ . \tag{GMP4}$$

The property (GMP1) insists that the obtained inference cannot be more specific than the rule consequent, while (GMP2) insists on the monotonicity in the inference. Properties (GMP3) and (GMP4) deal with the cases when the input is either the complement of the rule antecedent, in which case the output is unspecified, or identical to the rule antecedent itself in which case the output is expected to be identical to the rule consequent. Immediately, it can be observed that (GMP4) is the classical modus ponens.

In the following two subsections we discuss the validity of these GMP rules with respect to the two inference schemes introduced above, viz., CRI and SBR.

GMP Rules and CRI

GMP1: We consider two cases here.

(i) If the input A' is a fuzzy singleton (8.1) attaining normality at an $x_0 \in X$ then we see from (8.5) that for any $y \in Y$, $B'(y) = I(A(x_0), B(y))$. Hence (GMP1) is satisfied in this case, i.e., $B'(y) \geq B(y)$ only if I has the following property:

$$I(p,q) \geq q\,, \qquad p,q \in [0,1]\,, \tag{8.23}$$

which is always true for any fuzzy implication that satisfies (NP).

(ii) In the case the input A' is not a fuzzy singleton, even if I has the property (8.23), by the bounds on a t-norm T, i.e., $T \leq T_{\mathbf{M}}$, used in the composition it can be seen from (8.4) that B' is not always greater than B.

GMP2: Let $y \in Y$ and $A' \subset A''$, i.e., $A'(x) \leq A''(x)$ for all $x \in X$. Then by the monotonicity of T

$$\begin{aligned} B''(y) &= A'' \overset{T}{\circ} R(A,B) \\ &= \sup_{x\in X} T(A''(x), I(A(x), B(y))) \\ &\geq \sup_{x\in X} T(A'(x), I(A(x), B(y))) \\ &= B'(y)\,. \end{aligned}$$

GMP3: Let $A' = \overline{A}$ be the complement of A with respect to a negation N. If A' is normal, then the kernel of $\overline{A}$, given by $X' = \{x \in X | \overline{A}(x) = 1\}$, is non-empty. Obviously, $A(x) = 0$ for every $x \in X'$. Now, for any $y \in Y$, we have

$$\begin{aligned} B'(y) &= \sup_{x\in X} T(\overline{A}(x), I(A(x), B(y))) \\ &\geq \sup_{x\in X'} T(\overline{A}(x), I(A(x), B(y))) \\ &= \sup_{x\in X'} T(\overline{A}(x), I(0, B(y))) \\ &= T(1,1) = 1\,. \end{aligned}$$

GMP4: Property (GMP4) reduces to the following functional equation:

$$B'(y) = \sup_{x\in X} T(A(x), I(A(x), B(y)))\,, \qquad y \in Y\,. \tag{8.24}$$

Here two seminal works have been done in analyzing when the generalized modus ponens coincides with the classical modus ponens. Towards this end, TRILLAS and VALVERDE [240, 241], proposed the following as the essential properties of any binary function T on $[0,1]$ - which in the following is denoted as m - employed in (8.24) should satisfy, where $x, x', y \in [0,1]$:

$$m(x, I(x,y)) \leq y \tag{MP1}$$
$$m(1,1) = 1 \tag{MP2}$$
$$m(0,1) = 0 \tag{MP3}$$
$$x \leq x' \Longrightarrow m(x,y) \leq m(x',y) \tag{MP4}$$

The authors call these functions m as *modus ponens generating functions* (MPGF) for a given fuzzy implication I.

While TRILLAS and VALVERDE [240, 241] investigated (8.24) by fixing the fuzzy implication I and characterizing functions $T = m$ (not necessarily t-norms) that satisfy (8.24), DUBOIS and PRADE [88, 89] considered the dual situation. The results of both the works are more or less similar and hence here we only cite the ones from the former works, which can now be slightly generalized as follows:

Theorem 8.4.1. *(i) If I is an (S,N)-implication $I_{S,N}$ obtained from a right-continuous t-conorm S and a continuous negation N, then the corresponding MPGF is given by:*

$$m_{I_{S,N}}(x,y) = \inf\{t \in [0,1] \mid S(N(x),t) \geq y\} .$$

(ii) If I is an R-implication I_T obtained from a left-continuous t-norm T, then T itself is the corresponding MPGF, i.e., $m_{I_T} = T$.

It is easy to verify that the $m_{I_{S,N}}$ and m_{I_T} are indeed MPGF and satisfy all the properties (MP1)–(MP4) and also (8.24). Moreover, it should be emphasized that $m_{I_{S,N}}$ is not always a t-norm.

One can readily recognize that (MP1) is the same as (TC) with less conditions on T. Thus every t-norm that satisfies (TC) with a chosen fuzzy implication I ensures that the CRI inference employing them does satisfy (8.24).

GMP Rules and SBR

Theorem 8.4.2. *For an SBR inference scheme with a matching function M and $J = I$, a fuzzy implication, the following are valid:*

(i) If I satisfies (8.23) *then SBR satisfies* (GMP1).
(ii) If the matching function M has (MF1) *then SBR satisfies* (GMP2).
(iii) If the matching function M has (MF2) *then SBR satisfies* (GMP3).
(iv) If the matching function M has (MF3) *and I satisfies* (NP) *then SBR satisfies* (GMP4).

Proof. Let $s \in [0,1]$ be the degree of similarity between an input A' and the rule antecedent A, i.e., $s = M(A', A)$.

(i) Once again, if I satisfies (8.23), then it can be easily seen that for any $y \in Y$ we have $B'(y) = I(s, B(y)) \geq B(y)$, i.e., SBR does satisfy (GMP1).
(ii) Let $y \in Y$. Let the matching function M have the property (MF1). Then, by (I1) of I, we have

$$B_2(y) = I(M(A_2, A), B) \geq I(M(A_1, A), B) = B_1(y) .$$

(iii) Let $A' = \overline{A}$. Since M has (MF2), we get

$$B'(y) = I(M(A, \overline{A}), B(y)) = I(0, B(y)) = 1 ,$$

for any $y \in Y$, i.e., $B' = V$ and hence SBR does satisfy (GMP3).
(iv) If the matching function M satisfies (MF3) and if I satisfies (NP), then $B'(y) = I(M(A, A), B(y)) = I(1, B(y)) = B(y)$ for any $y \in Y$, i.e., $B' = B$ and hence SBR does satisfy (GMP4).

8.4.2 Function Approximation and AR

One of the best known applications of approximate reasoning is in the field of fuzzy control. The main job of a controller (either conventional or fuzzy) is to control a plant or a system under consideration. The output of the plant becomes the input for the controller and vice-versa. A fuzzy controller typically contains a rule base of fuzzy IF-THEN rules - which is supposed to have captured the working of the plant to be controlled - and given an input (usually crisp) employs any of the inference schemes of AR to obtain the corresponding output. This output in turn becomes the input for the plant, that effectively controls the outputs of the plant by controlling its inputs.

Hence, the capability of a fuzzy controller to uniformly approximate any given function depends on the following two factors:

(i) the rule base of fuzzy IF-THEN rules,
(ii) the fuzzy logic operators employed in the inference scheme.

In this section, we will be concerned only with the cases where the antecedents and consequents of the rules are related by fuzzy implications. We highlight how fuzzy implications can enrich inference schemes in AR to have good approximation capabilities under certain conditions. A controller using an inference scheme in AR based reasoning with fuzzy sets and fuzzy logic operators is usually referred to as a *fuzzy inference system* (FIS) and hence we shall also use the same term.

Remark 8.4.3. As Gottwald [114] states:

> "The main mathematical problem of fuzzy control, besides the engineering problem to get a suitable list of linguistic control rules for the actual control problem, is (therefore) the interpolation problem to find a function
>
> $$\Phi^* \colon \mathcal{F}(X) \to \mathcal{F}(Y) ,$$

which interpolates these data, i.e., which satisfies

$$\Phi^*(A_i) = B_i \ , \qquad i = 1, \ldots, n \ ,$$

and which, in this way, gives a fuzzy representation for the control function Φ.

Actually the standard approach is to look for one single function, more precisely: for some uniformly defined function, which should interpolate all these data, and which should be globally defined over $\mathcal{F}(X)$, or at least over a suitably chosen sufficiently large subclass of $\mathcal{F}(X)$".

The above remark is nothing but a different formulation of the classical modus ponens rule, namely, (GMP4). Hence it is clear that a fuzzy inference system is, in effect, a function approximator. Hence, unlike in a DSS or an ES where a fuzzy output is acceptable, the output of an FIS is usually defuzzified into a real number. Thus in fuzzy control, the goodness of inference of an FIS is measured based on how closely the FIS can approximate the control system.

The first such controller was proposed by MAMDANI [163] and later by MAMDANI ASSILIAN [165]. It employed $T_{\mathbf{M}}$ for relating the antecedents and consequents of the rules and in the presence of multiple rules, $S_{\mathbf{M}}$ was used as the aggregation operator. Later on it was shown that FIS using these operators (also called as min-max-FIS) with the antecedents and consequents of the rules chosen appropriately are universal approximators, i.e., they can approximate any continuous function arbitrarily closely (see, e.g., NGUYEN and WALKER [188], Chap. 10).

It was BALDWIN and GUILD [28] who pointed out that though the output of a min-max-FIS is a continuous function its slope is not as smooth as that of a min-I-FIS, i.e., an FIS employing a fuzzy implication I to relate the antecedents and consequents of the rules and $T_{\mathbf{M}}$ as the aggregation operator.

Let us consider an FIS consisting of the SISO rules of the type given in (8.13). Moreover, let the antecedent fuzzy sets A_i, $i \in \mathbb{N}_m = \{1, 2, \ldots, m\}$ satisfy the following requirements:

(A1) A_i is continuous for every $i \in \mathbb{N}_m$;
(A2) A_i is normal for every $i \in \mathbb{N}_m$;
(A3) The set $\{A_i\}_{i \in \mathbb{N}_m}$ is a complete covering, i.e., for any $x \in X$ there exists an $i \in \mathbb{N}_m$ such that $A_i(x) > 0$;
(A4) The set $\{A_i\}_{i \in \mathbb{N}_m}$ forms a Ruspini partition, i.e., $\sum_{i \in \mathbb{N}_m} A_i(x) = 1$ for every $x \in X$.

A defuzzifier, in general, is any operation that maps a fuzzy set to a single value. In our context, a defuzzifier converts the conclusions of the inference mechanism into a value that is suitably interpreted. For more information on defuzzifiers and defuzzification procedures, see, for example, DRIANKOV et al. [84], PASSINO and YURKOVICH [197]. Under the above assumptions LI et al. [152] have shown the following:

Theorem 8.4.4. *Consider a* min-I*-FIS consisting of the SISO rules of the type given in* (8.13), *whose antecedent fuzzy sets* A_i, $i \in \mathbb{N}_m = \{1, 2, \ldots, m\}$ *satisfy the assumptions (A1) - (A4). Then the FIS employing the Averaging of*

Maximum (MOA) defuzzifier, where the fuzzy implication used to relate the antecedents to the consequents is either an R-, (S,N)- or a QL-implication, is a universal approximator.

That is, for any continuous function $f\colon [a,b] \to \mathbb{R}$ and an arbitrary positive number ε, there is an FIS, its corresponding system function $y = G(x)$ based on an R-, (S,N)- or a QL-implication with MOA defuzzifier satisfies the inequality

$$\max_{x\in[a,b]} |f(x) - G(x)| < \varepsilon \,.$$

Remark 8.4.5. (i) Conditions (A1) - (A4) on the fuzzy sets are not very stringent. In practice, these are insisted as desirable properties, see, e.g., MARTIN-CLOUIARE [168], GOTTWALD [114, 115, 116] and hence the significance of the above result is immediate.
(ii) It should be noted that the above result is true under the same conditions on an FIS even when the rule base consists of MISO rules (see LI et al. [153]).

8.5 Efficiency of Inference Schemes in AR

The complexity of an inference algorithm stems mainly from two factors:

(i) *The process of inference itself.* The inferencing schemes in AR are generally resource consuming (both memory and time). Many of the inference schemes discretize the underlying domains and hence the process becomes computationally intensive.
(ii) *The structure, complexity and the number of rules.* Depending on the shape of the underlying fuzzy sets the number of parameters stored and processed varies. Similarly, the manner in which multiple antecedents are combined affects the processing complexity. Moreover, an increase in the number of rules only exacerbates the problem. As the number of input variables and/or input fuzzy sets increases, there is a combinatorial explosion of rules in multiple fuzzy rule based systems.

Remark 8.5.1. Despite the prevalent use of CRI the following are usually cited as its drawbacks (see, e.g., CORNELIS et al. [62], MARTIN-CLOUAIRE [168], DEMIRLI and TÜRKSEN [73]):

(i) *Computational complexity*: Although the computational complexity largely depends on the choice of operators employed, in general, for an n-input 1-output system $(A_1, A_2, \ldots, A_n) \longrightarrow B$ with the cardinality of the base sets X_i of each of the inputs A_i being n_i the complexity of a single inference is proportional to $\mathcal{O}\left(\prod_{i=1}^{n} n_i\right)$. If $n_i = m$, then it is $\mathcal{O}(m^n)$.
(ii) *Space complexity:* Again, for an n-input 1-output system we have an n-dimensional matrix having $\prod_{i=1}^{n} n_i$ entries. Therefore we need to store n-dimensional matrices for every fuzzy IF-THEN rule.

For example, consider inferencing with the CRI inference scheme (8.9) in the case of a 2-input fuzzy IF-THEN rule. Let the base sets X, Y, Z be discretized into m, n, l points, respectively. Then the memory requirements of the algorithm are as follows (see, for instance, DEMIRLI and TÜRKSEN [73]): $m \cdot n \cdot l$ for $(A(x) \odot B(y)) \longrightarrow C$, $m \cdot n$ for combining the given facts $A'(x) \odot B'(y)$ and l for the consequent. Overall, it is $m \cdot n \cdot l + m \cdot n + l$. In the case $m = n = l$, the memory requirements of the algorithm become $m^3 + m^2 + m$.

The many attempts towards reducing this complexity in the inference schemes in AR can be broadly classified along the following lines:

- Modification of the original inference algorithm,
- Transformation of the structure of the rules in the given rule base.

Remark 8.5.2. It is interesting to note that, on the one hand, in the case of CRI inference since we employ FATI procedure with implicative rules, the number of rules does not add to the complexity while, as shown in Remark 8.5.1, the multi-dimensionality of rules does influence the complexity of the algorithm. On the other hand, in SBR, even in the case of implicative rules, since the inference is done locally with each rule and finally aggregated the number of rules does influence the complexity, while the multi-dimensionality of rules is unlikely to add any additional complexity to what is inherent in the algorithm.

8.5.1 Modification of the CRI Inference Algorithm

Recently, JAYARAM [126] has introduced a modified CRI called Hierarchical CRI (HCRI), whose space complexity is lesser than the classical CRI, and shown that under certain conditions on the underlying operators the inference obtained from the Hierarchical CRI and the classical CRI are identical.

We first present the HCRI inference algorithm and discuss its complexity vis-á-vis the classical CRI's complexity (see Remark 8.5.1). Following this we give a sufficient condition under which the inference obtained from the Hierarchical CRI and the classical CRI are identical, which highlights the role played by the law of importation in AR.

Hierarchical CRI

In the field of fuzzy control, Hierarchical Fuzzy Systems (HFS) hold a centre stage, since, where applicable, they help to a large extent in breaking down the complexity of the system being modelled, both in terms of efficiency and understandability. For a good survey of HFS see TORRA [231] and the references therein.

JAYARAM [126] proposed a new, modified form of CRI termed 'Hierarchical CRI' owing to the way in which multiple antecedents of a fuzzy rule are operated upon in the inferencing.

Procedure for Hierarchical CRI

Step 1. Calculate $R' = B \longrightarrow C$.
Step 2. Calculate $C^* = B' \overset{T}{\circ} R' = B' \overset{T}{\circ} (B \longrightarrow C)$.
Step 3. Calculate $R'' = A \longrightarrow C^*$.
Step 4. Finally, calculate

$$\begin{aligned} C'' &= A' \overset{T}{\circ} R'' = A' \overset{T}{\circ} (A \longrightarrow C^*) \\ &= A' \overset{T}{\circ} \{A \longrightarrow (B' \overset{T}{\circ} (B \longrightarrow C))\} . \end{aligned} \tag{8.25}$$

We illustrate the gain in efficiency from using the Hierarchical CRI through the following example. For simplicity, we have considered the sup – min, i.e., sup $-T_{\mathbf{M}}$, composition in the example, though it can be substituted by a sup $-T$ composition, for any t-norm T.

Example 8.5.3. Let the fuzzy sets A, B, C be as in Example 8.3.5. Let I be the Łukasiewicz implication $I_{\mathbf{LK}}(x, y) = \min(1, 1 - x + y)$ and the antecedent combiner $\odot$ be the t-norm $T_{\mathbf{M}}(x, y) = \min(x, y)$.

Case 1: Inference with the classical CRI

Taking the Cartesian product of A and B with respect to $T_{\mathbf{M}}$, we have

$$T_{\mathbf{M}}(A, B) = \begin{pmatrix} 0.9 & 0.6 & 0.8 \\ 0.8 & 0.6 & 0.8 \\ 0.7 & 0.6 & 0.7 \\ 0.7 & 0.6 & 0.7 \end{pmatrix} .$$

Now, with $x \longrightarrow y = I_{\mathbf{LK}}(x, y)$, $T_{\mathbf{M}}(A, B) \longrightarrow C$ from (8.7) will be given by $R(A, B; C) = [R(z_1)\ R(z_2)\ R(z_3)]$, where

$$T_{\mathbf{M}}(A, B) \longrightarrow C = T_{\mathbf{M}}(A, B) \longrightarrow [0.1\ 0.1\ 0.2]$$

and $R(z_i) = T_{\mathbf{M}}(A, B) \longrightarrow z_i$. Thus

$$R(z_1) = R(z_2) = \begin{pmatrix} 0.2 & 0.5 & 0.3 \\ 0.3 & 0.5 & 0.3 \\ 0.4 & 0.5 & 0.4 \\ 0.4 & 0.5 & 0.4 \end{pmatrix} ,$$

$$R(z_3) = \begin{pmatrix} 0.3 & 0.6 & 0.4 \\ 0.4 & 0.6 & 0.4 \\ 0.5 & 0.6 & 0.5 \\ 0.5 & 0.6 & 0.5 \end{pmatrix} .$$

Let $A' = [0\ 0\ 1\ 0], B' = [0\ 1\ 0]$ be the given fuzzy singleton inputs. Then $T_{\mathbf{M}}(A', B')$ is still as given in (8.11). Taking the sup − min composition, we have

$$C' = T_{\mathbf{M}}(A', B') \overset{T_{\mathbf{M}}}{\circ} (T_{\mathbf{M}}(A, B) \longrightarrow C) = [0.5\ \ 0.5\ \ 0.6]\,. \tag{8.26}$$

Case 2: Inference with the Hierarchical CRI

Inferencing with the Hierarchical CRI, given an input (A', B') we have

$$\textbf{Step 1}\quad B \longrightarrow C = I_{\mathbf{LK}}(B, C) = \begin{pmatrix} 0.1 & 0.1 & 0.2 \\ 0.5 & 0.5 & 0.6 \\ 0.3 & 0.3 & 0.4 \end{pmatrix}$$

$$\begin{aligned} \textbf{Step 2}\qquad C^* &= B' \overset{T_{\mathbf{M}}}{\circ} (B \longrightarrow C) \\ &= [0\ 1\ 0] \overset{T_{\mathbf{M}}}{\circ} (B \longrightarrow C) \\ &= [0.5\ \ 0.5\ \ 0.6] \end{aligned}$$

$$\textbf{Step 3}\quad A \longrightarrow C^* = I_{\mathbf{LK}}(A, C^*) = \begin{pmatrix} 0.6 & 0.6 & 0.7 \\ 0.7 & 0.7 & 0.8 \\ 0.8 & 0.8 & 0.9 \\ 0.8 & 0.8 & 0.9 \end{pmatrix}$$

$$\begin{aligned} \textbf{Step 4}\qquad C'' &= A' \overset{T_{\mathbf{M}}}{\circ} (A \longrightarrow C^*) \\ &= [0\ 0\ 1\ 0] \overset{T_{\mathbf{M}}}{\circ} (A \longrightarrow C^*) \\ &= [0.8\ \ 0.8\ \ 0.9]\,. \end{aligned}$$

Remark 8.5.4. From the above example it is clear that we can convert a multi-input system employing CRI inference to a single-input hierarchical system employing CRI. The effect becomes more pronounced when we have more than two input variables. From the listed advantages of Hierarchical CRI in JAYARAM [126], the following is of interest in this context. From the above Example 8.5.3 it can be noticed that the most memory intensive step in the inference is the calculation of the 'current' output fuzzy set (Steps 2 & 4). Once again, considering the case of a k-input fuzzy IF-THEN rule, if the input universes of discourse $X_i, i = 1, 2, \ldots, k$ are discretized into n_i points and the output base set Z into l points, then the memory requirements of this step, and hence of the algorithm itself, can easily be seen to be $n^* \cdot l + l + n^*$, where $n^* = \max_{i=1}^{k} n_i$. In the case $m = n = l$ we have the overall memory requirements to be $2m + m^2$. It should also be emphasized that the memory requirements are independent of the number of input variables, as can be expected in any hierarchical setting.

Although the Example 8.5.3 illustrates the computational efficiency of Hierarchical CRI, we see that the inference obtained from the classical CRI is different from the one obtained from the proposed Hierarchical CRI, i.e., $C' \neq C''$. The following result proposes some sufficiency conditions under which the outputs of the classical and Hierarchical CRI schemes are identical, for the same inputs and thus highlights the importance of the law of importation (LI) discussed in Chap. 7, which is summarized in the following result.

Theorem 8.5.5. *Let the inputs to the fuzzy system be fuzzy singletons. If the t-norm T employed for the antecedent combiner and the implication I are such that* (LI) *holds, then* (8.8) *and* (8.25) *are equivalent, i.e.,*

$$(A', B') \overset{T_{\mathbf{M}}}{\circ} ((A, B) \longrightarrow C) \equiv A' \overset{T_{\mathbf{M}}}{\circ} (A \longrightarrow (B' \overset{T}{\circ} (B \longrightarrow C))) .$$

Example 8.5.6. Let the fuzzy sets A, B, C be as defined in Example 8.3.5. In Example 8.3.5 the antecedent combiner was taken to be the product t-norm $T_{\mathbf{P}}$ and the considered implication I was the Reichenbach implication $I_{\mathbf{RC}}$ which is an S-implication. From Theorem 7.3.2 and Table 7.1 we see that the pair $(T_{\mathbf{P}}, I_{\mathbf{RC}})$ does satisfy the conditions given in Theorem 8.5.5.

In this example we infer using the Hierarchical CRI for the identical fuzzy singleton inputs A', B' given in Example 8.3.5 and show that the output obtained is the same as that in Example 8.3.5, viz., (8.12). Inferencing with the Hierarchical CRI, given the input (A', B'), we have

$$\textbf{Step 1} \quad B \longrightarrow C = I_{\mathbf{RC}}(B, C) = \begin{pmatrix} 0.10 & 0.10 & 0.20 \\ 0.46 & 0.46 & 0.52 \\ 0.28 & 0.28 & 0.36 \end{pmatrix}$$

$$\begin{aligned} \textbf{Step 2} \qquad C^* &= B' \overset{T_{\mathbf{P}}}{\circ} [B \longrightarrow C] \\ &= [0\ 1\ 0] \overset{T_{\mathbf{P}}}{\circ} [B \longrightarrow C] \\ &= [0.46\ \ 0.46\ \ 0.52] . \end{aligned}$$

Now, after some, rather tedious calculations, it can be seen that

$$\begin{aligned} C'' &= A' \overset{T_{\mathbf{P}}}{\circ} (A \longrightarrow C^*) \\ &= [0\ 0\ 1\ 0] \overset{T_{\mathbf{P}}}{\circ} (A \longrightarrow C^*) \\ &= [0.622\ \ 0.622\ \ 0.664] . \end{aligned} \tag{8.27}$$

Quite evidently, the inference obtained from the Hierarchical CRI C'' (8.27) is equal to that obtained from the classical CRI C' (8.12), under the conditions of Theorem 8.5.5.

Remark 8.5.7. Similarly, in Example 8.5.3, the employed fuzzy implication I is the Łukasiewicz implication $I_{\mathbf{LK}}$ with minimum t-norm $T_{\mathbf{M}}$ as the antecedent combiner. Now, according to the conditions given in Theorem 8.5.5 if we consider the Łukasiewicz t-norm $T_{\mathbf{LK}}$ as the antecedent combiner it can be easily verified that the outputs obtained from the classical CRI and the Hierarchical CRI are indeed identical.

8.5.2 Transformation of the Structure of the Rules

In the case of a MISO rule, as pointed out in Remark 8.5.1, a number of multi-dimensional arrays need to be stored and processed in the CRI inference scheme. Techniques that transform the rules themselves have been proposed to reduce the complexity in the inference algorithm. DEMIRLI and TÜRKSEN [73], and later on COMBS and ANDREWS [56], have suggested to break up complex rules into a set of simpler rules, so that the inference can be perfomed with these simpler rules. On the other hand, JAYARAM [125] have proposed the merging of rules having identical consequents towards faster on-line inferencing.

Demirli and Türksen's Rule Break Up Method for CRI

From Remark 8.5.1(ii) we see that the memory requirements to infer a conclusion with a single MISO rule is m^3+m^2+m, where m is the discretisation granularity of the underlying sets. Evidently, this is mainly due to the manner of inference in CRI, which reasons with the multi-dimensional fuzzy sets obtained by combining the multiple facts and also from the multiple antecedents in a rule.

Towards reducing this complexity, DEMIRLI and TÜRKSEN [73] proposed to break an MISO rule of the form $(A, B) \longrightarrow C$ into two simpler SISO rules of the form $A \longrightarrow C$ and $B \longrightarrow C$. Subsequently the inference is done with these SISO rules independently and the obtained local conclusions are aggregated appropriately to give the final conclusion. The above discussion is summarized in the following result of DEMIRLI and TÜRKSEN [73] and is illustrated with an example.

Theorem 8.5.8. *Let a single MISO rule of the type* (8.6), *denoted* $(A, B) \longrightarrow C$, *be given along with a multiple-input* (A', B'), *where* A, A', B, B' *are fuzzy sets on appropriate non-empty sets. Consider the CRI inference scheme* (8.9) *where* $T = T_{\mathbf{M}}$ *and* I *is a fuzzy implication, i.e., the MISO rule is an implicative rule. If the inputs* A', B' *are normal, then the CRI inference obtained with the given 'complex' MISO rule is equivalent to the aggregation of the CRI inference obtained with the simpler rules, for an appropriate aggregation operator.*

In other words, for the above rule and inputs the fuzzy sets C, C' *obtained as follows are equal, i.e.,*

$$C = (A', B') \overset{T_{\mathbf{M}}}{\circ} ((A, B) \longrightarrow C) \,,$$

$$C' = (A' \overset{T_{\mathbf{M}}}{\circ} A \longrightarrow C) \cup (B' \overset{T_{\mathbf{M}}}{\circ} B \longrightarrow C) \,,$$

where $\cup$ *is interpreted as the t-conorm* $S_{\mathbf{M}}$.

Although the above result was originally proposed for S- and R-implications, from the proof it is obvious that it is true for any binary operation on $[0, 1]$ that is non-increasing in the first variable, i.e., satisfies (I1), and hence is valid for any fuzzy implication. It should be emphasized that if the inputs A', B' are not normalized, then the above equivalence is not guaranteed (see the working example in DICK and KANDEL [76]).

Example 8.5.9. Consider Example 8.5.3 once again where the antecedent combiner is the t-norm $T_{\mathbf{M}}$. For the same sets A, B, A', B' we have

$$\textbf{Step 1}\ A \longrightarrow C = \begin{pmatrix} 0.2 & 0.2 & 0.3 \\ 0.3 & 0.3 & 0.4 \\ 0.4 & 0.4 & 0.5 \\ 0.4 & 0.4 & 0.5 \end{pmatrix}$$

$$\textbf{Step 2}\quad C_1 \quad = A' \overset{T_{\mathbf{M}}}{\circ} (A \longrightarrow C) = [0.4\ \ 0.4\ \ 0.5]$$

$$\textbf{Step 3}\ B \longrightarrow C = \begin{pmatrix} 0.1 & 0.1 & 0.2 \\ 0.5 & 0.5 & 0.6 \\ 0.3 & 0.3 & 0.4 \end{pmatrix}$$

$$\textbf{Step 4}\quad C_2 \quad = B' \overset{T_{\mathbf{M}}}{\circ} (B \longrightarrow C) = [0.5\ \ 0.5\ \ 0.6]$$

$$\textbf{Step 5}\ C_1 \vee C_2 = [0.5\ \ 0.5\ \ 0.6] = C' \text{ from (8.26).}$$

Remark 8.5.10. The memory requirements for the Rule Break Up method can be easily seen to be $(m \cdot l + m)$ for $A' \overset{T}{\circ} (A \longrightarrow C)$, $(n \cdot l + n)$ for $B' \overset{T}{\circ} (B \longrightarrow C)$ and $2l$ for both the local conclusions C_1, C_2. In the case $m = n = l$ we have the overall memory requirements to be $2m^2 + 4m$ against $m^3 + m^2 + m$ for the original CRI inference with multiple antecedents (see Remark 8.5.1(ii)). In the case of k multiple antecedents the above procedure can be extended and the space requirements shown to be a linear function of k, in fact it is equal to $k \cdot m^2 + 2lm$ (compare also with Remark 8.5.4).

Combs and Andrews' Union Rule Configuration

Combs and Andrews [56], analogous to the Rule Break Up method, broke a complex MISO rule into multiple SISO rules, but suggested an alternate method of inferencing than that of Demirli and Türksen called the union rule configuration (URC) and the inference as URC inference.

To better appreciate the idea of URC, we briefly discuss what Combs and Andrews [56] term as the Intersection Rule Configuration (IRC) of fuzzy rules.

Fuzzy Inference Systems used in an Expert System or Decision Support Systems differ from the ones employed in Fuzzy Control, in that it is not uncommon to have an inconclusive output for some inputs (see (GMP3)). On the other hand, in FC, where the main task is to approximate a given function the above scenario is quite unacceptable and hence completeness of a rule base is mostly insisted, since for any combination of inputs an output is expected.

Consider an FIS with n-input variables and a single output variable. If there are n_i, for $i = 1, 2, \ldots, n$, fuzzy sets defined on each of the n input domains, then $\Pi_{i=1}^{n} n_i$ number of combinations are possible. In the case the antecedents of the rules are related by conjunction, a complete rule base has $\Pi_{i=1}^{n} n_i$ number of rules and is referred by Combs and Andrews as the IRC of fuzzy rules.

For instance, consider a system with 2-input variables and 1-output variable. Let the base sets be X, Y, Z with 2 fuzzy sets each on X, Y and 4 fuzzy sets on Z, which are denoted as follows: $A_1, A_2 \in \mathcal{F}(X)$, $B_1, B_2 \in \mathcal{F}(Y)$, $C, D, E, F \in \mathcal{F}(Z)$. Then for a complete rule base we need a total of $2 \times 2 = 4$ rules. Let the rules be as follows:

$$\begin{aligned} A_1, B_1 &\longrightarrow C\,, \\ A_1, B_2 &\longrightarrow D\,, \\ A_2, B_1 &\longrightarrow E\,, \\ A_2, B_2 &\longrightarrow F\,. \end{aligned} \tag{8.28}$$

This can also be written in a tabular form as follows:

Table 8.2. IRC table for the rule base in (8.28)

–	B_1	B_2
A_1	C	D
A_2	E	F

COMBS and ANDREWS [56] propose that the above MISO rule base (8.28) can be broken into the following simpler SISO rule base, much like the Rule Break Up method:

$$\begin{aligned} A_1 &\longrightarrow C\,, & A_1 &\longrightarrow D\,, \\ B_1 &\longrightarrow C\,, & B_1 &\longrightarrow E\,, \\ A_2 &\longrightarrow E\,, & A_2 &\longrightarrow F\,, \\ B_2 &\longrightarrow F\,, & B_2 &\longrightarrow D\,. \end{aligned} \tag{8.29}$$

If the aggregation over the rules in (8.28) is done disjunctively then as in the Rule Break Up method we can take the disjunction of the rules in (8.29). Here the authors suggest the following way of inferencing from the rules in (8.29):

$$\begin{aligned} C =& A' \overset{T}{\circ} ((A_1 \longrightarrow C) \vee (A_1 \longrightarrow D) \vee (A_2 \longrightarrow E) \vee (A_2 \longrightarrow F)) \\ &\vee\; B' \overset{T}{\circ} ((B_1 \longrightarrow C) \vee (B_1 \longrightarrow E) \vee (B_2 \longrightarrow F) \vee (B_2 \longrightarrow D))\,. \end{aligned} \tag{8.30}$$

Let us assume that the following subsethood relationship holds among fuzzy sets $C, D, E, F \in \mathcal{F}(Z)$:

$$C \prec D\,, C \prec E\,, D \prec F\,, E \prec F\,.$$

Once again using the distributivity of a fuzzy implication over max (see Proposition 7.2.26) the inference in (8.30) becomes

$$\begin{aligned} C =& A' \overset{T}{\circ} ((A_1 \longrightarrow D) \vee (A_2 \longrightarrow F)) \\ &\vee\; B' \overset{T}{\circ} ((B_1 \longrightarrow E) \vee (B_2 \longrightarrow F))\,, \end{aligned} \tag{8.31}$$

which can be obtained from the following rules, referred by Combs and Andrews as the Union Rule Configuration (URC) of fuzzy rules:

$$\begin{aligned} A_1 &\longrightarrow D\,, \qquad & B_1 &\longrightarrow E\,, \\ A_2 &\longrightarrow F\,, \qquad & B_2 &\longrightarrow F\,. \end{aligned} \tag{8.32}$$

and the corresponding Union Rule Matrix (URM) as given in Table 8.3

Table 8.3. URC table for the rule base in (8.32)

X :	$A_1 \longrightarrow D$	$A_2 \longrightarrow F$
Y :	$B_1 \longrightarrow E$	$B_2 \longrightarrow F$

So far, the method suggested by Combs and Andrews may seem like the Rule Break Up method packaged differently. In fact, to see the significance of the proposed method, consider the following situations.

Increase in the number of input fuzzy sets

Let us add one more fuzzy set to both the input domains X, Y to the rule base in (8.28), viz., $A_3 \in \mathcal{F}(X)$, $B_3 \in \mathcal{F}(Y)$. Now to obtain a complete rule base for the system with 3-input fuzzy sets on each of X, Y we need to define and add the following rules (this being a synthetic example, the consequents of these rules are arbitrarily assigned):

$$\begin{aligned} A_3, B_1 &\longrightarrow C\,, \qquad & A_1, B_3 &\longrightarrow C\,, \\ A_3, B_2 &\longrightarrow F\,, \\ A_3, B_3 &\longrightarrow E\,, \qquad & A_2, B_3 &\longrightarrow D\,, \end{aligned} \tag{8.33}$$

and the IRC matrix of the original system is as given in Table 8.4.

Table 8.4. IRC table for the rule base in (8.33)

–	B_1	B_2	B_3
A_1	C	D	C
A_2	E	F	D
A_3	C	F	E

Whereas, using the subsethood relations and breaking the rules as proposed above, the URC matrix for (8.33) is given in Table 8.5.

It can be easily seen that in the case of IRC rules we needed to add 5 rules while by employing URC we have reduced the number of additions to just two.

Table 8.5. URC table for the rule base in (8.32)

X :	$A_1 \longrightarrow D$	$A_2 \longrightarrow F$	$A_3 \longrightarrow F$
Y :	$B_1 \longrightarrow E$	$B_2 \longrightarrow F$	$B_3 \longrightarrow E$

Increase in the number of input variables /domains

On the other hand, let us consider the case when a third input variable is added to the system, say over the domain W, with 3 fuzzy sets $G_1, G_2, G_3 \in \mathcal{F}(W)$. Now, not only the number of rules to obtain a complete rule base for the system increases, but also the complexity of the rules also increases as given by the following rules (once again, the consequents of these rules are arbitrarily assigned):

$$\begin{array}{lll} A_1, B_1, G_1 \longrightarrow C\ , & A_1, B_1, G_2 \longrightarrow D\ , & A_1, B_1, G_3 \longrightarrow F\ , \\ A_1, B_2, G_1 \longrightarrow D\ , & A_1, B_2, G_2 \longrightarrow E\ , & A_1, B_2, G_3 \longrightarrow C\ , \\ A_2, B_1, G_1 \longrightarrow D\ , & A_2, B_1, G_2 \longrightarrow E\ , & A_2, B_1, G_3 \longrightarrow D\ , \\ A_2, B_2, G_1 \longrightarrow C\ , & A_2, B_2, G_2 \longrightarrow C\ , & A_2, B_2, G_3 \longrightarrow E\ . \end{array} \tag{8.34}$$

As mentioned earlier we have $2 \times 2 \times 3 = 12$ rules for a complete rule base. Once again, using the subsethood relations, the URC matrix for (8.34) is given in Table 8.6.

Table 8.6. URC table for the SISO rule base corresponding to the MISO rule base in (8.34)

X :	$A_1 \longrightarrow D$	$A_2 \longrightarrow F$	$A_3 \longrightarrow F$
Y :	$B_1 \longrightarrow E$	$B_2 \longrightarrow F$	$B_3 \longrightarrow E$
W :	$G_1 \longrightarrow D$	$G_2 \longrightarrow E$	$G_3 \longrightarrow F$

It can be easily seen that in the case of IRC rules we needed to add 8 rules while, by employing URC, we have reduced the number of additions to just 3.

Merger of rules with identical consequents

BALASUBRAMANIAM and RAO [23, 24] have proposed a simple rule reduction technique of combining the antecedents of rules with identical consequents. Such an approach can be taken in the case of both CRI and SBR. In this context, when we use implicative rules, since it is the FATI inference in CRI that is usually employed, the above approach is inconsequential. On the other hand, JAYARAM [125] has proposed some sufficient conditions on the different operators employed in SBR that ensure that the inferences obtained from the original rule base and the reduced rule base are identical.

In SBR the steps involved are the following:

1. Selection of a matching function M to match the antecedent A of the rule to the current input/observation A'.
2. Selection of the modification function J to modify the consequent B according to the degree of compatibility between A and A' to obtain B'.
3. In the case of MISO fuzzy rule bases, an additional step employing a commutative and associative operator K is required for combining the matching degrees of the antecedents A_i to the given inputs A_i'.
4. When there are more than one rule, an associative aggregation operator G is employed over the rules and the inference is obtained by (8.20) or (8.22), using J, M and K.

We denote the SISO-SBR inference scheme employed in the case of SISO fuzzy rule base by the triple (G, J, M), such that if $\mathcal{R}$ denotes the given SISO fuzzy rule base, then the inference is given by (8.20). Similarly, we denote the MISO-SBR inference scheme employed in the case of MISO fuzzy rule base $\mathcal{R}$ by the quadruple (G, J, K, M), where the inference is given by (8.22).

Definition 8.5.11. *Let X be any non-empty set, $A_1, A_2, A' \in \mathcal{F}(X)$ and $x, y, z \in [0,1]$. An MISO-SBR inference scheme (G, J, K, M) is said to be* consistent *if the operations G, J, K, M satisfy the following distributive equations:*

$$G(J(x,z), J(y,z)) = J(K(x,y), z) \, , \tag{C1}$$
$$M(K(A_1, A_2), A') = K(M(A_1, A'), M(A_2, A')) \, . \tag{C2}$$

Theorem 8.5.12. *Let a MISO fuzzy rule base $\mathcal{R}$ be given over some non-empty input sets X_i, for $i = 1, 2, \ldots, n$ and a non-empty output set Y. Let the inference be drawn using the MISO-SBR inference scheme (G, J, K, M), viz., (8.22). If the MISO-SBR inference scheme (G, J, K, M) is consistent, then inference invariant rule reduction is possible by combining antecedents of those rules in $\mathcal{R}$ whose consequents are identical.*

Example 8.5.13. In this example we show the efficiency and invariance in the inference obtained when the above rule reduction procedure is employed. Consider a rule base consisting of the following three rules:

$$\begin{aligned} &\text{IF } \tilde{x_1} \text{ is } A_1 \text{ AND } \tilde{x_2} \text{ is } B_1 \text{ THEN } \tilde{y} \text{ is } C\,, \\ &\text{IF } \tilde{x_1} \text{ is } A_2 \text{ AND } \tilde{x_2} \text{ is } B_2 \text{ THEN } \tilde{y} \text{ is } C\,, \\ &\text{IF } \tilde{x_1} \text{ is } A_3 \text{ AND } \tilde{x_2} \text{ is } B_3 \text{ THEN } \tilde{y} \text{ is } D, \end{aligned} \tag{$\mathcal{R}_O$}$$

where A_i, B_i for $i = 1, 2, 3$ are fuzzy sets defined on $X = \{x_1, x_2, x_3, x_4\}$, $Y = \{y_1, y_2, y_3\}$, respectively, while C, D are fuzzy sets defined on $Z = \{z_1, z_2, z_3, z_4\}$. They are given as follows:

$$\begin{aligned} A_1 &= [0.30 \;\; 0.50 \;\; 0.0 \;\; 1.00] \, , & B_1 &= [0.30 \;\; 0.40 \;\; 0.90] \, , \\ A_2 &= [0.36 \;\; 0.25 \;\; 0.30 \;\; 0.80] \, , & B_2 &= [0.12 \;\; 0.67 \;\; 0.99] \, , \\ A_3 &= [0.90 \;\; 0.0 \;\; 0.80 \;\; 0.50] \, , & B_3 &= [0.2 \;\; 0.7 \;\; 0.6] \, , \\ C &= [0.10 \;\; 0.25 \;\; 0.0 \;\; 0.70] \, , & D &= [0.10 \;\; 0.20 \;\; 0.0 \;\; 1.0] \, . \end{aligned}$$

We employ the CDR [187] inference scheme with $G = S_{\mathbf{M}}$, $J = I_{\mathbf{LK}}$, $K = T_{\mathbf{M}}$ and $M = M_{\mathbf{S}}$ (see Table 8.1). Let the given input be ($\tilde{x_1}$ is A'; $\tilde{x_2}$ is B') where

$$A' = [0.4\ 0.7\ 0.8\ 0.0]\ , \qquad B' = [0.2\ 0\ 1]\ .$$

In the following we infer both with the original rule base ($\mathcal{R}_{\mathbf{O}}$) and the reduced rule base ($\mathcal{R}_{\mathbf{R}}$) and show that the inferred output is identical in both the cases.

Inference with the original rule base ($\mathcal{R}_{\mathbf{O}}$)

Calculating the matching degrees

$$\begin{aligned}
M_{\mathbf{S}}(A', A_1) &= \min(0.6,\ 0.5,\ 0.2,\ 1.0) = 0.2\ ,\\
M_{\mathbf{S}}(A', A_2) &= \min(0.6,\ 0.3,\ 0.3,\ 1.0) = 0.3\ ,\\
M_{\mathbf{S}}(A', A_3) &= \min(0.9,\ 0.3,\ 0.8,\ 1.0) = 0.3\ ,\\
M_{\mathbf{S}}(B', B_1) &= \min(0.8,\ 1.0,\ 0.9) = 0.8\ ,\\
M_{\mathbf{S}}(B', B_2) &= \min(0.8,\ 1.0,\ 0.99) = 0.8\ ,\\
M_{\mathbf{S}}(B', B_3) &= \min(0.8,\ 1.0,\ 0.6) = 0.6\ .
\end{aligned}$$

Combining the matching degrees to obtain similarity values s_i
The similarity values are calculated using the operator $K = T_{\mathbf{M}}$ as follows:

$$\begin{aligned}
s_1 &= K(M_{\mathbf{S}}(A_1, A'), M_{\mathbf{S}}(B_1, B')) = \min(0.2,\ 0.8) = 0.2\ ,\\
s_2 &= \min(0.3,\ 0.8) = 0.3\ ,\\
s_3 &= \min(0.3,\ 0.6) = 0.3\ .
\end{aligned}$$

Modifying the consequents based on the similarity values s_i

$$\begin{aligned}
J(s_1, C) &= I_{\mathbf{LK}}(s_1, C) = [0.9\ 1\ 0.8\ 1] = C_1'\ ,\\
J(s_2, C) &= I_{\mathbf{LK}}(s_2, C) = [0.8\ 0.95\ 0.7\ 1] = C_2'\ ,\\
J(s_3, D) &= I_{\mathbf{LK}}(s_3, D) = [0.8\ 0.9\ 0.7\ 1] = C_3'\ .
\end{aligned}$$

Combining the obtained consequents for a conclusion

$$C' = G(C_1', C_2', C_3') = S_{\mathbf{M}}(C_1', C_2', C_3') = [0.9\ 1.0\ 0.8\ 1.0]\ . \tag{8.35}$$

Inference with the reduced rule base ($\mathcal{R}_{\mathbf{R}}$)

Firstly, note that (C1) is satisfied with $G = S_{\mathbf{M}}$, $J = I_{\mathbf{LK}}$ and $K = T_{\mathbf{M}}$ and subsequently, (C2) is satisfied with $K = T_{\mathbf{M}}$ and $M = M_{\mathbf{S}}$. Hence, the MISO-SBR inference scheme ($S_{\mathbf{M}}, I_{\mathbf{LK}}, T_{\mathbf{M}}, M_{\mathbf{S}}$) is consistent. Now, from Theorem 8.5.12 we see that inference invariant rule reduction is possible by combining the antecedents of the rules that have the same consequent fuzzy set.

As can be seen, the first two rules in the original rule base ($\mathcal{R}_{\mathbf{O}}$) have the same consequent fuzzy set C and hence can be reduced to the rule base consisting of the following two rules:

$$\begin{aligned} &\text{IF } \tilde{x_1} \; is \; A^* \text{ AND } \tilde{x_2} \; is \; B^* \text{ THEN } \tilde{y} \; is \; C, \\ &\text{IF } \tilde{x}_1 \; is \; A_3 \text{ AND } \tilde{x_2} \; is \; B_3 \text{ THEN } \tilde{y} \; is \; D, \end{aligned} \tag{$\mathcal{R}_{\mathbf{R}}$}$$

where

$$\begin{aligned} A^* &= K(A_1, A_2) = T_{\mathbf{M}}(A_1, A_2) = [0.3\ 0.25\ 0\ 0.8] \;, \\ B^* &= K(B_1, B_2) = T_{\mathbf{M}}(B_1, B_2) = [0.1\ 0.4\ 0.9] \;. \end{aligned}$$

Once again, calculating the matching degrees with respect to the same input pair $(A'; B')$ we obtain

$$\begin{aligned} M_{\mathbf{S}}(A^*, A') &= \min(0.6,\ 0.3,\ 0.2,\ 1.0) = 0.2 \;, \\ M_{\mathbf{S}}(B^*, B') &= \min(0.8,\ 1.0,\ 0.9) = 0.8 \;. \end{aligned}$$

Hence the similarity value of the input to the new rule is $s_1^* = \min(0.2, 0.8) = 0.2$, which modifies the consequent C as

$$C_1^* = I_{\mathbf{LK}}(s_1^*, C) = [0.9\ 1.0\ 0.8\ 1.0] \;.$$

Combining the obtained consequents for a conclusion we obtain

$$C'' = G(C_1^*, C_3') = S_{\mathbf{M}}(C_1^*, C_3') = [0.9\ 1.0\ 0.8\ 1.0] \;,$$

i.e., $C'' = C'$ in (8.35). In other words, we have shown that the inference obtained for the same inputs from the original and reduced rule bases is identical.

8.6 Bibliographical Remarks

Most of the earliest works on fuzzy implications dealt predominantly with their suitability in approximate reasoning. For example, see Fukami et al. [109], Mizumoto [185, 186], Mizumoto and Zimmermann [184], Weber [255], Yager [259], Kiszka et al. [139, 138], Trillas et al. [239, 240, 241, 242], Wangming [254], Ahlquist [2, 3], Cao and Kandel [47], Hall [118], Ying [265], Cao et al. [48], Kandel et al. [136], Park et al. [196], Ruan and Kerre [213], Cárdenas et al. [51]. In fact, this line of approach still continues as is evident from, for instance, Yager [260, 261], Cordon et al. [59, 60], Cai [46] and Pei [206].

It should be noted that, other than the CRI and SBR inference schemes, many inference schemes employing fuzzy sets have been proposed but which do not strictly fall under these two categories. For example, the inference scheme of Raha et al. [207] that is a combination of both the above approaches, Fuzzy Truth Value Modification inference of Baldwin [26], the scheme for implication based rules proposed by Ughetto et al. [246], the perception based logical deduction of Novák [189] and Novák and Perfilieva [192], which uses evaluative linguistic expressions Novák [190], Novák and Perfilieva [191]. See also earlier papers of Mizumoto [186], Mizumoto and Zimmermann [184], Roger [212] and Dubois and Prade [92], which is still considered as a very good survey of various AR schemes.

Instead of the sup $-T$ composition one could also use the alternative inf $-S$ composition in the CRI inference scheme as follows:

$$B'(y) = A'(x) \overset{S}{\oslash} R(x,y) = \inf_{x \in X} S(A'(x), R(x,y)) ,$$

where S is a t-conorm. Unfortunately, this dual composition has not found as much acceptance as sup $-T$ composition in the community. Yet another method of inferencing employing fuzzy relations to represent fuzzy rules is the triangle product of Bandler and Kohout [29, 30], Boixader and Jacas [37]. This type of inference, usually called the *Bandler-Kohout* product, is an inf $-I$ composition where I is a fuzzy implication. For more recent work on this topic see Štěpnička et al. [229] and the references therein.

There are many other matching functions than those given in Sect. 8.3.4. Zwick et al. [270] have compared 19 such similarity measures based on a few parameters. See also Bien and Chun [35], Pedrycz [204], Pappis and Karacapilidis [195], Wang [251], Chen [52] for more such measures. For a comparative study of many SBR inference schemes see, for example, Yeung and Tsang [264], Wang et al. [252].

The matching function $M_{\mathbf{S}}$ is a specific example of fuzzy set inclusion grade indicator, denoted by *Inc* in Cornellis et al. [63], where the authors discuss Inclusion-Based Reasoning (IBR). Obviously, SBR subsumes IBR. In the same work the authors revise the 8 axioms that *Inc* is expected to satisfy according to the Sinha-Dougherty principle in [225]. The aforesaid list is quite different from, and in some cases more general than, the desirable properties (MF1) – (MF4) of a matching function in an SBR, since they are primarily meant for the measure *Inc* which employs a fuzzy implication operator. See also the related work of Burillo et al. [43].

For a detailed discussion on inclusion grade indicators and the properties they are expected to satisfy see Cornellis et al. [63]. As observed therein, Kitainik [140, 141] had developed similar axioms for an inclusion operator which was later characterized by Fodor and Yager [107]. Briefly, the result states that any inclusion operator satisfies Kitainik's axioms iff it is the $M_{\mathbf{S}}$ matching function where the fuzzy implication operator I has the contrapositive symmetry with a strong negation N, i.e., I has CP(N). The above result has been further generalized in Cornellis et al. [63] to an inclusion operator satisfying the S-D axioms for which a necessary and sufficient condition is that the I in $M_{\mathbf{S}}$ satisfy the ordering property (OP) and the following condition:

$$I(x,y) = 0 \Longleftrightarrow x = 1 \text{ or } y = 1 , \qquad x, y \in [0,1] .$$

The property (GMP4) has been considered only in the context of a single rule, while one typically deals with a set of rules, in which case the above property translates to proving the validity of the following equation:

$$A_j \circ \widetilde{R} = B_j \,, \qquad j \in \mathbb{N}_n \,. \tag{8.36}$$

In the case of implicative rules the relation $\widetilde{R}$ in (8.36) is given by (8.17).

SANCHEZ [220] was the earliest to investigate (8.36) and showed that (8.36) is solvable if and only if $\widetilde{R}$ as given by (8.17) is a solution of it. Although this result was at first obtained for the case when $T = T_{\mathbf{M}}$ and $I = I_{\mathbf{GD}}$, it has been generalized to the case of the R-implications obtained from left continuous t-norms by GOTTWALD [112]. A complete solvability condition is still elusive and is considered an open problem (see GOTTWALD [116]). However, under some reasonable assumptions on the fuzzy sets A_j, B_j, a necessary and sufficient condition for conjuncitve rules to be a solution of a system of fuzzy relation equation under the sup $-T$ composition has been proposed in DE BAETS [66], KLAWONN [142], NOVÁK and PERFILIEVA [193]. See also similar investigations by DI NOLA et al. [79].

Although there are many more GMP rules proposed and investigated (see FUKAMI et al. [108, 109], BALDWIN and PILSWORTH [27], MIZUMOTO [186], MAGREZ and SMETS [162], DUBOIS and PRADE [92], KLIR and YUAN [147]) they can all be seen as specific cases of the four rules (GMP1) – (GMP4) considered in Sect. 8.4. See also JENEI [130] for an investigation of these properties for CRI in the context of continuity of their inference.

For the motivation behind the properties (MP1)-(MP4), we refer the readers to the original works of TRILLAS and VALVERDE [240, 241].

As noted earlier, the methods considered in Sect. 8.5 are so chosen to highlight the significant role played by the properties a fuzzy implication possesses. The other significant works on this topic include, for example, the following: MARTIN-CLOUIARE [168, 169] on Fast GMP and UGHETTO et al. [246] (see also their related works [247, 248]).

A distinction should be made between the method proposed in JAYARAM [126] (see Sect. 8.5.1) - which uses the given MISO fuzzy rules without any modification to them - and typical inferencing in HFS which is dictated by the hierarchical structure that exists among the modeled system variables. It should be emphasized that the term 'Hierarchical CRI', as employed here, refers to the modifications in the inferencing procedure of CRI and does not impose a hierarchical architecture on the MISO fuzzy rules.

Although it was mentioned that a complete rule base is necessary in the case of fuzzy control, KÓCZY and his group have pioneered the cause of 'Sparse' rule bases, i.e., rule bases which do not encode the resultant outputs for every combination of inputs. Many investigations have been carried out to plug these 'gaps' in the knowledge by various methods, chief among them being interpolative techniques. See, e.g., the works of KÓCZY et al. [148, 149], DUBOIS and PRADE [93], BOUCHON-MEUNIER et al. [39, 40, 41] and that of JENEI [133].

It is desirable that the inference obtained from the original/untransformed inference schemes should be retained and not lost in the efforts to reduce the complexity of the algorithm under consideration. It was pointed out by Dick and Kandel [76] that the equivalence between the original CRI inference (8.8) and the inference in (8.30) obtained by URC may not always hold. It should be remarked that in the case the conjunction used is the t-norm $T_{\mathbf{M}}$ and the inputs A', B' are normal, then the inference in (8.30) and the original CRI inference (8.8) are identical, and hence is equivalent to the inference in (8.32). The conditions under which equivalence between IRC and URC can be guaranteed is studied by Weinschenk et al. [256].

One of the earliest works that considered combining antecedents in fuzzy rules with identical consequents was by Dubois and Prade [90]. The focus of their study was the conditions on the underlying possibility distributions that enabled meaningful combination, whereas the agenda in Jayaram [125] is to study the conditions on the operators used in the SBR inference mechanisms that allows combining antecedents without losing the obtained inference.

Appendix

A Some Results on Real Functions

By the facts from the theory of real functions (cf. [155], §0.1), we obtain

Lemma A.0.1. *Let* $T \neq \emptyset$. *If* $(f_t)_{t\in T}$ *is one parameter family of monotonic, bounded functions of the same kind* ($f_t\colon [a,b] \to [c,d], t \in T$), *then* $\sup_{t\in T} f_t$ *and* $\inf_{t\in T} f_t$ *are monotonic, bounded functions of this kind.*

Proof. Let us assume that f_t are increasing for all $t \in T$. If $s \in T$ is fixed, then we obtain

$$(\inf_{t\in T} f_t)(x) = \inf_{t\in T} f_t(x) \le f_s(x) \le f_s(y) \,.$$

Therefore, for every $s \in T$ we have $(\inf_{t\in T} f_t)(x) \le f_s(y)$. From the last inequality we get

$$(\inf_{t\in T} f_t)(x) \le (\inf_{s\in T} f_s)(y) \,.$$

This implies that $\inf_{t\in T} f_t$ is increasing. Similarly we can prove that $\sup_{t\in T} f_t$ is increasing. In the same way, we get verification for decreasing or constant functions. □

Definition A.0.2. *Let* A *be any non-empty set. A function* $f\colon A \to \mathbb{R}$ *has the Darboux property if whenever* $a, b \in A$ *and* γ *is between* $f(a)$ *and* $f(b)$, *then there is a* c *between* a *and* b *such that* $f(c) = \gamma$.

Proposition A.0.3. *If* $f\colon [a,b] \to \mathbb{R}$ *is continuous, then* f *has the Darboux property.*

Heine-Cantor Theorem. *If* (X, d_X) *is a compact metric space, then every continuous function* $f\colon X \to Y$, *where* (Y, d_Y) *is a metric space, is uniformly continuous, this means that for a given* $\varepsilon > 0$, *there exists* $\delta > 0$ *such that for all* $x, y \in X$, *if* $d_X(x,y) < \delta$, *then* $d_Y(f(x), f(x)) < \varepsilon$.

For example, if $f\colon [a,b] \to \mathbb{R}$ is a continuous function, then it is uniformly continuous.

Dini's Theorem. *Let* X *be a compact topological space. Suppose we have a sequence of continuous functions* $f_n\colon X \to \mathbb{R}$, $n \in \mathbb{N}$ *such that*

(a) $(f_n)_{n\in\mathbb{N}}$ *is an increasing sequence i.e.,* $f_n(x) \le f_{n+1}(x)$ *for all* $x \in X$ *and* $n \in \mathbb{N}$,

M. Baczyński and B. Jayaram: Fuzzy Implications, STUDFUZZ 231, pp. 281–287, 2008.
springerlink.com

(b) $(f_n)_{n\in\mathbb{N}}$ *converges pointwise to a continuous function* f.

Then $(f_n)_{n\in\mathbb{N}}$ *converges uniformly to* f, *i.e., for a given* $\varepsilon > 0$, *there exists* $n_0 \in \mathbb{N}$ *such that for all natural* $n \geq n_0$ *and* $x \in X$ *we have* $|f_n(x) - f(x)| \leq \varepsilon$.

Instead of an increasing sequence $(f_n)_{n\in\mathbb{N}}$, one may assume a decreasing sequence.

We would like to underline that the following result concerning the continuity of functions of two variables is well-known in the theory of real functions and often used in the theory of fuzzy logic (see [222], Theorem 3.1.3; [146], Proposition 1.19; [113], Proposition 5.5.1).

Theorem A.0.4. *For a function* $F\colon [0,1]^2 \to [0,1]$ *which is monotonic with respect to one variable the following statements are equivalent:*

(i) F *is continuous.*
(ii) F *is continuous in each variable.*

Proof. *(i)* $\Longrightarrow$ *(ii)* It is obvious that a continuous function of two variables is continuous with respect to each variable.

(ii) $\Longrightarrow$ *(i)* Let $F\colon [0,1]^2 \to [0,1]$ be continuous in each variable and without loss of generality assume that F is increasing with respect to the second variable. Fix arbitrarily $x_0, y_0 \in [0,1]$ and take $\varepsilon > 0$. Assume firstly, that $y_0 > 0$ and take any strictly increasing sequence $(y_n)_{n\in\mathbb{N}}$ converging to y_0. Let us define functions $f_n\colon [0,1] \to [0,1]$ by $f_n(x) := F(x, y_n)$ for all $n \in \mathbb{N}_0$. By the monotonicity of F,

$$f_n(x) = F(x, y_n) \leq F(x, y_{n+1}) = f_{n+1}(x)\,, \qquad x \in [0,1]\,.$$

Hence, from the continuity of F with respect to the second variable, f_n converges pointwise to f_0, which is a continuous function. Therefore, by Dini's Theorem, f_n converges uniformly to f_0. Thus, there exists $n_0 \in \mathbb{N}$ such that for all $n \geq n_0$

$$f_0(x) - f_n(x) < \frac{\varepsilon}{3}\,, \qquad x \in [0,1]\,.$$

Using the definition of f_n we get

$$F(x, y_0) - F(x, y_{n_0}) < \frac{\varepsilon}{3}\,, \qquad x \in [0,1]\,. \tag{A.1}$$

Moreover, by the monotonicity

$$F(x, y) - F(x, y_{n_0}) < \frac{\varepsilon}{3}\,, \qquad x \in [0,1]\,, \quad y \in [y_{n_0}, y_0]\,, \tag{A.2}$$

where $y_{n_0} < y_0$. From our assumption, f_0 is a continuous function on $[0,1]$, so because of Heine-Cantor Theorem it is uniformly continuous, i.e., there exists $\delta > 0$ such that

$$|x - x_1| < \delta \quad \Longrightarrow \quad |f_0(x) - f_0(x_1)| < \frac{\varepsilon}{3}\,, \tag{A.3}$$

for all $x, x_1 \in [0,1]$. Now, taking any $x' \in [0,1]$ such that $|x_0 - x'| < \delta$ and $y' \in [y_{n_0}, y_0]$, from the triangle inequality, (A.3), (A.1) and (A.2) we get

$$\begin{aligned}
|F(x_0,y_0) - F(x',y')| &= |F(x_0,y_0) - F(x',y_0) + F(x',y_0) - F(x',y_{n_0}) \\
&\quad + F(x',y_{n_0}) - F(x',y')| \\
&\le |F(x_0,y_0) - F(x',y_0)| + |F(x',y_0) - F(x',y_{n_0})| \\
&\quad + |F(x',y_{n_0}) - F(x',y')| \\
&< \frac{\varepsilon}{3} + \frac{\varepsilon}{3} + \frac{\varepsilon}{3} = \varepsilon .
\end{aligned}$$

Therefore, if $y_0 = 1$, then putting $\delta_1 := y_0 - y_{n_0}$ we get

$$|x_0 - x'| < \delta, |1 - y'| < \delta_1 \implies |F(x_0,1) - F(x',y')| < \varepsilon ,$$

so F is continuous at the point $(x_0, 1)$.

Similarly, if we assume that $y_0 < 1$, then by considering a strictly decreasing sequence $(y'_n)_{n\in\mathbb{N}}$ converging to y_0 we can prove that there exists $\delta' > 0$ and $y_{n_1} > y_0$ such that $|F(x_0,y_0) - F(x',y')| < \varepsilon$, whenever $|x_0 - x'| < \delta'$ and $y' \in [y_0, y_{n_1}]$. Therefore, if $y_0 = 0$, then putting $\delta_2 := y_{n_1} - y_0$ we get

$$|x_0 - x'| < \delta, |y'| < \delta_2 \implies |F(x_0,0) - F(x',y')| < \varepsilon ,$$

so F is continuous at the point $(x_0, 0)$.

Finally, if $y_0 \in (0,1)$, with $\delta'' := \min\{\delta, \delta'\}$ and $\delta''' := \min\{y_0 - y_{n_0}, y_{n_1} - y_0\}$, we get

$$|x_0 - x'| < \delta'', |y_0 - y'| < \delta''' \implies |F(x_0,y_0) - F(x',y')| < \varepsilon ,$$

so F is continuous at the point (x_0, y_0). Since the above point was fixed arbitrarily from the unit interval, F is continuous as a function of two variables and this ends the proof. □

Theorem A.0.5. *(i) The family of all fuzzy negations is a complete, completely distributive lattice with lattice operations*

$$(N_1 \vee N_2)(x) := \max(N_1(x), N_2(x)) , \qquad x \in [0,1] , \tag{A.4}$$
$$(N_1 \wedge N_2)(x) := \min(N_1(x), N_2(x)) , \qquad x \in [0,1] . \tag{A.5}$$

(ii) The family of all strict negations is a distributive lattice.
(iii) The family of all strong negations is a distributive lattice.

Proof. (i) Let N_t be fuzzy negations for every $t \in T$, where $T \neq \emptyset$. By Lemma A.0.1, functions $\sup_{t\in T} N_t$, $\inf_{t\in T} N_t$ are also decreasing. Moreover,

$$\sup_{t\in T} N_t(0) = \sup_{t\in T} 1 = 1 , \qquad \sup_{t\in T} N_t(1) = \sup_{t\in T} 0 = 0 .$$

The last equalities are also true for the infimum, so $\sup_{t\in T} N_t, \inf_{t\in T} N_t$ are fuzzy negations. Therefore, the family of all fuzzy negations is a complete lattice. According to BIRKHOFF [36], Chap. V, the family $[0,1]^X$ of all functions $f\colon X \to [0,1]$ forms a completely distributive lattice. Thus the family of all fuzzy negations is a complete sublattice for $X = [0,1]$, and we get that the family of all fuzzy negations is a completely distributive lattice.

(ii) Since the composition of continuous functions is continuous, the family of all strict negations is a distributive lattice. It is not complete, since it is well known that certain sequences of continuous functions have limits which are not continuous.

(iii) Let us take any two strong negations N_1, N_2 and consider a fuzzy negation $N := N_1 \vee N_2$. We show that N is an involution. Let $x \in [0,1]$ and without loss of generality, let $N_1(x) \le N_2(x)$. Thus $N_1(N_2(x)) \le N_1(N_1(x)) = x$ and $N(x) = N_2(x)$. By easy calculations, we get

$$\begin{aligned} N(N(x)) &= \max(N_1(N(x)), N_2(N(x))) \\ &= \max(N_1(N_2(x)), N_2(N_2(x))) \\ &= \max(N_1(N_2(x)), x) = x \,. \end{aligned}$$

Similar reasoning can be carried out for $N = N_1 \wedge N_2$. □

Lemma A.0.6. *If N is a continuous fuzzy negation, then the function $\mathfrak{N}$: $[0,1] \to [0,1]$ defined by*

$$\mathfrak{N}(x) = \begin{cases} N^{(-1)}(x), & \textit{if } x \in (0,1] \,, \\ 1, & \textit{if } x = 0 \,, \end{cases} \qquad x \in [0,1] \,, \tag{A.6}$$

is a strictly decreasing fuzzy negation. Moreover,

$$\mathfrak{N}^{(-1)} = N \,, \tag{A.7}$$

$$N \circ \mathfrak{N} = \mathrm{id}_{[0,1]} \,, \tag{A.8}$$

$$\mathfrak{N} \circ N \mid_{\mathrm{Ran}(\mathfrak{N})} = \mathrm{id}_{\mathrm{Ran}(\mathfrak{N})} \,. \tag{A.9}$$

Proof. Let N be a continuous fuzzy negation. By virtue of Corollary 3.3(ii) from [146], we get that the pseudo-inverse $N^{(-1)}$ is a decreasing function. Moreover,

$$N^{(-1)}(1) = \sup\{x \in [0,1] \mid N(x) > 1\} = \sup \emptyset = 0 \,,$$

but

$$N^{(-1)}(0) = \sup\{x \in [0,1] \mid N(x) > 0\} \,,$$

can be less than 1. Let us define a function $\mathfrak{N}$: $[0,1] \to [0,1]$ by (A.6). Firstly, see that this function is now a well defined fuzzy negation. Further, since N is continuous, from the above mentioned corollary we have

$$\left(N^{(-1)}\right)^{(-1)} = N \,, \tag{A.10}$$

which implies (A.7). Indeed, let us define the following two sets

$$\begin{aligned} A(y) &= \{x \in [0,1] \mid \mathfrak{N}(x) > y\} \,, \\ B(y) &= \{x \in [0,1] \mid N^{(-1)}(x) > y\} \,, \end{aligned}$$

for every $y \in [0,1]$. We consider the following three cases.

- If $A(y) = \emptyset$, then $y = 1$, since $\mathfrak{N}(0) = 1$. Then $\mathfrak{N}^{(-1)}(1) = \sup \emptyset = 0 = N(1)$, i.e., we get (A.7) for $y = 1$.
- If $A(y) = \{0\}$, then $\mathfrak{N}^{(-1)}(y) = \sup\{0\} = 0$. Moreover, only for $x = 0$ we have $\mathfrak{N}(0) > y$, i.e., for all $x \in (0,1]$ we have $\mathfrak{N}(x) \le y$. Hence, by the definition of $\mathfrak{N}$, we get that $N^{(-1)}(x) \le y$ for all $x \in (0,1]$. Therefore, $B(y) = \{0\}$ or $B(y) = \emptyset$. In both cases we get $\left(N^{(-1)}\right)^{(-1)}(y) = \sup B(y) = 0$, for such y. By (A.10), we have that (A.7) is true in this case.
- If $A(y) \neq \emptyset$ and $A(y) \neq \{0\}$, then there exists $x_0 \in (0,1]$ such that $x_0 \in A(y)$. Since for all $x \in [x_0, 1]$ we have $\mathfrak{N}(x) = N^{(-1)}(x)$, again from (A.10) we get $\sup A(y) = \sup B(y)$ i.e., (A.7) is also true.

By Remark 3.4(ii) from [146] (for $f = \mathfrak{N}$), the fuzzy negation $\mathfrak{N}$ is strictly monotonic on $N([0,1))$. Continuity of N implies that $(0,1] \subset N([0,1))$, so $\mathfrak{N}$ is strictly decreasing on $[0,1]$.

Finally, by virtue of Remark 3.4 (vi) from [146] (for $f = \mathfrak{N}$), since $\mathfrak{N}$ is a strictly decreasing fuzzy negation, we obtain

$$N \circ \mathfrak{N}(x) = \mathfrak{N}^{(-1)} \circ \mathfrak{N}(x) = x\ , \qquad x \in [0,1]\ ,$$
$$\mathfrak{N} \circ N(x) = \mathfrak{N} \circ \mathfrak{N}^{(-1)}(x) = x\ , \qquad x \in \mathrm{Ran}(\mathfrak{N})\ .$$

Therefore, (A.8) and (A.9) are true. □

Theorem A.0.7. *Let T be any t-norm and N_T its natural negation.*

(i) If N_T is continuous, then it is strong.
(ii) If N_T is discontinuous, then it is not strictly decreasing.

Proof. (i) Firstly, we show that N_T is strict. Assume to the contrary that N_T is not strict, i.e., it is constant on some interval $[x,y]$, where without loss of the generality, we assume that $0 < x < y < 1$. Therefore, there exists $p \in [0,1]$ such that

$$N_T(x) = N_T(y) = p\ .$$

If $p = 0$, then $N_T(x) = 0$, which implies that $T(\varepsilon, x) = T(x, \varepsilon) > 0$ for an arbitrary small $\varepsilon > 0$. Therefore, $N_T(\varepsilon) < x$. Since N_T is continuous, as $\varepsilon \to 0$, we have that $N_T(0) \le x$. However, $N_T(0) = 1$, a contradiction.

If $p = 1$, then $T(1-\varepsilon, x) = T(x, 1-\varepsilon) = 0$ for an arbitrary small $\varepsilon > 0$. Thus, $N_T(1-\varepsilon) \ge x$. Since N_T is continuous, as $\varepsilon \to 0$, we have that $N_T(1) \ge x$. However, $N_T(1) = 0$, a contradiction.

Hence, we consider now the situation, when $p \in (0,1)$. Since $N_T(z) = p$ for any $z \in (x,y)$, by the definition of N_T we have

$$T(z, p-\varepsilon) = T(p-\varepsilon, z) = 0\ ,$$
$$T(z, p+\varepsilon) = T(p+\varepsilon, z) > 0\ ,$$

for any arbitrary small $\varepsilon > 0$. Thus

$$N_T(p+\varepsilon) < z \le N_T(p-\varepsilon)\ .$$

Since N_T is continuous, as $\varepsilon \to 0$, we have that $N_T(p) = z$. Now this happens for every $z \in (x, y)$, which once again contradicts the fact that N_T is a function itself, or the fact that N_T is continuous. Hence N_T is strict.

To show that N_T is strong, we show that $N_T(N_T(N_T(x))) = N_T(x)$ for all $x \in [0,1]$. The above is clear for any $x \in \{0,1\}$. Since N_T is strict, for any $x \in (0,1)$ and an arbitrary small $\varepsilon > 0$, we have the following inequalities:

$$\begin{aligned} &x - \varepsilon < x < x + \varepsilon \\ &\qquad\qquad \Longrightarrow N_T(x-\varepsilon) > N_T(x) > N_T(x+\varepsilon) \\ &\qquad\qquad \Longrightarrow N_T(N_T(x-\varepsilon)) < N_T(N_T(x)) < N_T(N_T(x+\varepsilon)) . \end{aligned}$$

By the definition of N_T, we have

$$T(x, N_T(x+\varepsilon)) = T(N_T(x+\varepsilon), x) = 0 ,$$

thus $x \le N_T(N_T(x+\varepsilon))$, which implies that

$$N_T(x) \ge N_T(N_T(N_T(x+\varepsilon))) .$$

Once again, by the continuity of N_T, as $\varepsilon \to 0$, we have

$$N_T(x) \ge N_T(N_T(N_T(x))) .$$

By the definition of N_T, if $z < N_T(x)$, then $T(z,x) = 0$. Now, since

$$N_T(N_T(x-\varepsilon)) < N_T(N_T(x)) ,$$

we also have

$$T(N_T(N_T(x-\varepsilon)), N_T(x)) = 0 .$$

Once again, by the definition of N_T, we have $N_T(x) \le N_T(N_T(N_T(x-\varepsilon)))$, and as $\varepsilon \to 0$, we have

$$N_T(x) \le N_T(N_T(N_T(x))) .$$

From the above inequalities, we have

$$N_T(x) = N_T(N_T(N_T(x))), \qquad x \in [0,1] .$$

Now, by the continuity of N_T one can easily see that N_T is involutive.

(ii) Let N_T be discontinuous at some $p \in [0,1]$. By the monotonicity of N_T, there exist constants $x, y \in [0,1]$ such that

$$x = \begin{cases} \lim\limits_{t \to p^+} N_T(t), & \text{if } p < 1 , \\ 0, & \text{if } p = 1 , \end{cases} \qquad\qquad y = \begin{cases} \lim\limits_{t \to p^-} N_T(t), & \text{if } p > 0 , \\ 1, & \text{if } p = 0 . \end{cases}$$

From the discontinuity of decreasing negation N_T at p we have $x < y$. Now, we consider the following two cases:

a) Let $N_T(p) = x$. In particular this implies that $p > 0$. Let us fix arbitrarily $z \in (x, y)$. It is obvious that $N_T(p) < z$, so by the definition of N_T we get $T(p, z) > 0$, which implies $N_T(z) \leq p$. On the other side, $N_T(p - \varepsilon) \geq y$ for an arbitrary small $\varepsilon > 0$, thus $T(p-\varepsilon, z) = 0$, therefore $N_T(z) \geq p-\varepsilon$. Taking the limit $\varepsilon \to 0$ we get $N_T(z) \geq p$. Since z was arbitrarily fixed, the above implies that $N_T(z) = p$ for every $z \in (x, y)$ and hence N_T is constant on this interval (x, y).
b) Let $N_T(p) = z'$, where $x < z' \leq y$. In particular this implies that $p < 1$. We now claim that N_T is a constant on the interval (x, z'). Let us fix arbitrarily $z \in (x, z')$. In this case we have that $T(p, z) = 0$, so $N_T(z) \geq p$. Once again, we claim that $N_T(z) = p$. Instead, if $N_T(z) > p$, then $T(p + \varepsilon, z) = 0$ for some $\varepsilon > 0$. Thus, by the definition of N_T, we have that $N_T(p + \varepsilon) \geq z$, and from the decreasing nature of N_T and the definition of x we obtain

$$x \geq N_T(p + \varepsilon) \geq z \ ,$$

a contradiction to the fact that $x < z$. Therefore, $N_T(z) = p$ for every $z \in (x, z')$ and hence N_T is a constant on this interval (x, z'). □

References

1. Aczél, J.: Lectures on functional equations and their applications. Academic Press, New York (1966)
2. Ahlquist, J.E.: Application of fuzzy implication to probe nonsymmetric relations: Part I. Fuzzy Sets and Systems 22, 229–244 (1987)
3. Ahlquist, J.E.: Application of fuzzy implication to probe nonsymmetric relations: Part II. Fuzzy Sets and Systems 25, 87–95 (1988)
4. Alcalde, C., Burusco, A., Fuentes-González, R.: A constructive method for the definition of interval-valued fuzzy implication operators. Fuzzy Sets and Systems 153, 211–227 (2005)
5. Alsina, C., Trillas, E.: When (S,N)-implications are (T, T_1)-conditional functions? Fuzzy Sets and Systems 134, 305–310 (2003)
6. Baczyński, M.: Characterization of Dienes implication. In: [211], pp. 299–305 (1999)
7. Baczyński, M.: On a class of distributive fuzzy implications. Internat. J. Uncertain. Fuzziness Knowledge-Based Systems 9, 229–238 (2001)
8. Baczyński, M.: Contrapositive symmetry of distributive fuzzy implications. Internat. J. Uncertain. Fuzziness Knowledge-Based Systems 10, 135–147 (2002)
9. Baczyński, M.: Residual implications revisited. Notes on the Smets-Magrez Theorem. Fuzzy Sets and Systems 145, 267–277 (2004)
10. Baczyński, M.: On the distributivity laws for fuzzy implications. In: Atanassov, K.T., Hryniewicz, O., Kacprzyk, J. (eds.) Soft Computing. Foundations and Theoretical Aspects, EXIT, Warsaw, pp. 53–66 (2004)
11. Baczyński, M., Drewniak, J.: Conjugacy classes of fuzzy implication. In: [211], pp. 287–298 (1999)
12. Baczyński, M., Drewniak, J.: Monotonic fuzzy implication. In: Szczepaniak, P.S., Lisboa, P.J.G., Kacprzyk, J. (eds.) Fuzzy Systems in Medicine. Studies in Fuzzines and Soft Computing, vol. 41, pp. 90–111. Physica-Verlag, Heidelberg (2000)
13. Baczyński, M., Drewniak, J., Sobera, J.: Semigroups of fuzzy implications. Tatra Mt. Math. Publ. 21, 61–71 (2001)
14. Baczyński, M., Drewniak, J., Sobera, J.: On sup $-*$ compositions of fuzzy implications. In: [218], pp. 274–279 (2003)
15. Baczyński, M., Jayaram, B.: On the characterizations of (S,N)-implications. Fuzzy Sets and Systems 158, 1713–1727 (2007)
16. Baczyński, M., Jayaram, B.: Yager's classes of fuzzy implications: some properties and intersections. Kybernetika 43, 157–182 (2007)

17. Baczyński, M., Jayaram, B.: (S,N)- and R-implications: a state-of-the-art survey. Fuzzy Sets and Systems 159, 1836–1859 (2008)
18. Baczyński, M., Jayaram, B.: On the distributivity of fuzzy implications over nilpotent or strict triangular conorms. IEEE Trans. Fuzzy Systems (in press) doi:10.1109/TFUZZ.2008.924201
19. Baczyński, M., Jayaram, B.: On the distributivity of fuzzy implications over representable uninorms. In: Magdalena, L., Ojeda-Aciego, M., Verdegay, J.L. (eds.) Proc. of IPMU 2008, Torremolinos (Malaga), June 22-27, 2008, pp. 1318–1325 (2008)
20. Balaji, P., Rao, C.J.M., Balasubramaniam, J.: On the suitability of Yager's implication for fuzzy systems. In: [208], pp. 354–357 (2003)
21. Balasubramaniam, J.: Contrapositive symmetrization of fuzzy implications – revisited. Fuzzy Sets and Systems 157, 2291–2310 (2006)
22. Balasubramaniam, J.: Yager's new class of implications J_f and some classical tautologies. Inform. Sci. 177, 930–946 (2007)
23. Balasubramaniam, J., Rao, C.J.M.: R-implication operators and rule detection in Mamdani-type fuzzy systems. In: [50], pp. 82–84 (2002)
24. Balasubramaniam, J., Rao, C.J.M.: A lossless rule reduction technique for a class of fuzzy system. In: Mastorakis, N.E. (ed.) Recent Advances in Simulation, Computational Methods and Soft Computing, Proc. 3rd WSEAS International Conference on Fuzzy Sets and Fuzzy Systems, Interlaken, Switzerland, February 2002, pp. 228–233. WSEAS press (2002)
25. Balasubramaniam, J., Rao, C.J.M.: On the distributivity of implication operators over T- and S-norms. IEEE Trans. Fuzzy Systems 12, 194–198 (2004)
26. Baldwin, J.F.: A new approach to approximate reasoning using a fuzzy logic. Fuzzy Sets and Systems 2, 309–325 (1979)
27. Baldwin, J.F., Pilsworth, B.W.: Axiomatic approach to implication for approximate reasoning with fuzzy logic. Fuzzy Sets and Systems 3, 193–219 (1980)
28. Baldwin, J.F., Guild, N.C.F.: Modelling controllers using fuzzy relations. Kybernetes 9, 223–229 (1980)
29. Bandler, W., Kohout, L.J.: Fuzzy relational products and fuzzy implication operators. In: International Workshop on Fuzzy Reasoning Theory and Applications, London, September 1978, Queen Mary College, University of London (1978)
30. Bandler, W., Kohout, L.J.: Semantics of implication operators and fuzzy relational products. Internat. J. Man-Mach. Stud. 12, 89–116 (1979)
31. Bandler, W., Kohout, L.: Fuzzy power sets and fuzzy implication operators. Fuzzy Sets and Systems 4, 13–30 (1980)
32. Bertoluzza, C.: On the distributivity between t-norms and t-conorms. In: Proc. 2nd IEEE Int. Conf. on Fuzzy Systems (FUZZ-IEEE 1993), San Francisco, USA, March 1993, pp. 140–147 (1993)
33. Bertoluzza, C., Doldi, V.: On the distributivity between t-norms and t-conorms. Fuzzy Sets and Systems 142, 85–104 (2004)
34. Bezdek, J.C., Dubois, D., Prade, H. (eds.): Fuzzy sets in approximate reasoning and information systems. Kluwer, Dordrecht (1999)
35. Bien, Z., Chun, M.G.: An inference network for bidirectional approximate reasoning based on equality measures. IEEE Trans. Fuzzy Systems 2, 177–180 (1994)
36. Birkhoff, G.: Lattice theory, 3rd edn. American Mathematical Society, Providence, Rhode Island (1967)
37. Boixader, D., Jacas, J.: Extensionality based approximate reasoning. Internat. J. Approx. Reason. 19, 221–230 (1998)

38. Borkowski, L. (ed.): Jan Łukasiewicz – Selected works. Studies in Logic and the Foundations of Mathematics. North-Holland, Amsterdam-London; PWN-Polish Scientific Publishers, Warsaw (1970)
39. Bouchon-Meunier, B., Delechamp, J., Marsala, C., Mellouli, N., Rifqi, M., Zerrouki, L.: Analogy and fuzzy interpolation in the case of sparse rules. In: De Baets, B., Fodor, J.C. (eds.) Proc. Joint Eurofuse-Soft and Intelligent Computing Conference (EUROFUSE-SIC 1999), Budapest, Hungary, May 1999, pp. 132–136 (1999)
40. Bouchon-Meunier, B., Marsala, C., Rifqi, M.: Interpolative reasoning based on graduality. In: Proc. 9th IEEE International Conference on Fuzzy Systems (FUZZ-IEEE 2000), San Antonio, USA, pp. 483–487 (2000)
41. Bouchon-Meunier, B., Dubois, D., Marsala, C., Prade, H., Ughetto, L.: A comparative view of interpolation methods between sparse fuzzy rules. In: Proc. Joint 9th IFSA World Congress and 20th NAFIPS International Conference, Vancouver, Canada, pp. 2499–2504 (2001)
42. Bouchon-Meunier, B., Kreinovich, V.: Axiomatic description of implication leads to a classical formula with logical modifiers (In particular, Mamdani's choice of "AND" as implication is not so weird after all), University of Texas at El Paso, Computer Science Department, Tech. Rep. No. UTEP-CS-96-23 (1996)
43. Burillo, P., Frago, N., Fuentes, R.: Inclusion grade and fuzzy implication operators. Fuzzy Sets and Systems 114, 417–429 (2000)
44. Bustince, H., Barrenechea, E., Mohedano, V.: Intuitionistic fuzzy implication operators – an expression and main properties. Internat. J. Uncertain. Fuzziness Knowledge-Based Systems 12, 387–406 (2004)
45. Bustince, H., Burillo, P., Soria, F.: Automorphisms, negation and implication operators. Fuzzy Sets and Systems 134, 209–229 (2003)
46. Cai, K.-Y.: Robustness of fuzzy reasoning and δ-equalities of fuzzy sets. IEEE Trans. Fuzzy Syst. 9, 738–750 (2001)
47. Cao, Z., Kandel, A.: Applicability of some fuzzy implication operators. Fuzzy Sets and Systems 34, 135–144 (1989)
48. Cao, Z., Park, D., Kandel, A.: Investigations on the applicability of fuzzy inference. Fuzzy Sets and Systems 49, 151–169 (1992)
49. Carbonell, M., Mas, M., Suner, J., Torrens, J.: On distributivity and modularity in De Morgan triplets. Internat. J. Uncertain. Fuzziness Knowledge-Based Systems 4, 351–368 (1996)
50. Caulfield, H.J., et al. (eds.): Proc. 6th Joint Conference on Information Science, JCIS 2002, Durham, USA. Research Triangle Park, North Carolina (2002)
51. Cárdenas, E., Castillo, J.C., Cordón, O., Herrera, F., Peregrín, A.: Influence of fuzzy implication functions and defuzzification methods in fuzzy control. BUSEFAL 57, 69–79 (1994)
52. Chen, S.-M.: A new approach to handling fuzzy decision-making problems. IEEE Trans. on Syst. Man and Cyber. 18, 1012–1016 (1988)
53. Cignoli, R., Esteva, F., Godo, L., Torrens, A.: Basic Fuzzy Logic is the logic of continuous t-norms and their residua. Soft Comp. 4, 106–112 (2002)
54. Cignoli, R., Esteva, F., Godo, L., Montagna, F.: On a class of left-continuous t-norms. Fuzzy Sets and Systems 131, 283–296 (2002)
55. Cignoli, R., D'Ottaviano, I.M.L., Mundici, D.: Algebraic foundations of many-valued reasoning. Kluwer, Dordrecht (1999)
56. Combs, W.E., Andrews, J.E.: Combinatorial rule explosion eliminated by a fuzzy rule configuration. IEEE Trans. Fuzzy Systems 6, 1–11 (1998)

57. Combs, W.E.: Author's reply. IEEE Trans. Fuzzy Systems 7, 371–373 (1999)
58. Combs, W.E.: Author's reply. IEEE Trans. Fuzzy Systems 7, 477–478 (1999)
59. Cordón, O., Herrera, F., Peregrin, A.: Applicability of the fuzzy operators in the design of fuzzy logic controllers. Fuzzy Sets and Systems 86, 15–41 (1997)
60. Cordón, O., Herrera, F., Peregrin, A.: Searching for basic properties obtaining robust implication operators in fuzzy control. Fuzzy Sets and Systems 111, 237–251 (2000)
61. Cornelis, Ch., Deschrijver, G., Kerre, E.E.: Classification of intuitionistic fuzzy implicators: An algebraic approach. In: [50], pp. 105–108 (2002)
62. Cornelis, Ch., DeCock, M., Kerre, E.E.: Efficient approximate reasoning with positive and negative information. In: Negoita, M.G., Howlett, R.J., Jain, L.C. (eds.) KES 2004. LNCS (LNAI), vol. 3214, pp. 779–785. Springer, Heidelberg (2004)
63. Cornelis, Ch., Van der Donck, C., Kerre, E.E.: Sinha-Dougherty approach to the fuzzification of set inclusion revisited. Fuzzy Sets and Systems 134, 283–295 (2003)
64. Cornelis, Ch., Deschrijver, G., Kerre, E.E.: Implication in intuitionistic fuzzy and interval-valued fuzzy set theory: construction, classification, application. Internat. J. Approx. Reason. 35, 55–95 (2004)
65. Cross, V., Sudkamp, T.: Fuzzy implication and compatibility modification. In: Proc. 2nd IEEE International Conference on Fuzzy Systems (FUZZ-IEEE 1993), San Francisco, USA, pp. 219–224 (1993)
66. De Baets, B.: A note on Mamdani controllers. In: Da, R., et al. (eds.) Proc. 2nd Internat. FLINS Workshop, Mol, Belgium, pp. 22–28. World Scientific, Singapore (1996)
67. De Baets, B.: Idempotent uninorms. European J. Oper. Res. 118, 631–642 (1999)
68. De Baets, B., Kerre, E.E.: Fuzzy inclusion and the inverse problem. In: [267], pp. 940–945 (1994)
69. De Baets, B., Fodor, J.C.: On the structure of uninorms and their residual implicators. In: Gottwald, S., Klement, E.P. (eds.) Proc. 18th Linz Seminar on Fuzzy Set Theory, Linz, Austria, pp. 81–87 (1997)
70. De Baets, B., Fodor, J.C.: Residual operators of representable uninorms. In: [268], pp. 52–56 (1997)
71. De Baets, B., Fodor, J.C.: Residual operators of uninorms. Soft Comput. 3, 89–100 (1999)
72. De Baets, B., Mesiar, R.: Residual implicators of continuous t-norms. In: [268], pp. 27–31 (1996)
73. Demirli, K., Türksen, I.B.: Rule Break up with Compositional Rule of Inference. In: Proc. FUZZ-IEEE 1992, pp. 749–756 (1992)
74. Deschrijver, G., Kerre, E.E.: Smets-Magrez axioms for R-implicators in interval-valued and intuitionistic fuzzy set theory. Internat. J. Uncertain. Fuzziness Knowledge-Based Systems 13, 453–464 (2005)
75. Deschrijver, G., Kerre, E.E.: Implicators based on binary aggregation operators in interval-valued fuzzy set theory. Fuzzy Sets and Systems 153, 229–248 (2005)
76. Dick, S., Kandel, A.: Comments on Combinatorial rule explosion eliminated by a fuzzy rule configuration [and reply]. IEEE Trans. Fuzzy Systems 7, 475–478 (1999)
77. Dienes, Z.P.: On an implication function in many-valued systems of logic. J. Symb. Logic 14, 95–97 (1949)
78. Di Nola, A.: The role of Łukasiewicz logic in fuzzy set theory. In: [269], pp. 120–134 (1980)

79. Di Nola, A., Pedrycz, W., Sessa, S.: An aspect of discrepancy in the implementation of modus ponens in the presence of fuzzy quantities. Internat. J. Approx. Reas. 3, 259–265 (1989)
80. Drewniak, J.: Selfconjugate fuzzy implications. In: De Baets, B., Fodor, J.C. (eds.) Proc. Joint Eurofuse-Soft and Intelligent Computing Conference (EUROFUSE-SIC 1999), Budapest, Hungary, May 1999, pp. 351–353 (1999)
81. Drewniak, J.: Invariant fuzzy implications. Soft Comput. 10, 506–513 (2006)
82. Drewniak, J., Kula, K.: Generalized compositions of fuzzy relations. Internat. J. Uncertain. Fuzziness Knowledge-Based Systems 10, 149–164 (2002)
83. Drewniak, J., Sobera, J.: Composition of invariant fuzzy implications. Soft Comput. 10, 514–520 (2006)
84. Driankov, D., Hellendoorn, H., Reinfrank, M.: An introduction to fuzzy control, 2nd edn. Springer, Berlin (1993)
85. Dubois, D., Prade, H.: Fuzzy-set theoretic differences and inclusions and their uses in fuzzy arithmetic and analysis. In: Klement, E.P. (ed.) Proc. 5th Internat. Seminar on Fuzzy Set Theory, Johannes Kepler Universität, Linz, Austria, September 5-9 (1983)
86. Dubois, D., Prade, H.: A theorem on implication functions defined from triangular norms. Stochastica 8, 267–279 (1984)
87. Dubois, D., Prade, H.: A theorem on implication functions defined from triangular norms. BUSEFAL 18, 33–41 (1984)
88. Dubois, D., Prade, H.: Fuzzy logic and the generalized modus ponens revisited. Internat. J. Cybernetics and Systems 15, 293–331 (1984)
89. Dubois, D., Prade, H.: The generalized modus ponens under sup-min composition - a theoretical study. In: Gupta, M.M., Kandel, A., Bandler, W., Kiszka, J.B. (eds.) Approximate Reasoning in Expert Systems, pp. 157–166. Elsevier, North Holland (1985)
90. Dubois, D., Prade, H.: On the combination of uncertain or imprecise pieces of information in rule-based systems - A discussion in the framework of possibility theory. Internat. J. Approx. Reas. 2, 65–87 (1988)
91. Dubois, D., Martin-Clouaire, R., Prade, H.: Practical computing in fuzzy logic. In: Gupta, M.M., Yamakawa, T. (eds.) Fuzzy Computing, pp. 11–34. Elsevier, North Holland (1988)
92. Dubois, D., Prade, H.: Fuzzy sets in approximate reasoning. Part 1: Inference with possibility distributions. Fuzzy Sets and Systems 40, 143–202 (1991)
93. Dubois, D., Prade, H.: On fuzzy interpolation. Internat. J. Gen. Systems 28, 103–114 (1999)
94. Dubois, D., Prade, H. (eds.): Fundamentals of fuzzy sets. Kluwer, Boston (2000)
95. Durante, F., Klement, E.P., Mesiar, R., Sempi, C.: Conjunctors and their residual implicators: characterizations and construction methods. Mediterr. J. Math. 4, 343–356 (2007)
96. Esteva, F., Gispert, J., Godo, L., Montagna, F.: On the standard and rational completeness of some axiomatic extensions of the monoidal t-norm logic. Studia Logica 71, 199–226 (2002)
97. Esteva, F., Godo, L.: Monoidal t-norm based logic: towards a logic for left-continuous t-norms. Fuzzy Sets and Systems 124, 271–288 (2001)
98. Esteva, F., Godo, L., Hájek, P., Navara, M.: Residuated fuzzy logic with an involutive negation. Arch. Math. Logic 39, 103–124 (2000)
99. Fodor, J.C.: Some remarks on fuzzy implication operations. BUSEFAL 38, 42–46 (1989)

100. Fodor, J.C.: On fuzzy implication operators. Fuzzy Sets and Systems 42, 293–300 (1991)
101. Fodor, J.C.: A new look at fuzzy connectives. Fuzzy Sets and Systems 57, 141–148 (1993)
102. Fodor, J.C.: On contrapositive symmetry of implications in fuzzy logic. In: Proc. 1st European Congress on Fuzzy and Inteligent Technologies (EUFIT 1993), pp. 1342–1348. Verlag der Augustinus Buchhandlung, Aachen (1993)
103. Fodor, J.C.: Contrapositive symmetry of fuzzy implications. Fuzzy Sets and Systems 69, 141–156 (1995)
104. Fodor, J.C., Keresztfalvi, T.: Nonstandard conjunctions and implications in fuzzy logic. Internat. J. Approx. Reason. 12, 69–84 (1995)
105. Fodor, J.C., Roubens, M.: Fuzzy preference modelling and multicriteria decision support. Kluwer, Dordrecht (1994)
106. Fodor, J.C., Yager, R.R., Rybalov, A.: Structure of uninorms. Internat. J. Uncertain. Fuzziness Knowledge-Based Systems 5, 411–427 (1997)
107. Fodor, J., Yager, R.R.: Fuzzy set - theoretic operators and quantifiers. In: [94], pp. 125–193 (2000)
108. Fukami, S., Mizumoto, M., Tanaka, K.: On fuzzy reasoning. Systems-Comput.-Controls 9, 44–53 (1980)
109. Fukami, S., Mizumoto, M., Tanaka, K.: Some considerations on fuzzy conditional inference. Fuzzy Sets and Systems 4, 243–273 (1980)
110. Gaines, B.R.: Foundations of fuzzy reasoning. Int. J. Man-Machine Studies 8, 623–668 (1976)
111. Goguen, J.A.: The logic of inexact concepts. Synthese 19, 325–373 (1969)
112. Gottwald, S.: Characterizations of the solvability of fuzzy equations. Elektron. Informationsverarb. Kybernet. 22, 67–91 (1986)
113. Gottwald, S.: A treatise on many-valued logics. Research Studies Press, Baldock (2001)
114. Gottwald, S.: An abstract approach toward the evaluation of fuzzy rule systems. In: Wang, P.P., Ruan, D., Kerre, E.E. (eds.) Fuzzy Logic. A Spectrum of Theoretical & Practical Issues. Studies in Fuzziness and Soft Computing, vol. 215, pp. 291–306. Springer, Berlin (2007)
115. Gottwald, S.: Fuzzy control as a general interpolation problem. In: Reusch, B. (ed.) Computational Intelligence, Theory and Applications - Advances in Soft Computing, pp. 197–204. Springer, Berlin (2005)
116. Gottwald, S.: Mathematical fuzzy control. A survey of recent results. Logic J. IGPL 13, 525–541 (2005)
117. Gödel, K.: Zum intuitionistischen Aussagenkalkül. Auzeiger der Akademie der Wissenschaften in Wien, Mathematisch, naturwissenschaftliche Klasse 69, 65–66 (1932)
118. Hall, L.O.: The choice of ply operator in fuzzy intelligent systems. Fuzzy Sets and Systems 34, 135–144 (1990)
119. Hájek, P.: Basic fuzzy logic and BL-algebras. Soft Comp. 2, 124–128 (1998)
120. Hájek, P.: Metamatematics of fuzzy logic. Kluwer, Dordrecht (1998)
121. Hájek, P., Kohout, L.: Fuzzy implications and generalized quantifiers. Internat. J. Uncertain. Fuzziness Knowledge-Based Systems 4, 225–233 (1996)
122. Heyting, A.: Die formalen Regeln der intuitionistischen Logik. I, II, III. Sitzungsberichte Akad., 42–56, 57–71, 158–169 (1930)
123. Igel, Ch., Temme, K.-H.: The chaining syllogism in fuzzy logic. IEEE Trans. Fuzzy Syst. 12, 849–853 (2004)

124. Ivanek, J.: Construction of implicational quantifiers from fuzzy implications. Fuzzy Sets and Systems 151, 381–391 (2005)
125. Jayaram, B.: Rule reduction for efficient inferencing in similarity based reasoning. Internat. J. Approx. Reason. 48, 156–173 (2008)
126. Jayaram, B.: On the law of importation $(x \wedge y) \longrightarrow z \equiv (x \longrightarrow (y \longrightarrow z))$ in fuzzy logic. IEEE Trans. Fuzzy Systems 16, 130–144 (2008)
127. Jayaram, B., Baczyński, M.: (U, N)-implications and their characterizations. In: Štěpnička, M., Novák, V., Bodenhofer, U. (eds.) New Dimensions in Fuzzy Logic and Related Technologies. Proc. 5th EUSFLAT Conference, Ostrava, Czech Republic, Czech Republic, September 2007, vol. I, pp. 105–110. University of Ostrava (2007)
128. Jayaram, B., Baczyński, M.: Intersections between basic families of fuzzy implications (S,N)-, R- and QL-implications. In: Štěpnička, M., Novák, V., Bodenhofer, U. (eds.) New Dimensions in Fuzzy Logic and Related Technologies. Proc. 5th EUSFLAT Conference, Ostrava, Czech Republic, September 2007, vol. I, pp. 111–118. University of Ostrava (2007)
129. Jayaram, B., Baczyński, M.: Algebra of (S,N)-implications. In: Klement, E.P., et al. (eds.) Proc. 8th International Conference FSTA 2008, Liptovský Ján, The Slovak Republic, pp. 61–62 (2008)
130. Jenei, S.: Continuity of Zadeh's compositional rule of inference. Fuzzy Sets and Systems 99, 333–339 (1999)
131. Jenei, S.: New family of triangular norms via contrapositive symmetrization of residuated implications. Fuzzy Sets and Systems 110, 157–174 (2000)
132. Jenei, S.: Structure of left-continuous t-norms with strong associated negations. I. Rotation construction. J. Appl. Non-Classical Logics 10, 83–92 (2000)
133. Jenei, S.: A new approach for interpolation and extrapolation of compact fuzzy quantities. The one dimensional case. In: Klement, E.P., Stout, L.N. (eds.) Proc. 21th Linz Seminar on Fuzzy Set Theory, Linz, Austria, pp. 13–18 (2000)
134. Jenei, S.: Structure of left-continuous t-norms with strong associated negations. II. Rotation-annihilation construction. J. Appl. Non-Classical Logics 11, 351–366 (2001)
135. Jenei, S.: How to construct left-continuous triangular norms – state of the art. Fuzzy Sets and Systems 143, 27–45 (2004)
136. Kandel, A., Li, L., Cao, Z.: Fuzzy inference and its applicability to control systems. Fuzzy Sets and Systems 48, 99–111 (1992)
137. Khaledi, Gh., Mashinchi, M., Ziaie, S.A.: The monoid structure of e-implications and pseudo-e-implications. Inform. Sci. 174, 103–122 (2005)
138. Kiszka, J.B., Gupta, M.M., Trojan, G.M.: Multivariable fuzzy controller under Gödel's implication. Fuzzy Sets and Systems 34, 301–321 (1990)
139. Kiszka, J.B., Kochańska, M.E., Śliwińska, D.S.: The influence of some fuzzy implication operators on the accuracy of a fuzzy model. Fuzzy Sets and Systems 15 (Part 1), 111–128, (Part 2), 223–240 (1985)
140. Kitainik, L.: Fuzzy inclusions and fuzzy dichotomous decision procedures. In: Kacprzyk, J., Orlovski, S. (eds.) Optimization models using fuzzy sets and possibility theory, pp. 154–170. D. Reidel, Dordrecht (1987)
141. Kitainik, L.: Fuzzy decision procedures with binary relations. Kluwer, Dordrecht (1993)
142. Klawonn, F.: Fuzzy points, fuzzy relations and fuzzy functions. In: Novák, V., Perfilieva, I. (eds.) Discovering the World with Fuzzy Logic. Studies in Fuzzines and Soft Computing, vol. 41, pp. 431–453. Springer, Physica-Verlag, Heidelberg (2000)

143. Kleene, S.C.: On a notation for ordinal numbers. J. Symb. Log. 3, 150–155 (1938)
144. Klement, E.P., Mesiar, R.: Logical, algebraic, analytic and probabilistic aspects of triangular norms. Elsevier, Amsterdam (2005)
145. Klement, E.P., Mesiar, R., Pap, E.: On the relationship of associative compensatory operators to triangular norms and conorms. Internat. J. Uncertain. Fuzziness Knowledge-Based Systems 4, 129–144 (1996)
146. Klement, E.P., Mesiar, R., Pap, E.: Triangular norms. Kluwer, Dordrecht (2000)
147. Klir, G.J., Yuan, B.: Fuzzy sets and fuzzy logic. Theory and applications. Prentice Hall, New Jersey (1995)
148. Kóczy, L., Hirota, K.: Interpolative reasoning with insufficient evidence in sparse fuzzy rule bases. Inform. Sci. 71, 169–201 (1993)
149. Kóczy, L., Hirota, K.: Approximate reasoning by linear rule interpolation and general approximation. Internat. J. Approx. Reas. 9, 197–225 (1993)
150. Kuczma, M.: Functional equations in a single variable. PWN – Polish Scientific Publishers, Warszawa (1968)
151. Larsen, P.M.: Industrial applications of fuzzy logic control. Int. J. Man Mach. Studies 12, 3–10 (1980)
152. Li, Y.-M., Shi, Z.-K., Li, Z.-H.: Approximation theory of fuzzy systems based upon genuine many-valued implications - SISO cases. Fuzzy Sets and Systems 130, 147–157 (2002)
153. Li, Y.-M., Shi, Z.-K., Li, Z.-H.: Approximation theory of fuzzy systems based upon genuine many-valued implications - MISO cases. Fuzzy Sets and Systems 130, 159–174 (2002)
154. Ling, C.H.: Representation of associative functions. Publ. Math. Debrecen. 12, 189–212 (1965)
155. Łojasiewicz, S.: An introduction to the theory of real functions. Wiley, New York (1988)
156. Łukasiewicz, J.: O logice trójwartościowej. Ruch Filozoficzny 5, 92–93 (1920) [English translation in [38]]
157. Łukasiewicz, J.: Interpretacja liczbowa teorii zdań. Ruch Filozoficzny 7, 92–93 (1923) [English translation in [38]]
158. Maes, K.C., De Baets, B.: A contour view on uninorm properties. Kybernetika 42, 303–318 (2006)
159. Maes, K.C., De Baets, B.: Rotation-invariant t-norm solutions of a system of functional equations. Fuzzy Sets and Systems 157, 373–397 (2006)
160. Maes, K.C., De Baets, B.: On the structure of left-continuous t-norms that have a continuous contour line. Fuzzy Sets and Systems 158, 843–860 (2007)
161. Maes, K.C., De Baets, B.: The triple rotation method for constructing t-norms. Fuzzy Sets and Systems 158, 1652–1674 (2007)
162. Magrez, P., Smets, P.: Fuzzy Modus Ponens: a new model suitable for applications in Knowledge-based systems. Internat. J. Intell. Systems 4, 181–200 (1989)
163. Mamdani, E.H.: Application of fuzzy algorithms for control of simple dynamic plant. Proc. IEEE 121, 1585–1588 (1974)
164. Mamdani, E.H.: Application of fuzzy logic to approximate reasoning using linguistic synthesis. IEEE Trans. Comput. 26, 1182–1191 (1977)
165. Mamdani, E.H., Assilian, S.: An experiment in linguistic synthesis with a fuzzy logic controller. Int. J. Man-Machine Studies 7, 1–13 (1975)
166. Mantaras, R.L.: Approximate Reasoning Models. Ellis Horwood Series in Artifical Intelligence. Wiley, New York (1990)
167. Martin, J., Mayor, G., Torrens, J.: On locally internal monotonic operators. Fuzzy Sets and Systems 137, 27–42 (2003)

168. Martin-Clouaire, R.: A fast generalized modus ponens. BUSEFAL 18, 75–82 (1984)
169. Martin-Clouaire, R.: Semantics and computation of the generalized modus ponens: the long paper. Internat. J. Approx. Reas. 3, 195–217 (1989)
170. Mas, M., Mayor, G., Torrens, J.: t-operators. Internat. J. Uncertain. Fuzziness Knowledge-Based Systems 7, 31–50 (1999)
171. Mas, M., Mayor, G., Torrens, J.: The distributivity condition for uninorms and t-operators. Fuzzy Sets and Systems 128, 209–225 (2002)
172. Mas, M., Monserrat, M., Torrens, J.: S-implications and R-implications on a finite chain. Kybernetika 40, 3–20 (2004)
173. Mas, M., Monserrat, M., Torrens, J.: On two types of discrete implications. Internat. J. Approx. Reas. 40, 262–279 (2005)
174. Mas, M., Monserrat, M., Torrens, J.: QL-implications versus D-implications. Kybernetika 42, 351–366 (2006)
175. Mas, M., Monserrat, M., Torrens, J.: Two types of implications derived from uninorms. Fuzzy Sets and Systems 158, 2612–2626 (2007)
176. Mas, M., Monserrat, M., Torrens, J., Trillas, E.: A survey on fuzzy implication functions. IEEE Trans. Fuzzy Systems 15, 1107–1121 (2007)
177. Mas, M., Monserrat, M., Torrens, J.: Modus Ponens and Modus Tollens in discrete implications. Internat. J. Approx. Reas. (submitted)
178. Mayor, G., Torrens, J.: Triangular norms in discrete settings. In: [144], pp. 193–236 (2005)
179. Mendel, J.M., Liang, Q.: Comments on Combinatorial rule explosion eliminated by a fuzzy rule configuration [and reply]. IEEE Trans. Fuzzy Systems 7, 369–373 (1999)
180. Menger, K.: Statistical metrics. Proc. Nat. Acad. Sci. USA 28, 535–537 (1942)
181. Mesiar, R.: Fuzzy difference posets and MV-algebras. In: Bouchon-Meunier, B., Yager, R.R., Zadeh, L.A. (eds.) IPMU 1994. LNCS, vol. 945, pp. 208–212. Springer, Heidelberg (1995)
182. Mesiar, R., Mesiarová, A.: Residual implications and left-continuous t-norms which are ordinal sums of semigroups. Fuzzy Sets and Systems 143, 47–57 (2004)
183. Miyakoshi, M., Shimbo, M.: Solutions of composite fuzzy relational equations with triangular norms. Fuzzy Sets and Systems 16, 53–63 (1985)
184. Mizumoto, M., Zimmermann, H.J.: Comparison of fuzzy reasoning methods. Fuzzy Sets and Systems 8, 253–283 (1982)
185. Mizumoto, M.: Fuzzy conditional inference under $\max - \odot$ composition. Inform. Sci. 27, 183–209 (1982)
186. Mizumoto, M.: Fuzzy controls under various fuzzy reasoning methods. Inform. Sci. 45, 129–151 (1988)
187. Morsi, N.N., Fahmy, A.A.: On generalized modus ponens with multiple rules and a residuated implication. Fuzzy Sets and Systems 129, 267–274 (2002)
188. Nguyen, H.T., Walker, E.A.: A first course in fuzzy logic, 2nd edn. CRC Press, Boca Raton (2000)
189. Novák, V.: Perception-based logical deduction. In: Reusch, B. (ed.) Computational intelligence, Theory and Applications. Advances in Soft Computing, pp. 237–250. Springer, Heidelberg (2005)
190. Novák, V.: A comprehensive theory of trichotomous evaluative linguistic expressions. Fuzzy Sets and Systems (in press) doi:10.1016/j.fss.2008.02.023
191. Novák, V., Perfilieva, I.: Evaluating linguistic expressions and functional fuzzy theories in fuzzy logic. In: Zadeh, L.A., Kacprzyk, J. (eds.) Computing with Words in Information/Intelligent Systems 1, pp. 383–406. Springer, Heidelberg (1999)

192. Novák, V., Perfilieva, I.: On the semantics of perception-based fuzzy logic deduction. Int. J. Intell. Syst. 19, 1007–1031 (2004)
193. Novák, V., Perfilieva, I.: System of fuzzy relations as a continuous model of IF-THEN rules. Inform. Sci. 177, 3218–3227 (2007)
194. Ovchinnikiov, S.V., Roubens, M.: On fuzzy strict preference, indifference, and incomparability relations. Fuzzy Sets and Systems 47, 49, 313–318, 15–20 (1992)
195. Pappis, C.P., Karacapilidis, N.I.: A comparative assessment of measures of similarity of fuzzy values. Fuzzy Sets and Systems 56, 171–174 (1993)
196. Park, D., Cao, Z., Kandel, A.: Investigations on the applicability of fuzzy inference. Fuzzy Sets and Systems 49, 151–169 (1992)
197. Passino, K.M., Yurkovich, S.: Fuzzy control. Addison-Wesley-Longman, Menlo Park (1998)
198. Pavelka, J.: On fuzzy logic. I. Many-valued rules of inference. Z. Math. Logik Grundlagen Math. 25, 45–52 (1979)
199. Pavelka, J.: On fuzzy logic. II. Enriched residuated lattices and semantics of propositional calculi. Z. Math. Logik Grundlagen Math. 25, 119–134 (1979)
200. Pavelka, J.: On fuzzy logic. III. Semantical completeness of some many-valued propositional calculi. Z. Math. Logik Grundlagen Math. 25, 447–464 (1979)
201. Pedrycz, W.: Fuzzy relational equations with triangular norms and their resolutions. BUSEFAL 11, 24–32 (1982)
202. Pedrycz, W.: Numerical and applicational aspects of fuzzy relational equations. Fuzzy Sets and Systems 11, 1–18 (1983)
203. Pedrycz, W.: On generalized fuzzy relational equations and their applications. J. Math. Anal. Appl. 107, 520–536 (1985)
204. Pedrycz, W.: Neurocomputations in relational systems. IEEE Trans. Pattern Anal. Mach. Intell. 13, 289–297 (1991)
205. Pei, D.: R_0 implication: characteristics and applications. Fuzzy Sets and Systems 131, 297–302 (2002)
206. Pei, D.: Fuzzy implications suitable for triple I based fuzzy reasoning. J. Fuzzy Math. 11, 817–826 (2003)
207. Raha, S., Pal, N.R., Ray, K.S.: Similarity based approximate reasoning: methodology and application. IEEE Transactions on Systems, Man and Cybernetics 32, 541–547 (2002)
208. Rajendra, A. (ed.): Proc. 2nd International Conference on Applied Artificial Intelligence (ICAAI 2003), Pune, India, December 15-16 (2003)
209. Reichenbach, H.: Wahrscheinlichkeitslogik. Erkenntnis 5, 37–43 (1935)
210. Rescher, N.: Many-valued logic. McGraw-Hill, New York (1969)
211. Reusch, B. (ed.): Fuzzy Days 1999. LNCS, vol. 1625. Springer, Heidelberg (1999)
212. Martin-Clouaire, R.: Semantics and computation of the generalized modus ponens: the long paper. Internat. J. Approx. Reason. 3, 195–217 (1989)
213. Ruan, D., Kerre, E.E.: Fuzzy implication operators and generalized fuzzy method of cases. Fuzzy Sets and Systems 54, 23–37 (1993)
214. Ruiz, D., Torrens, J.: Distributive idempotent uninorms. Internat. J. Uncertain. Fuzziness Knowledge-Based Systems 11, 413–428 (2003)
215. Ruiz, D., Torrens, J.: Residual implications and co-implications from idempotent uninorms. Kybernetika 40, 21–38 (2004)
216. Ruiz-Aguilera, D., Torrens, J.: Distributivity of strong implications over conjunctive and disjunctive uninorms. Kybernetika 42, 319–336 (2006)
217. Ruiz-Aguilera, D., Torrens, J.: Distributivity of residual implications over conjunctive and disjunctive uninorms. Fuzzy Sets and Systems 158, 23–37 (2007)

218. Rutkowski, L., Kacprzyk, J. (eds.): Neural networks and soft computing. Advances in Soft Computing. Physica-Verlag, Heidelberg (2003)
219. Sainio, E., Turunen, E., Mesiar, R.: A characterization of fuzzy implications generated by generalized quantifiers. Fuzzy Sets and Systems 159, 491–499 (2008)
220. Sanchez, E.: Resolution of composite fuzzy relation equations. Info. Control 30, 38–48 (1976)
221. Schweizer, B., Sklar, A.: Associative functions and statistical triangle inequalities. Publ. Math. Debrecen. 8, 169–186 (1961)
222. Schweizer, B., Sklar, A.: Probabilistic metric spaces. North-Holland, New York (1983)
223. Shi, Y., Ruan, D., Kerre, E.E.: On the characterization of fuzzy implications satisfying $I(x, y) = I(x, I(x, y))$. Info. Sci. 177, 2954–2970 (2007)
224. Shi, Y., Ruan, D., Van Gasse, B., Kerre, E.E.: On the first place antitonicity in QL-implications. Fuzzy Sets and Systems (in press) doi:10.1016/j.fss.2008.04.012
225. Sinha, D., Dougherty, E.R.: Fuzzification of set inclusion: Theory and applications. Fuzzy Sets and Systems 55, 15–42 (1991)
226. Smets, P., Magrez, P.: Implication in fuzzy logic. Internat. J. Approx. Reason. 1, 327–347 (1987)
227. Smogura, R.: A note on the article: Residual implications revisited. Notes on the Smets-Magrez theorem. Fuzzy Sets and Systems 157, 466–469 (2006)
228. Sokhansanj, B.A., Fitch, J.P.: URC fuzzy modeling and simulation of gene regulation. In: Proc. 23rd Annual International Conference of the IEEE Engineering in Medicine and Biology Society, Instanbul, Turkey, October 25-28, pp. 2918–2921 (2001)
229. Štěpnička, M., De Baets, B., Nosková, L: On additive and multiplicative fuzzy models. In: Štěpnička, M., Novák, V., Bodenhofer, U. (eds.) New Dimensions in Fuzzy Logic and Related Technologies. Proc. 5th EUSFLAT Conference, Ostrava, Czech Republic, Czech Republic, September 2007, vol. 2, pp. 95–102. University of Ostrava (2007)
230. Thiele, H.: Aximoatic considerations of the concepts of R-Implication and t-norm. University of Dortmund, Reihe Computational Intelligence, Tech. Rep. No. CI-35/98 (1988)
231. Torra, V.: A review of the construction of hierarchical fuzzy systems. Int. J. Intell. Syst. 17, 531–543 (2002)
232. Trillas, E.: Sobre funciones de negación en la teoria de conjuntos difusos. Stochastica 3, 47–59 (1979); [English translation In: Barro, S., Bugarín, A., Sobrino, A. (eds.) Advances in fuzzy logic, Public. Univ. Santiago de Compostela, Spain, pp. 31–45 (1998)]
233. Trillas, E., Alsina, C.: On the law $[p \wedge q \to r] = [(p \to q) \vee (p \to r)]$ in fuzzy logic. IEEE Trans. Fuzzy Systems 10, 84–88 (2002)
234. Trillas, E., Alsina, C., Pradera, A.: On MPT-implication functions for fuzzy logic. Rev. R. Acad. Cien. Serie A Mat. 98, 259–271 (2004)
235. Trillas, E., Alsina, C., Renedo, E., Pradera, A.: On contra-symmetry and MPT-conditionality in fuzzy logic. Internat. J. Intell. Systems 20, 313–326 (2005)
236. Trillas, E., Campo, C., del Cubillo, S.: When QM-operators are implication functions and conditional fuzzy relations. Internat. J. Intell. Systems 15, 647–655 (2000)
237. Trillas, E., de Soto, A.R., del Cubillo, S.: A glance at implication and T-conditional functions. In: Novák, V., Perfilieva, I. (eds.) Discovering the World with Fuzzy Logic. Studies in Fuzzines and Soft Computing, vol. 41, pp. 126–147. Springer, Physica-Verlag, Heidelberg (2000)

238. Trillas, E., Valverde, L.: On some functionally expressable implications for fuzzy set theory. In: Klement, E.P. (ed.) Proc. 3rd Inter. Seminar on Fuzzy Set Theory, Linz, Austria, pp. 173–190 (September 1981)
239. Trillas, E., Valverde, L.: On implication and indistinguishability in the setting of fuzzy logic. In: Kacprzyk, J., Yager, R.R. (eds.) Management decision support systems using fuzzy sets and possibility theory, TÜV-Rhineland, Cologne, pp. 198–212 (1985)
240. Trillas, E., Valverde, L.: On mode and implication in approximate reasoning. In: Gupta, M.M., Kandel, A., Bandler, W., Kiszka, J.B. (eds.) In approximate reasoning in expert systems, pp. 157–166. Elsevier, North Holland (1985)
241. Trillas, E., Valverde, L.: On modus ponens in fuzzy logic. In: Proc. 15th International Symposium on Multiple-Valued Logic, Kingston, Canada, pp. 294–301 (1985)
242. Trillas, E., Valverde, L.: On inference in fuzzy logic. In: Proc. 2nd IFSA World Congress, Gakushu-in University, Japan, pp. 294–297 (1987)
243. Turunen, E.: Mathematics Behind Fuzzy Logic. Springer, Heidelberg (1999)
244. Türksen, I.B., Kreinovich, V., Yager, R.R.: A new class of fuzzy implications. Axioms of fuzzy implication revisited. Fuzzy Sets and Systems 100, 267–272 (1998)
245. Türksen, I.B., Zhong, Z.: An approximate analogical reasoning approach based on similarity measures. IEEE Trans. on Syst. Man and Cyber. 18, 1049–1056 (1988)
246. Ughetto, L., Dubois, D., Prade, H.: Efficient inference procedures with fuzzy inputs. In: Proc. 6th IEEE International Conference on Fuzzy Systems (FUZZ-IEEE 1997), Barcelona, Spain, July 1997, pp. 567–572 (1997)
247. Ughetto, L., Dubois, D., Prade, H.: Implicative and conjunctive fuzzy rules – a tool for reasoning from knowledge and examples. In: Hendler, J., Subramanian, D., Uthurusamy, R., Hayes-Roth, B. (eds.) Proc. 16th National Conference on Artificial Intelligence and the 11th Innovative Aplications of Artificial Antelligence conference, Orlando, USA, July 18-22, pp. 214–219. American Association for Artificial Intelligence, Menlo Park (1999)
248. Ughetto, L., Dubois, D., Prade, H.: A new perspective on reasoning with fuzzy rules. In: Pal, N.R., Sugeno, M. (eds.) AFSS 2002. LNCS (LNAI), vol. 2275, pp. 1–11. Springer, Heidelberg (2002)
249. Viceník, P.: Additive generators and discontinuity. BUSEFAL 76, 25–28 (1998)
250. Villar, J., Sanz-Bobi, M.A.: Semantic analysis of fuzzy models, application to Yager models. In: Mastorakis, N. (ed.) Advances in Fuzzy Systems and Evolutionary Computation. Proc. 2nd WSES International Conference on Fuzzy Sets and Fuzzy Systems (FSFS 2001), Puerto de la Cruz, Tenerife, February 11-15, pp. 82–87. WSES Press (2001)
251. Wang, W.-J.: New similarity measures on fuzzy sets and on elements. Fuzzy Sets and Systems 85, 305–309 (1997)
252. Wang, X., De Baets, B., Kerre, E.: A comparative study of similarity measures. Fuzzy Sets and Systems 73, 259–268 (1995)
253. Wang, Z., Fang, J.: Residual operations of left and right uninorms on a complete lattice. Fuzzy Sets and System (in press) doi:10.1016/j.fss.2008.03.001
254. Wangming, W.: Fuzzy reasoning and fuzzy relational equations. Fuzzy Sets and Systems 20, 67–78 (1986)
255. Weber, S.: A general concept of fuzzy connectives, negations and implications based on t-norms and t-conorms. Fuzzy Sets and Systems 11, 115–134 (1983)
256. Weinschenk, J., Combs, W., Marks, R.: Avoidance of rule explosion by mapping fuzzy systems to a union rule configuration. In: Proc. IEEE International Conference on Fuzzy Systems, May 25-28, pp. 43–48 (2003)

257. Willmott, R.: Two fuzzier implication operators in the theory of fuzzy power sets. Fuzzy Sets and Systems 4, 31–36 (1980)
258. Yager, R.R.: An approach to inference in approximate reasoning. Internat. J. Man-Machine Studies 13, 323–338 (1980)
259. Yager, R.R.: On the implication operator in fuzzy logic. Inform. Sci. 31, 141–164 (1983)
260. Yager, R.R.: On global requirements for implication operators in fuzzy modus ponens. Fuzzy Sets and Systems 106, 3–10 (1999)
261. Yager, R.R.: On some new classes of implication operators and their role in approximate reasoning. Inform. Sci. 167, 193–216 (2004)
262. Yager, R.R.: Modeling holistic fuzzy implication using co-copulas. Fuzzy Optim. and Decis. Making 5, 207–226 (2006)
263. Yager, R.R., Rybalov, A.: Uninorm aggregation operators. Fuzzy Sets and Systems 80, 111–120 (1996)
264. Yeung, D.S., Tsang, E.C.C.: A comparative study on similarity-based fuzzy reasoning methods. IEEE Trans. on Syst. Man and Cyber. 27, 216–227 (1997)
265. Ying, M.: Reasonableness of the compositional rule of fuzzy inference. Fuzzy Sets and Systems 36, 305–310 (1990)
266. Zadeh, L.A.: Outline of a new approach to the analysis of complex systems and decision processes. IEEE Trans. on Syst. Man and Cyber. 3, 28–44 (1973)
267. Zimmermann, H.J. (ed.): Proc. 2nd European Congress on Intelligent Techniques and Soft Computing, EUFIT 1994, Aachen, Germany, September 1994. ELITE, Aachen (1994)
268. Zimmermann, H.J. (ed.): Proc. 4th European Congress on Intelligent Techniques and Soft Computing, EUFIT 1996, Aachen, Germany, September 2-6. ELITE, Aachen (1996)
269. Zimmermann, H.J. (ed.): Proc. 6th European Congress on Intelligent Techniques and Soft Computing, EUFIT 1998, Aachen, Germany, September 7-10. ELITE, Aachen (1998)
270. Zwick, R., Carlstein, E., Budescu, D.V.: Measures of similarity among fuzzy concepts: a comparative analysis. Internat. J. Approx. Reason. 1, 221–242 (1987)

List of Figures

List of Tables

Index

MIX
Papier aus verantwortungsvollen Quellen
Paper from responsible sources
FSC® C105338

If you have any concerns about our products,
you can contact us on
ProductSafety@springernature.com

In case Publisher is established outside the EU,
the EU authorized representative is:
Springer Nature Customer Service Center GmbH
Europaplatz 3, 69115 Heidelberg, Germany

Printed by Libri Plureos GmbH
in Hamburg, Germany